Mobile and Wireless: Communication and Technology

Mobile and Wireless: Communication and Technology

Edited by Grant Harper

CLANRYE INTERNATIONAL
www.clanryeinternational.com

Clanrye International,
750 Third Avenue, 9th Floor,
New York, NY 10017, USA

ISBN: 978-1-63240-793-1

Cataloging-in-Publication Data

Mobile and wireless : communication and technology / edited by Grant Harper.
p. cm.
Includes bibliographical references and index.
ISBN 978-1-63240-793-1
1. Mobile communication systems. 2. Wireless communication systems.
3. Communication and technology. I. Harper, Grant.
TK5103.2 .M63 2019
621.384--dc23

For information on all Clanrye International publications visit our website at www.clanryeinternational.com

CLANRYE INTERNATIONAL

Printed in China.

Contents

Preface

Every book is a source of knowledge and this one is no exception. The idea that led to the conceptualization of this book was the fact that the world is advancing rapidly; which makes it crucial to document the progress in every field. I am aware that a lot of data is already available, yet, there is a lot more to learn. Hence, I accepted the responsibility of editing this book and contributing my knowledge to the community.

Wireless communication technology is used for easy transfer of data between two or more remote points of exchange without the use of wires. It is characterised by the use of radio waves for fast long range informational exchange and communication. Modern technological systems also employ electro-magnetic radiations like light or vibrations for wireless transmission. Telecommunication industry is the major field in which wireless communication is used for varied purposes. This book elucidates concepts and innovative models around prospective developments with respect to technology of mobile and wireless communication. It strives to provide a fair idea about this discipline and to help develop a better understanding of the latest advances within this field. Information technology experts, researchers and students will find this book beneficial.

While editing this book, I had multiple visions for it. Then I finally narrowed down to make every chapter a sole standing text explaining a particular topic, so that they can be used independently. However, the umbrella subject sinews them into a common theme. This makes the book a unique platform of knowledge.

I would like to give the major credit of this book to the experts from every corner of the world, who took the time to share their expertise with us. Also, I owe the completion of this book to the never-ending support of my family, who supported me throughout the project.

Editor

1

Collaborative neighbor discovery in directional wireless sensor networks: algorithm and analysis

Fernaz Narin Nur[1†], Selina Sharmin[1†], Md. Ahsan Habib[1†], Md. Abdur Razzaque[1*] [†],
Md. Shariful Islam[2], Ahmad Almogren[3], Mohammad Mehedi Hassan[3] and Atif Alamri[3]

Abstract

In directional wireless sensor networks (DSNs), sensor nodes with directional antennas provide extended network lifetime and better coverage performance. However, one of the key challenges of directional nodes is to discover their neighbors due to difficulty in achieving synchronization among their directed transmissions and receptions. Existing solutions suffer from high discovery latency and poor percentage of neighbor discovery either due to lack of proper coordination or centralized management of the discovery operation. In this work, we develop a collaborative neighbor discovery (COND) mechanism for DSNs. Each COND node polls to directly discover its neighbors in a distributed way and collaborates with the already discovered nodes so as to allow indirect discovery. It helps to increase the neighbor discovery performance significantly. A Markov chain-based analytical model is developed to quantify theoretical performances of the proposed COND system. The performance of the COND system is evaluated in Network Simulator Version 3, and simulation results reveal that it greatly reduces the discovery latency and increases neighbor discovery ratio compared to state-of-the-art approaches.

Keywords: Directional sensor networks, Directional antenna, Synchronization, Indirect neighbor discovery, Discovery latency, Discovery ratio, Performance analysis

1 Introduction

In directional wireless sensor networks (DSNs), sensor devices with directional antennas sense data and deliver them in a multihop fashion toward the sink node. Nowadays, DSNs are being used for the implementation of many real-time applications including infrastructure monitoring, health-care monitoring, robotic exploration, battlefield surveillance, target tracking, and disaster response [1, 2]. The directional sensors have been proved to provide better energy efficiency, sensing quality, bandwidth utilization, etc. [3, 4]. However, the directional transmissions and receptions of the sensor nodes have made the development of data communication and networking protocols for DSNs more challenging.

Neighbor discovery is defined as the problem of identifying all nodes within the communication range of a sensor device. The problem of neighbor discovery in directional sensor network not only is more challenging than that in its omnidirectional counterpart but also requires complete redesign and implementation. The key fact behind this requirement is that the directional nodes have limited width of communication sectors, and thus, two nearby nodes must steer their communication sectors to each other for a successful communication [5–7]. Efficient strategies need to be employed so that a pair of neighbor nodes eventually beamform their antennas to each other at a certain time instance and can discover themselves. The primary challenge lies in increasing the number of nodes discovered in bounded delay through better coordination in a distributed way, when the nodes are not clock-synchronized.

*Correspondence: razzaque@du.ac.bd
Special thanks to the Information and Communication Technology Department of the Ministry of Posts, Telecommunications and Information Technology, Peoples Republic of Bangladesh, for the student research fellowship.
†Equal contributors
[1]Green Networking Research (GNR) Group, Department of Computer Science and Engineering, University of Dhaka, Dhaka 1000, Bangladesh
Full list of author information is available at the end of the article

In the literature, the problem of neighbor discovery with omnidirectional antenna has well been explored [1, 8]. Nevertheless, recently, a few neighbor discovery mechanisms are developed that are based on directional communications. A neighbor discovery algorithm based on contention-based and contention-free approaches have been developed in [8], where only the sink node uses directional antenna, whereas the other nodes use omnidirectional one, resulting in asymmetry-in-gain problem [9, 10]. A scan-based asynchronous neighbor discovery algorithm (SBAN) is proposed in [11], where the neighbor discovery is based on fully directional transmissions and receptions. The probabilistic HELLO message broadcasting with fast reply mechanism reduces collisions among the sensors. A randomized two-way neighbor discovery algorithm has been developed in [12]. The authors have presented an asymptotic analysis of one-way and two-way directional neighbor discovery algorithms and have designed a two-way neighbor discovery mechanism with selective feedback that reduces the neighbor discovery time. However, both the works [11, 12] lack in guaranteeing responder the reception of reply packets which increases the discovery latency for a dense network. A fully directional and centralized sectored antenna neighbor discovery (SAND) algorithm is developed in [13], where a token is passed among the nearby nodes to successfully learn each other. However, for token management, a central node is required that is not feasible for many applications of DSNs, especially in achieving scalability. Moreover, the rotational latency of a token affects the neighbor discovery accuracy and it increases the overall neighbor discovery latency. None of these works have well-thought-out the directional coordination between two nodes.

In this paper, a collaborative neighbor discovery (COND) algorithm is proposed, where the directional sensors discover their neighboring nodes collaboratively by sharing the discovered neighborhood information with the surrounding nodes. The key idea behind the COND mechanism is that each directional node performs neighbor discovery in each sector using a contention-based polling mechanism. The goal of neighbor discovery mechanism is to increase the number of discovered nodes with minimum delay using a two-way collaborative polling mechanism. The preliminary concept of this work has been published in [14]. In this paper, we augment the detail operation method of indirect neighbor discovery through collaboration. We carry out theoretical analysis using Markov chain model that quantifies expected number of discovered neighbors and the corresponding delay incurred. We also present performance analysis and results for average discovery latency per node, sensing wastage, and energy cost per neighbor discovery. The results show that the proposed COND mechanism greatly outperforms the state-of-the-art systems—randomized two-way [12] and SAND [13].

The main contributions of this work are summarized as follows:

- A fully distributed novel neighbor discovery mechanism, collaborative neighbor discovery (COND) algorithm is developed for sensor nodes in DSNs.
- A two-way polling-based collaborative mechanism is designed to decrease the possibility of conflict and to speed up the handshaking process. In this mechanism, each node learns about its neighbors collaboratively by direct and indirect neighbor discovery through sharing information of already discovered nodes to and from its neighbors.
- We develop a Markov chain model to analyze theoretically the expected number of neighbors discovered and the number of iterations required for discovery.
- Finally, the results of performance evaluations, carried out in Network Simulator Version 3 (NS-3) [15], show that our proposed COND mechanism outperforms SAND and randomized two-way mechanisms in terms of neighbor discovery latency and ratio, energy consumption, etc.

The rest of the paper is organized as follows. We give a detailed description of related state-of-the-art works in Section 2. In Section 3, we describe the network model and assumptions. The proposed algorithm for the neighbor discovery using directional antenna is discussed in detail in Section 4, and the theoretical analysis and performance studies for the given algorithm are discussed in Sections 5 and 6, respectively. Finally, we conclude the paper in Section 7.

2 Related work

Directional sensor networks (DSNs) are gaining much popularity in the recent years, due to their improved performance gains through directional antennas [7, 16–21]. Even though the problem of neighbor discovery was well investigated in the literature for omnidirectional and wireless ad hoc networks, that for directional sensor networks kept less focused. However, in recent years, we find a few of neighbor discovery algorithms that are based on directional antennas. The existing works can broadly be categorized in two types: one-way broadcasting mechanisms [22–27] and two- or three-way handshaking mechanisms [11, 13, 14, 28–30].

In one-way broadcast-based neighbor discovery mechanisms, nodes periodically broadcast their presence. On reception of at least one message successfully, a nearby node discovers the sender node. A probability-based neighbor discovery algorithm, namely I-SBA is proposed in [27], where nodes transmit, receive, and remain idle with certain probabilities. The mechanism decreases number of collisions among the neighboring nodes when the probabilities are selected properly. Nevertheless, with the growing number of nodes in the network, the possibility of directional synchronizations among the nodes is gradually decreased. In [23], authors develop a directional neighbor discovery mechanism where the number of slots for neighbor discovery is dependent on the order of the average number of nodes in the network. The base station sends a beacon signal toward all the nodes in the network to start the discovery process, and all the nodes are time-synchronized with each other. Their neighbor discovery mechanisms exploit hybrid usage of omnidirectional and directional antennas; hence, with the increasing number of nodes in the network, the mechanism initiates extra overhead for the node synchronization. A directional neighbor discovery protocol was developed in [25], which is based on one-way broadcast mechanism. A HELLO message is broadcasted by each node, and after hearing the message, the neighboring nodes learn about information of that node.

In none of the above one-way broadcast-based neighbor discovery mechanisms, the sender nodes know whether the neighboring nodes receive their message successfully or not. Thus, the one-way neighbor discovery algorithms might be suitable for omnidirectional networks but not for directional sensor networks, because the later requires synchronization between the transmitting and receiving nodes to make the communication a successful one. Therefore, two- or three-way handshaking protocols are more desirable for DSNs.

In two-way handshaking-based neighbor discovery mechanisms, a receiver node, after reception of a message, sends back a REPLY message toward the sender node, facilitating both nodes to discover each other in one event of communications. In directional transmission, a receiver node can only hear the message if it faces its antenna toward the sender and thus a feedback message is very important for neighbor discovery among the nodes. Hence, at least two-way handshaking is required for effective directional neighbor discovery as the nodes can agree on a future period for communication after the discovery. A fully directional scan-based asynchronous neighbor discovery algorithm is proposed in [11], where each node senses the medium and transmits HELLO message in a random direction and gathers location of the neighbor nodes, reducing the number of collisions among reply packets.

The urgent requirement of synchronization in between a transmitter and a receiver is kept unexplored in the above works. Moreover, nodes wait at each sector for a fixed period of time for neighbor discovery, which is highly inefficient. In [29], a directional neighbor discovery mechanism has been developed, where only the sink node of the network uses directional antenna, whereas the other sensors use omnidirectional antennas. The sink node applies contention-based and cotention-free methods to medium access and sequentially scans all its sectors to discover its neighbors. To minimize the medium access delay, the authors presented an optimization method for selecting the beam width and the persistence probability of neighbor discovery. This mechanism results asymmetry-in-gain problem [9] due to the difference between the antenna gain patterns of the sink node and other sensors. A randomized two-way neighbor discovery mechanism is proposed in [12], where the authors have presented an asymptotic analysis of one-way and two-way directional neighbor discovery algorithms. Their two-way neighbor discovery mechanism exploits selective feedback policy so as to reduce collisions among the reply packets and thus to decrease neighbor discovery time. However, the reception of a reply packet is not certain in the above works due to lack of synchronization between the sender and receiver nodes.

A fully directional and centralized three-way sectored antenna neighbor discovery (SAND) algorithm is proposed in [13], where a token is passed among network nodes sequentially to discover their surrounding neighbors. However, for token management, it requires a central node, which is not only often infeasible for many DSN applications but also limited by scalability. Furthermore, the rotational latency of a token affects the neighbor discovery accuracy and increases the overall neighbor discovery latency. The SAND mechanism also has not considered the directional synchronization issue among the nodes, causing deafness problem, and as a central controller controls the whole mechanism, the discovery latency sharply increases.

Our work is different from the above studies in that we develop a fully distributed neighbor discovery mechanism, where both the transmissions and receptions are directional and nodes collaborate with their (already discovered) neighbors to find other neighbors, reducing neighbor discovery latency significantly. This indirect neighbor discovery and selective feedback policies jointly help our algorithm to decrease the collisions in the network. In addition to that, a COND node employs dynamic waiting time at different directions while discovering nodes in the neighbor that greatly helps to mitigate the deafness problem.

3 Network model and assumption

We assume a directional sensor network of an area A where a set S of directional sensor nodes are deployed with uniform random distribution, resulting in deployment density,

$$\rho = \frac{|\mathcal{S}|}{A}. \quad (1)$$

Each node is designed with $\mathcal{M}$ sectored switched-beam directional antennas, as shown in Fig. 1. To cover the whole azimuth, fixed number of fixed beam width antenna elements is attached in each antenna of a sensor. Each node can activate one sector at a particular time. Nodes can switch from one sector to another by activating different antenna elements, and each element has high antenna gain to a particular sector. Basically, they direct a major or almost their entire signal in just one direction instead of dispersing all around a 360° circle. Typically, the switching delay from one sector to another is 5×10^{-6} μs for a traditional switched beam directional antenna [13, 31].

Each node has one transceiver and can form one directional beam in a sector at a given time. For the network nodes, all transmissions and receptions are directional and they do not require any additional omnidirectional antenna. Note that the usage of both directional and omnidirectional antennas causes additional hardware cost, complexity, and two other major problems. First, the transmission ranges of the omnidirectional antenna and directional antenna are different, which introduces asymmetry gain problem in communication links. Again, the spatial reuse benefits are greatly reduced since omnidirectional communication inhibits more simultaneous transmissions.

The communication radius (r) of all nodes in the network is homogeneous. Nodes in the network are not clock-synchronized with each other. Each node independently discovers its neighbors. The success event of a transmission is probabilistic; however, our dynamic polling period and selected response REPLY to HELLO messages decrease the probability of collision and increase the number of successful transmission events. If the discovery message is exchanged between two nodes for at least half of a slot, two nodes can discover at the same time slot [7]. Subsequently, in a distributed system, without any clock synchronization, two sensor nodes can discover each other within a bounded delay.

Initially, a node learns its position using a localization method [32]. We assume that the time is divided into contiguous slots and all nodes independently discover their neighbors within a bounded time. The total time frame is logically separated into transmit and receive time slots. Each slot is further divided into multiple mini-slots. A node either transmits *HELLO* message or listen to the medium for any message from its neighbor(s) in an individual slot. While communicating with each other, a node can identify the angular position of its neighbor by capturing its spatial signature vector [29]. After the deployment of the sensor nodes in the network, nodes discover their neighbors using COND mechanism, and during the neighbor discovery period, all the nodes will be active to discover most of the neighbors within short period of time. Later, the sensors collect sense data and start data transmission toward the sink node. An underlying medium access control protocol (e.g., S-MAC [33], T-MAC [34], B-MAC [35], DMAC [36], SAMAC [18]) can determine and control the energy-saving states of a sensor node, which is beyond the scope of this work. The notations used in this paper are summarized in Table 1.

4 Collaborative neighbor discovery mechanism

In this section, we present detail operation method of the proposed two-way collaborative neighbor discovery algorithm, COND, for DSNs. Each COND node tries

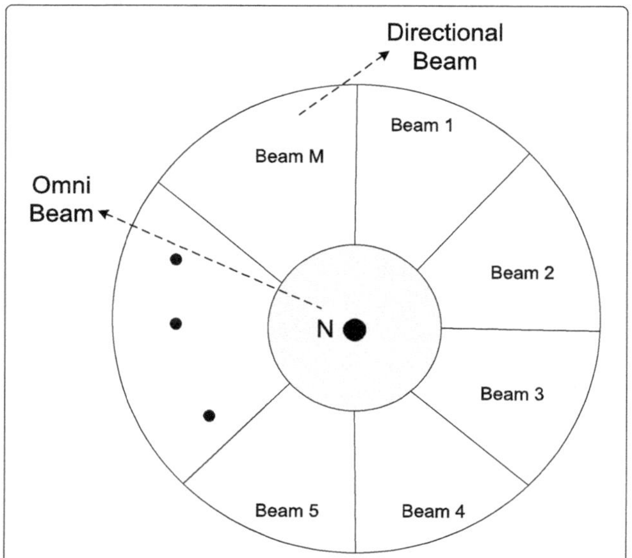

Fig. 1 Antenna model for each node

Table 1 List of notations

Symbol	Definition
$\mathcal{S}$	Set of all sensor nodes
$\mathcal{M}$	Set of sectors of directional antenna
P_t	Transmission probability
r	Communication radius of each sensor
$\mathcal{K}$	Delay tuning parameter
A	Deployment area of the network
ρ	Network density
$\mathcal{T}_a$	Neighbor table of node $a \in \mathcal{S}$
E_n	Expected number of nodes in the neighborhood
$\mathcal{N}_e^m$	Number of expected neighbor nodes in a sector $m \in \mathcal{M}$
$\mathcal{N}_d^m$	Number of discovered neighbor nodes after an iteration
$\mathcal{F}$	Flag indicating a neighbor is discovered directly or indirectly
$\mathcal{R}$	Iteration for neighbor discovery
x'	Mini sub-slot of a slot x
δ_m	Parameter that tunes $\mathcal{K}$

to explore maximum number of neighboring nodes in different sectors. A COND node stays in each of its sectors for a particular time duration and performs direct and indirect neighbor discovery processes using polling mechanism. Unlike [11, 13, 30], we allow COND nodes to dynamically vary the time duration operating in a sector based on percentage of its neighbors already discovered. Furthermore, unlike in our previous work [14], we allow COND nodes to explicitly track the neighbors discovered indirectly and take actions differently, facilitating to increase the neighbor discovery performance. In summary, the COND mechanism is condensed into the following three steps:

- Initialization of neighbor discovery with appropriate value of a delay tuning parameter ($\mathcal{K}$)
- Discovery of neighbor nodes in different sectors using contention-based polling mechanism
- Maintaining and updating the neighbor tables of COND nodes in collaborative fashion

What follows next, we present the detail operations of the COND steps.

4.1 Initialization

Each node in the network first calculates the approximate number of neighbors around them after the deployment of the nodes in the network. The expected number of neighbors of each node can be calculated as

$$E_n = \rho \times \pi \times r^2, \tag{2}$$

where r is the communication radius of a sensor node and ρ is the node deployment density. Then, the expected number of neighbors in a particular sector $m \in \mathcal{M}$ of a node is given by

$$\mathcal{N}_e^m = \frac{E_n}{|\mathcal{M}|}. \tag{3}$$

Initially, each sensor node $s \in \mathcal{S}$ starts to explore neighbors in all of its sectors $m \in \mathcal{M}$ by switching its direction clockwise after a certain interval of T seconds. T is the time interval for staying in a particular sector of each node. As we are assuming that each node switches its sector one by one for neighbor discovery, it is necessary to stay in a particular sector for a certain period so that two nodes can steer their antennas to each other in that period. Hence, T depends on the number of sectors of a node and the delay incurred for switching a direction, and it is determined as follows:

$$T = T_{\text{switch}} \times (|\mathcal{M}| - 1) \times \mathcal{K}, \tag{4}$$

where T_{switch} is the switching delay and $\mathcal{M}$ is the set of sectors. Furthermore, when a good number of nodes are discovered in a particular sector of a node, it must stay less period of time over there so as to reduce the discovery overhead. On the other hand, when a node has discovered very less number of neighbors in a sector, it must stay longer period in that particular sector in order to reduce the discovery latency. Thus, for determining the value of time period T, we introduce a delay tuning parameter $\mathcal{K}$ that is updated after each iteration depending on the number of discovered neighbor nodes in a particular sector as follows:

$$\mathcal{K} = \delta_m \times \mathcal{K}, \tag{5}$$

where the value of δ_m is chosen in such a way that the discovery latency in each sector $m \in \mathcal{M}$ is minimized. If a node discovers most of its neighbors in a given sector $m \in \mathcal{M}$, then it is wise for the node to stay for less period of time there and vice-versa; otherwise, discovery overhead will unnecessarily be increased. Thus, the value of δ_m helps us to dynamically control the staying period of a sensor node in a certain sector $m \in \mathcal{M}$ for neighbor discovery. Let N_d^m represents the percentage of nodes that are already discovered in sector $m \in \mathcal{M}$, then the value of δ_m is determined as follows:

$$\delta_m = \begin{cases} 2.0, & N_d^m \leq 30\%, \\ 1.5, & N_d^m \leq 50\%, \\ 1.0, & N_d^m \leq 80\%, \\ 0.5, & N_d^m > 80\%. \end{cases} \tag{6}$$

Note that the numerical values in Eq. (6) are not strict choices; rather, the values are depicted through numerous simulation experiments for a given network environment (stated in Section 6), and they are tunable by the network administrator. In the case the number of already discovered nodes in a certain sector crosses the expected number of neighbors, the node may stop staying in that sector. Hence, it may happen that a node misses a good number of neighbors to discover if the calculation of expected number of neighbors is incorrect (due to distortions in deployment). Therefore, we allow a COND node to continue the discovery process until it fails repeatedly (say, up to two iterations) to find a new neighbor in a particular sector.

4.2 Contention-based polling mechanism

The COND nodes contend among them to poll their neighbors in different sectors so as to discover those. The frame structure followed by the polling mechanism contains 'H' (Hello) and 'L' (Listen) slots, as shown in Fig. 2a. In H slots, a node broadcasts HELLO messages in random time slots with persistent probability P_t. In the remaining slots, the node listens to the medium for any HELLO message from the neighboring nodes.

Each slot is further divided into several mini sub-slots, as shown in Fig. 2a. If a node decides to transmit a HELLO

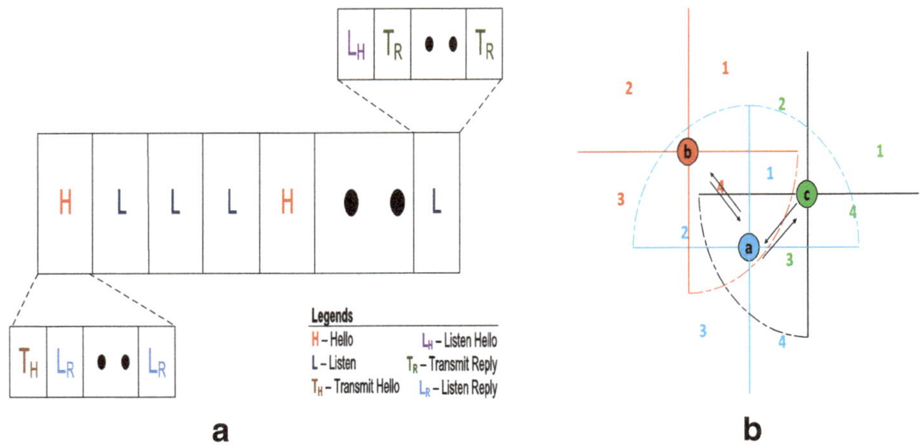

Fig. 2 Frame structure and neighbor discovery in COND system. **a** Frame structure. **b** Neighbor discovery

message in a slot, it does so in the first mini-slot and listens to the channel for any REPLY message(s) from its neighboring node(s). The HELLO message contains the node's ID, (*x*,*y*) coordinate, sector number, and neighbor table, i.e., $\Gamma_H = \{ID, \text{coord}(x,y), m, \mathcal{T}\}$. On reception of HELLO message, the surrounding nodes update their neighbor tables and relay back their information in a randomly chosen mini-slot. Thus, the nodes develop their neighbor tables for each sector with so far discovered nodes. The structure of the neighbor table is represented by the tuple, $< ID, \text{coord}(x,y), m, \mathcal{F} >$, where the fourth field is a flag that contains 1 if the designated neighbor is discovered through direct exchange of messages and 0 if it is discovered indirectly. The details of *direct* and *indirect* discovery processes are discussed in the following section.

4.3 Collaborative discovery of neighbor nodes

The key challenge in reducing neighbor discovery latency in DSNs is the synchronization of HELLO-REPLY transmissions among the nodes in a neighborhood. Therefore, in addition to discover neighbors through direct exchange of messages, we allow COND nodes to share their neighbor tables with each other and to update their tables collaboratively so as to further decrease the discovery latency. Thus, some neighbor nodes are discovered indirectly (via other nodes).

4.3.1 Direct discovery

When two nodes are beamformed to each other and successfully exchange HELLO and REPLY messages, they can directly discover each other. After the HELLO-REPLY handshaking, two nodes update their neighbor tables accordingly. Thus, it is highly dependent on the synchronization of corresponding sectors of two neighbors. In Fig. 2b, when the node *a* discovers a neighbor node *b* through direct handshaking, the flag entry for *b* in *a*'s neighbor table is set to 1. Unlike existing works in the literature that depend only on direct discovery, we allow indirect discovery of neighbor nodes through collaboration among them.

4.3.2 Indirect discovery

In omnidirectional communication, only direct discovery is sufficient for neighbor discovery. For directional nodes, it is the most important that two nodes are beamforming their antennas to each other; otherwise, the handshaking between the nodes would never occur. So, it may happen that when a node is broadcasting HELLO message in a particular sector, some of the neighbor nodes cannot receive the message because of the deafness problem [9]. Hence, the strict requirement of directional synchronization among nodes upsurges the discovery latency among the sensor nodes.

Consider a scenario in Fig. 2b, where pairs *a* and *b* and *a* and *c* have already discovered themselves in the present or previous time frame, but node *b* has not yet been discovered by *c* or vice-versa, as their antenna directions did not match during the direct neighbor discovery process. In order to ease the problem and to accelerate the discovery process, we allow COND nodes to share their neighbor tables with already discovered nodes. That is, node *a* shares its table with *b* and *c* so that they can discover each other as a neighbor of the respective sectors by checking their coordinates. We term this update of neighbor table as *indirect discovery*, and the flag entry for such discovered node is set to 0.

The proposed collaborative neighbor discovery (COND) mechanism has been summarized in Algorithm 1. Initially, the neighbor table of each node remains empty. After getting a HELLO/REPLY message from any neighbor, a node updates its own table (lines 6–23). The node that receives a HELLO message checks its neighbor table to find whether the node is discovered earlier or not. If there exists a record, the node checks the flag value and if the value is 1, it does not give any REPLY message to that node assuming that it has exchanged messages in the earlier slots with that node; otherwise, it updates the information in the neighbor table and sends a REPLY message in a random sub-slot to the direction of the sender node (lines 15–22). If the sender node receives a REPLY message (without any collision) from any neighbor, it adds the node in the neighbor table (lines 8–12). In this way, two nodes discover each other using two-way handshaking. In the case, a node cannot discover any new node in a certain sector in consecutive two iterations, we stop COND process in that sector (lines 25–27).

The time complexity of Algorithm 1 for one iteration is quite straightforward to follow. The lines 1–31, enclosed in a loop, iterate $|\mathcal{M}|$ times having a complexity of $\mathcal{O}(|\mathcal{M}|)$ in the worst case. Again, the statements 4–23 iterate $|T|$ times in a nested loop. The rest of the statements have constant unit time complexities. Therefore, the overall time complexity of the proposed COND algorithm is $\mathcal{O}(|\mathcal{M}| \times |T|)$.

Algorithm 1 Collaborative neighbor discovery (COND) at each node $a \in \mathcal{S}$

INPUT: $\rho, E_n, \mathcal{M}, \delta_m$
OUTPUT: Neighbor Table $\mathcal{T}_{\dashv}fornodea$

1: **for** each sector $m \in \mathcal{M}$ **do**
2:   **set** $\mathcal{R}_m \leftarrow 1$
3:   Calculate $\mathcal{N}_m$, T and $\mathcal{K}$ using Eqs. 3 -6
4:   **for** each time slot $x \in T$ **do**
5:     **if** x is a HELLO slot **then**
6:       a broadcasts Γ_H message in the first mini-slot
7:       **if** a receives Γ_R from any node b **then**
8:         **add** b in neighbor table $\mathcal{T}_a$ and **set** flag $\mathcal{F} \leftarrow 1$
9:         **add** $w \in \mathcal{T}_b$ that falls in sector m of node a to $\mathcal{T}_a$
10:         **set** flags $\mathcal{F} \leftarrow 0$ for the added nodes of $\mathcal{T}_b$
11:       **end if**
12:     **else**
13:       a **listens** the medium
14:       **if** a receives a Γ_H from any neighbor b **then**
15:         **if** $(b \notin \mathcal{T}_a)$ **OR** $(b \in \mathcal{T}_a$ **AND** $\mathcal{F} = 0)$ **then**
16:           **add** b in neighbor table $\mathcal{T}_a$ and **set** flag $\mathcal{F} \leftarrow 1$
17:           **add** $w \in \mathcal{T}_b$ that falls in sector m of node a to $\mathcal{T}_p$
18:           **set** flags $\mathcal{F} \leftarrow 0$ for the added nodes of $\mathcal{T}_b$
19:           a **replies** back Γ_R to b
20:         **end if**
21:       **end if**
22:     **end if**
23:   **end for**
24:   **update** $\mathcal{N}_d^m$, *i.e.*, the number of nodes discovered in sector m during current iteration
25:   **if** $\left(\mathcal{N}_d^m(\mathcal{R}_m - 1) == 0 \;\&\&\; \mathcal{N}_d^m(\mathcal{R}_m) == 0\right)$ **then**
26:     $\mathcal{M} \leftarrow \mathcal{M} \backslash m$
27:   **end if**
28:   **if** $\mathcal{M} = \emptyset$ **then**
29:     exit
30:   **end if**
31:   $\mathcal{R}_m \leftarrow \mathcal{R}_m + 1$
32: **end for**

5 Theoretical analysis of COND

At each iteration, a COND node performs neighbor discovery around its all sectors, and on completion of one iteration, the node gets an updated neighbor table with the newly discovered neighbors. In this section, we formulate a theoretical model for analyzing the neighbor discovery delay, average number of discovered nodes per iteration, and corresponding energy consumption of the COND system.

5.1 Discrete time Markov chain model

We have developed a discrete time Markov chain model for analyzing the performance of the proposed algorithm. In this model, a node can stay in one of the following six states: *idle* (I), *Transmitting HELLO message* (T), *Listening REPLY message* (L), *Receiving HELLO message* (R), *Transmitting REPLY message* (A), or *Collided* (C).

Here, L is the set of REPLY message listening states, i.e., $L = \{L_0^2, L_1^2, \ldots, L_v^2 \ldots L_0^u, L_1^u, \ldots, L_v^u\}$ and each $L_v^u \in L$ denotes that the v number of neighbor nodes is discovered in reply mini-slot u. Similarly, R is the set of states after receiving a HELLO message, i.e., $R = \{R_0^1, R_1^1, \ldots, R_v^1\}$, where each state R_v^1 represents that the v number of neighbors is discovered during the first mini-slot. The value of u is from 2 to the number of mini-slots (x') available in a slot and the maximum value of v can be $\mathcal{N}_e^m$, the expected number of neighbors in a certain sector m.

Figure 3 shows the state transition diagram of the COND mechanism. Given that a node transmits HELLO message with probability P_t, we can calculate the *idle* state (i.e., listening for any HELLO message) probability for a node as follows:

$$P_I = 1 - P_t, \tag{7}$$

where P_t is the probability that a node transmits HELLO message in a given slot. Suppose, a COND node a has sent a HELLO message in sector m. Another node b can only receive this HELLO message if it beamforms the antenna toward a in sector m', where $m \neq m'$ and $|m - m'| = 1$ for three sector nodes, $|m - m'| = 2$ for four sector nodes,

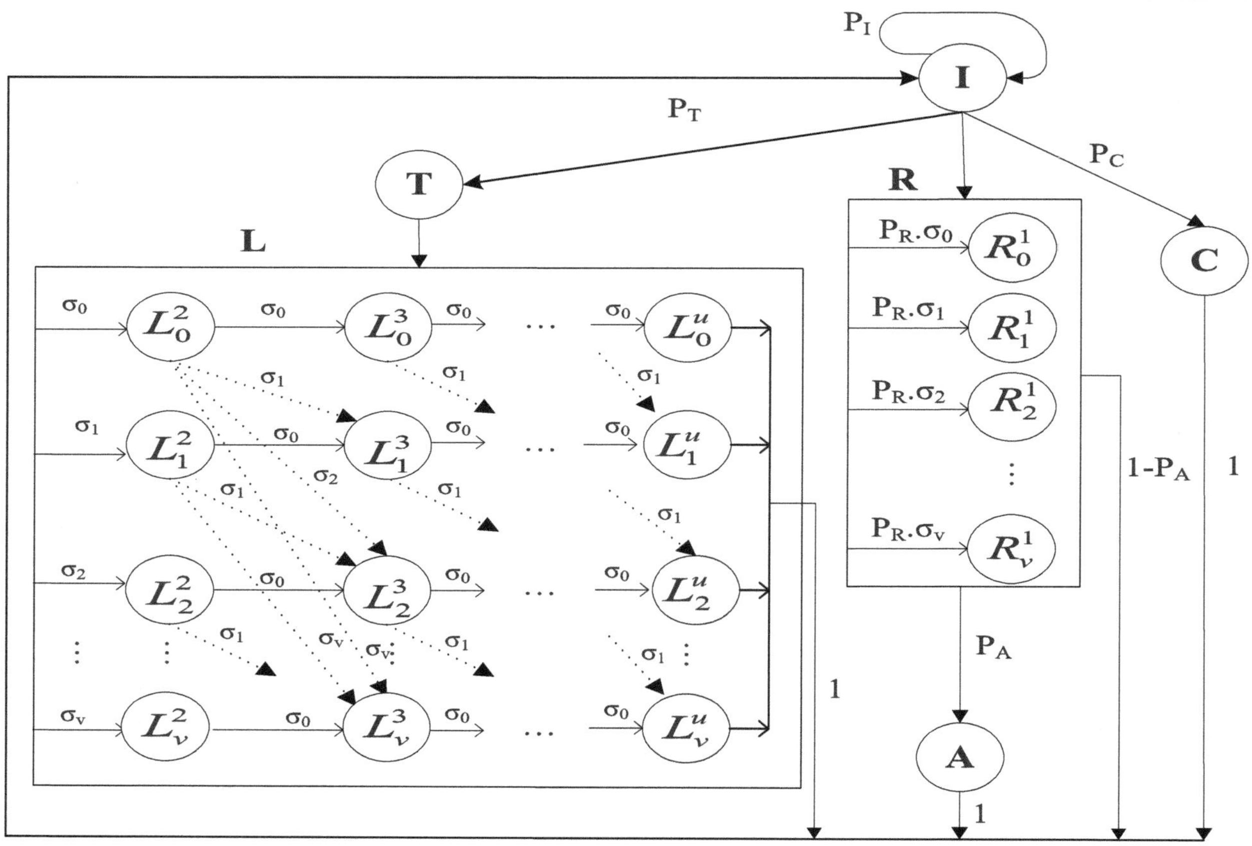

Fig. 3 State transition diagram for the COND mechanism

etc. The joint probability that the two nodes a and b will beamform to each other is thus calculated as follows:

$$P((b,m'),(a,m)) = P((a,m)\times P((b,m')|(a,m)) = \frac{1}{m^2}, \quad (8)$$

where the event (a,m) represents that the node a beam forms its antenna in sector m. Hence, in a given slot, a node a changes its state from I to T with probability,

$$P_T = P((b,m'),\ (a,m))\times P_t^a\times(1-P_t^b)\times(1-P_t)^{\mathcal{N}_e^{b,m'}-1}, \quad (9)$$

where the last part $(1-P_t)^{\mathcal{N}_e^{b,m'}-1}$ denotes the probability that no other nodes except a in the sector area of node b is transmitting any HELLO message at that time slot. Similarly, the receiving node b changes its state from I to R with probability,

$$P_R = P((a,m'),\ (b,m))\times P_t^a\times(1-P_t^b)\times(1-P_t)^{\mathcal{N}_e^{b,m}-1}. \quad (10)$$

Therefore, the probability of occurring a collision event is given by

$$P_C = 1-(P_I+P_T+P_R). \quad (11)$$

Note that, on successful reception of HELLO or REPLY messages, a node discovers neighbor(s) *directly* or *indirectly* in one of the L or R states. The probability of discovering v number of nodes in a mini-slot is

$$\sigma_v = \sigma^d\times\sigma_{v-1}^i+\sigma_v^i, \quad (12)$$

where $\sigma^d = \mu\times P_R$ is the probability that the sender node is directly discovered; μ is a binary variable representing whether the sender node is already discovered ($\mu = 0$) or not ($\mu = 1$). The receiver node indirectly discovers a neighbor with probability, $\sigma^i = \eta\times P_R$, where binary variable $\eta = 1$ if there remains a node in the sender's table that is not yet discovered and is located in the sector area of the receiver node. Thus, the probability that the k number of neighbors are indirectly discovered in a mini-slot is given by

$$\sigma_v^i = \binom{\mathcal{N}_e^m}{v}\times\left(\sigma^i\right)^v\times\left(1-\sigma^i\right)^{\mathcal{N}_e^m-v}. \quad (13)$$

Now, the probability that no neighbor is discovered in a mini-slot is given by

$$\sigma_0 = 1-\left(\sigma^d+\sigma^i\right). \quad (14)$$

Note that, after receiving a HELLO message successfully from a sender node a, the receiver node b sends back a REPLY message to the sender a if any of the following three conditions is met: (i) the information of a is not present in b's neighbor table, (ii) the node a is discovered indirectly, and (iii) the node b finds that $\mathcal{T}_b \setminus \mathcal{T}_a \neq \phi$. Let P_A denotes the probability of occurring any of the above events. Thus, the receiver node b transmits a REPLY message with probability P_A; otherwise, it returns to idle state. The probability distribution vector, S, at a given time slot x at equilibrium is expressed as

$$\mathbf{S} = \begin{bmatrix} S_I & S_T & S_v^u & S_v^1 & S_A & S_C \end{bmatrix}^x. \tag{15}$$

We can find the distribution vector at equilibrium by solving the equations $PS = S$ and $\sum \mathbf{S} = 1$.

The solution of the above two equations is given below,

$$s = \frac{1}{\xi} \times \begin{bmatrix} 1 \\ P_T \\ \sigma_0 \times P_T \\ \vdots \\ \sigma_v \times P_T \\ (\sigma_0)^2 \times P_T \\ 2 \times \sigma_0 \times \sigma_1 \times P_T \\ \vdots \\ P_T(2 \times \sigma_v \times \sigma_1 + (\sigma_2)^2 + \sigma_0 \times \sigma 4 + \sigma 4) \\ \sigma_0 \times P_R \\ \vdots \\ \sigma_v \times P_R \\ P_A \times P_R(\sigma_0 + \sigma_1 + \sigma_2 + \sigma_3 + \sigma_4 + \ldots \sigma_v) \\ P_C \end{bmatrix}, \tag{16}$$

where $\xi = 1 + P_T + (\sigma_0 + \sigma_1 + \sigma_2 + \sigma_3 + \cdots + \sigma_v)(P_T + P_R + P_A P_R) + P_T[(\sigma_0)^2 + 2\sigma_0\sigma_1 + 2\sigma_0(\sigma_1)^2 + 2\sigma_0\sigma_1\sigma_2\sigma_3 + 2\sigma_0\sigma_1(\sigma_2)^2\sigma_3(\sigma_4)^2 + \cdots + (P_T)^2] + P_C$.

The performance of the proposed COND mechanism greatly depends on the number of discovered neighbors in a time frame. From the Markov chain analysis we find the total number of neighbors discovered per slot. Thus, the expected number of discovered neighbors by a particular node is

$$E_v = \sum_{x=1}^{T} \left[v \times \sigma_u^v(x) + v \times \sigma_1^v(x) \right]. \tag{17}$$

Now, the neighbor discovery rate ζ of a particular node is

$$\zeta = \frac{E_v}{T}. \tag{18}$$

Figure 4a, b shows the number discovered nodes and required number of iterations for the varying number of nodes in the network when there are 15 slots in a time frame and each slot has three mini-slots.

5.2 Comparison of analytical and simulation results

In this section, we compare the simulation experiment results with those of theoretical analysis. We perform our simulation experiments in NS-3 for varying number of sensors, ranging from 5 to 50. We have fixed the transmission range of each node at 100 m and the number of sectors at 4. With the growing number of sensors in the network, the number of discovered nodes increases linearly in both theoretical and simulation results, as shown in Fig. 5a. Unsurprisingly, as depicted in graphs of Fig. 5b, the number of iterations also increases with growing number of nodes both in theoretical and simulation results. However, the rate of increase is not that

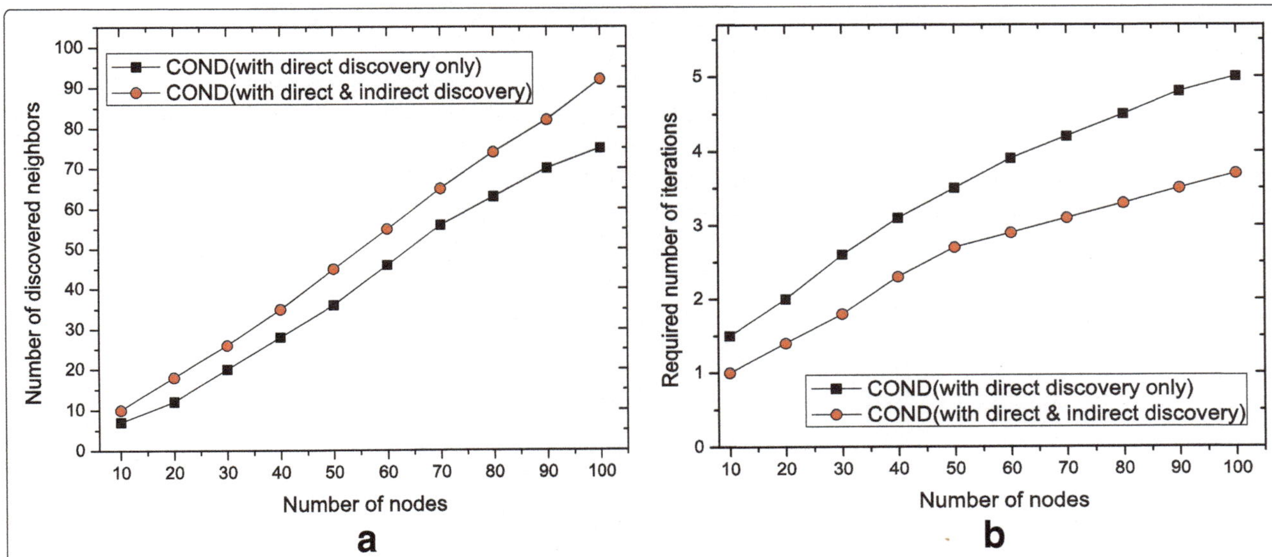

Fig. 4 Theoretical analysis for varying number of sensor nodes. **a** Neighbor discovery. **b** Required iterations

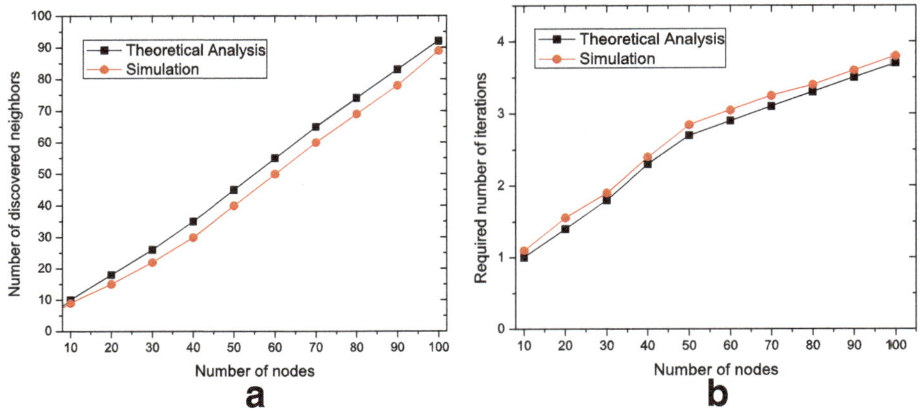

Fig. 5 Comparison of results from theoretical analysis and simulation experiments. **a** Neighbor discovery. **b** Required iterations

much higher because of the incorporation of indirect discovery in COND. Finally, the graphs of Fig. 5a, b show that the analytical results are quite closer to simulation outcomes in estimating the number of discovered neighbors and the required number of iterations for neighbor discovery.

6 Performance evaluation

We have implemented our proposed COND system in NS-3 [15], a discrete-event network simulator to verify the effectiveness of proposed algorithm. We have done a comparative study of COND performances with those of two state-of-the-art algorithms—randomized two-way neighbor discovery mechanism [12] and SAND [13].

6.1 Simulation environment

We use wifiSimpleAdhocGrid model for sensor nodes and YansWifiPhy model to define the channel properties such as propagation and loss characteristics. Each node has the information about network density, and it calculates the expected number of neighbor nodes in each sector based on that information. The COND mechanism stops when no neighbor is discovered in two subsequent rounds. The simulation configuration parameters are shown in Table 2. We run the simulation experiments 30 times with different random seed values and take the average of the results for each graph data points.

6.2 Performance metrics

- *Average discovery latency per node*: A round of node discovery is completed when a node finishes the operation in all sectors (either clockwise or anti-clockwise). Average discovery latency per node is the ratio of total discovery time to the number of discovered nodes.
- *Sensing wastage* is the count of wastage of time slots for neighbor discovery. A sensing time slot is regarded as wastage if no neighbor is discovered in it.
- *Average discovery ratio* is the ratio of the number of neighbor nodes successfully discovered to the total number of nodes in the network.
- *Control byte overhead* is the ratio of total number of control bytes (HELLO and REPLY messages) exchanged during the simulation period to the total number of neighbors discovered.
- *Average energy cost per node* discovery is the measure of average amount of energy consumed during the simulation period for discovering a single node in the network.

6.3 Simulation results

The comparative performance results are plotted for varying number of sensors deployed in the network, number of sectors a node has, and the size of the sensing time frame.

Table 2 Simulation parameters

Parameters	Value
Area deployment	500 m × 500 m
Type of deployment	Uniform random
Number of nodes deployed	100 ∼ 1000
Number of sectors	2 ∼ 6
Transmission range	100 m
Slot duration	1 ms
Packet size	40 bytes
Initial energy	3 J
Data transmission rate	1 Mbps
Bandwidth	5 MHz
Routing protocol	AODV
Channel model	Yans WiFi model
Propagation model	Friis propagation loss model
Simulation time	1000 s

6.3.1 Impacts of varying number of nodes

We have varied the number of nodes from 100 to 1000 and compared the performances of the proposed COND mechanism with those of other systems. The number of sectors for each node is fixed at 4, and we call a round of sensing is completed when a node is switched to all of its sectors for neighbor discovery. Each time frame contains 25 slots and each slot has 4 mini-slots.

The graph of Fig. 6a shows that the discovery latency increases with the growing number of nodes in all the studied systems. However, the rate of increase in our proposed COND mechanism is significantly lower than those in other two systems. This is achieved because of incorporating indirect discovery in COND and reduction of transmission collisions through randomized and controlled accesses of medium by its nodes. The graphs also depict that the latency is increased linearly in SAND with growing number of nodes as a central node manages the neighbor discovery.

As we have fixed the number of slots in each time frame, the sensing wastage is very high for a low dense network in all the studied approaches, causing poor utilization of time slots. The wastage of slots reduces with the growing number of nodes up to approximately 600 nodes in COND due to collision and conflicts of nodes in competing environment of node discovery. The COND has the minimum wastage of slots because of allowing indirect node discovery, whereas many slots are wasted in the other algorithms especially for the lack of sector synchronization, as shown in Fig. 6b.

The graphs in Fig. 6c depict the discovery ratio of the neighboring nodes during the simulation period. The discovery ratio is 100% for all the proposed algorithms when the network density is low. The discovery ratio decreases with the growing number of nodes. Our in-depth look into the simulation trace files depicts that the last 10% of nodes take the longer time to discover. The COND mechanism performs better than the other two mechanisms with higher network density as the indirect discovery of COND mechanism increases the neighbor discovery ratio. In the two-way randomized protocol and SAND, the overheads are higher than that of COND because of additional control packet exchanges among nodes at random times due to collision and loss of token, as depicted in Fig. 6d. The SAND has higher overhead among all the algorithms as

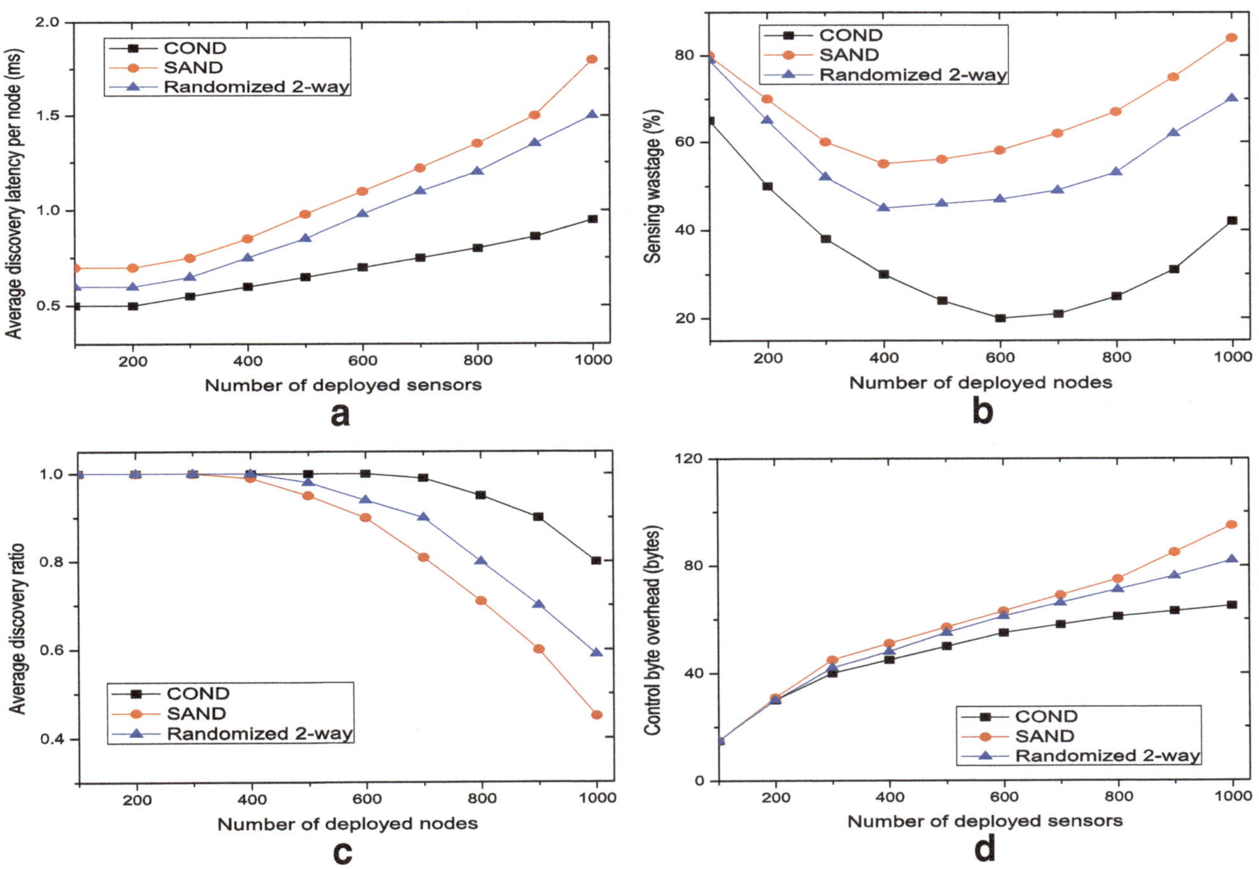

Fig. 6 Impacts of number of sensor nodes on performances of the studied neighbor discovery algorithms. **a** Average discovery latency. **b** Sensing wastage of discovery slots. **c** Average discovery ration. **d** Control byte overhead

it uses tokens to control and manage transmissions in the network.

6.3.2 *Impacts of varying number of sectors*

In this experiment, we allow each node to have two to six sectors. Here, the number of sensor nodes is fixed at 500 and the frame size at 25 time slots. The graphs of Fig. 7a depict that the discovery latencies of the evaluated algorithms decrease with growing number of sectors from two to four. This result is achieved due to increased spatial reusability with higher number of sectors. However, our COND mechanism achieves minimum latency, and it becomes stable gradually. On the contrary, in other mechanisms, the latency increases again with beamforming angle less than 90°. The reason is that, in this situation, nodes need longer period to discover their neighbors as synchronization of antenna beams is quite challenging for narrow beam widths. For the same reason, the wastage of time slots is also minimum in COND compared to the other mechanisms, as shown in Fig. 7b.

Directional antenna reduces the chance of collision among the neighboring nodes. The Fig. 7c shows that the neighbor discovery ratio gradually increases with increasing sectors in all the mechanisms, which is as expected theoretically. Nevertheless, excessive reduction of sensing angle reduces the chance of direction synchronization among nodes, and thus, it decreases the neighbor discovery ratio. Yet again, higher number of sectors causes the nodes to switch from one sector to another frequently, resulting in transmission of more HELLO/REPLY messages. Hence, the control byte overhead is increased with the growing number of sectors, as shown in Fig. 7d. Our COND mechanism outperforms the state-of-the-art mechanisms as it needs less control byte overheads.

6.3.3 *Impacts of varying number of time slots*

The number of time slots per round are varied from 10 to 40 for the simulation. Here, we assume fixed number of 500 sensor nodes and each node has 4 sensing sectors.

The graphs of Fig. 8a show that the average latency for all the mechanisms decrease progressively with the growing number of time slots as each node gets longer time to discover its neighbors in each sector. The COND utilizes the time slots more efficiently and has the

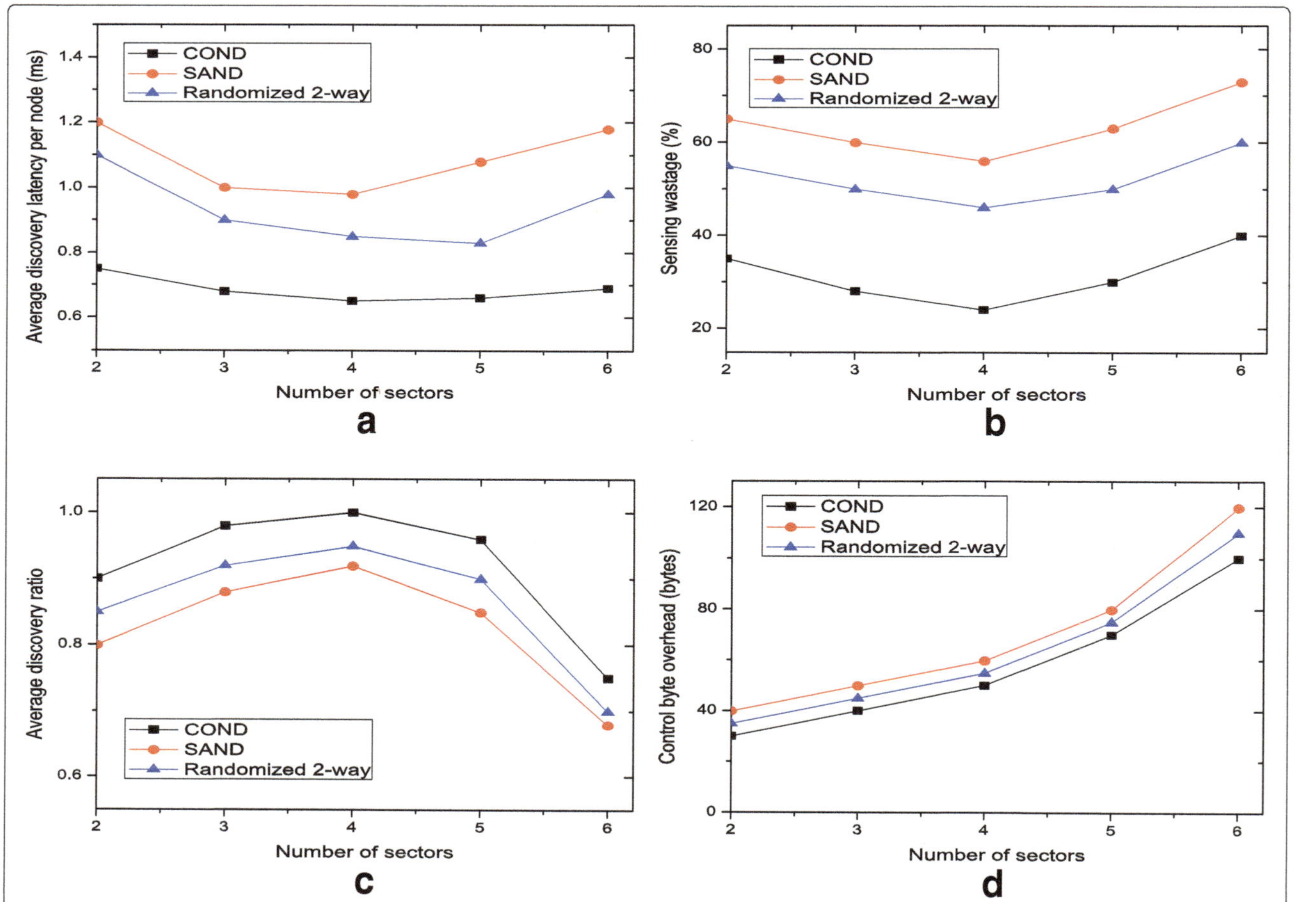

Fig. 7 Impacts of number of sectors on performances of the studied neighbor discovery algorithms. **a** Average discovery latency. **b** Sensing wastage of discovery slots. **c** Average discovery ration. **d** Control byte overhead

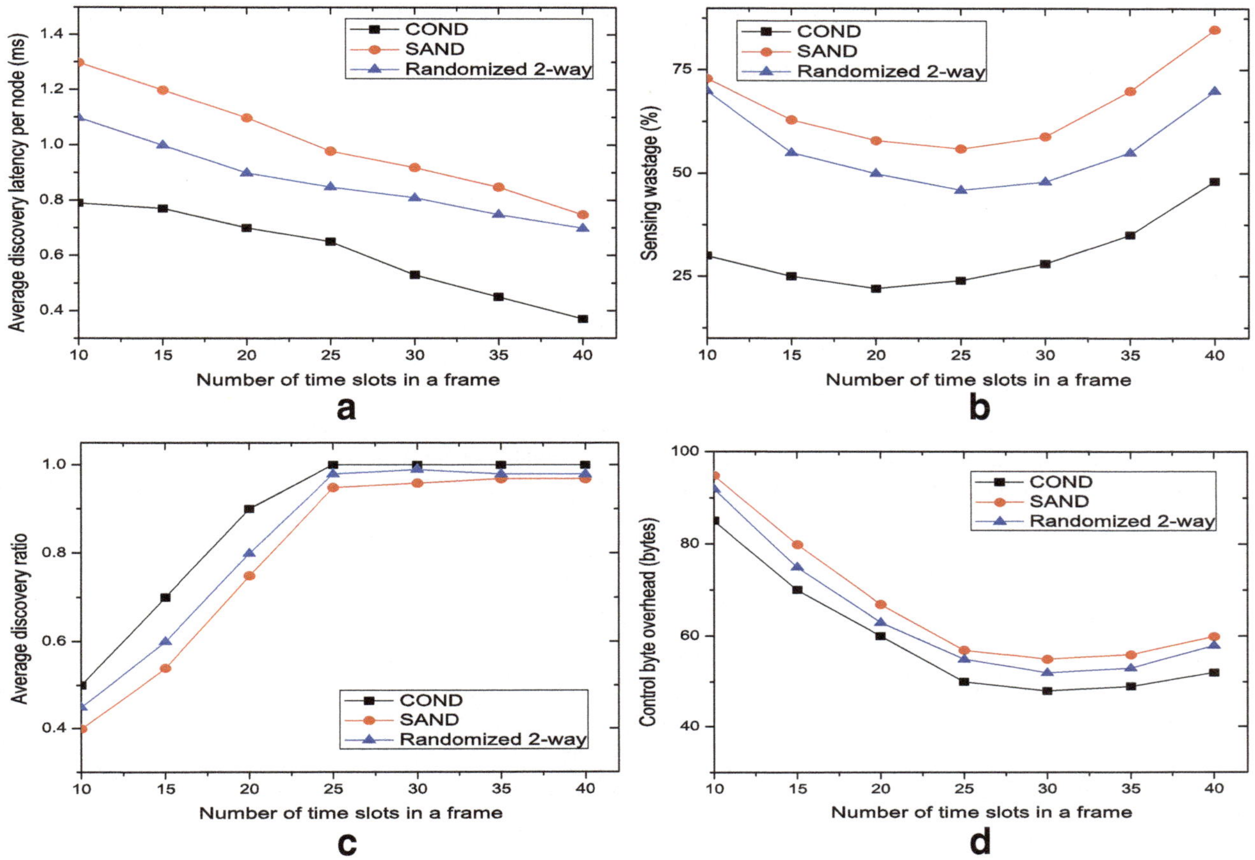

Fig. 8 Impacts of number of time slots in a frame on performances of the studied neighbor discovery algorithms. **a** Average discovery latency. **b** Sensing wastage of discovery slots. **c** Average discovery ration. **d** Control byte overhead

minimum latency by the way of indirect neighbor discovery. But unfortunately, with the growing number of slots, the sensing wastage increases as shown in Fig. 8b. As the number of nodes are fixed at 500 and there are many time slots for neighbor discovery, the wastage increases. For the same reason, the neighbor discovery ratio increases with the growing number of time slots. However, when the number of slots are greater than 25, it becomes stable as large number of time slots are not required for neighbor discovery in a low dense network. The COND has maximum discovery ratio compared to the other systems, as shown in Fig. 8c. The control byte overhead also decreases

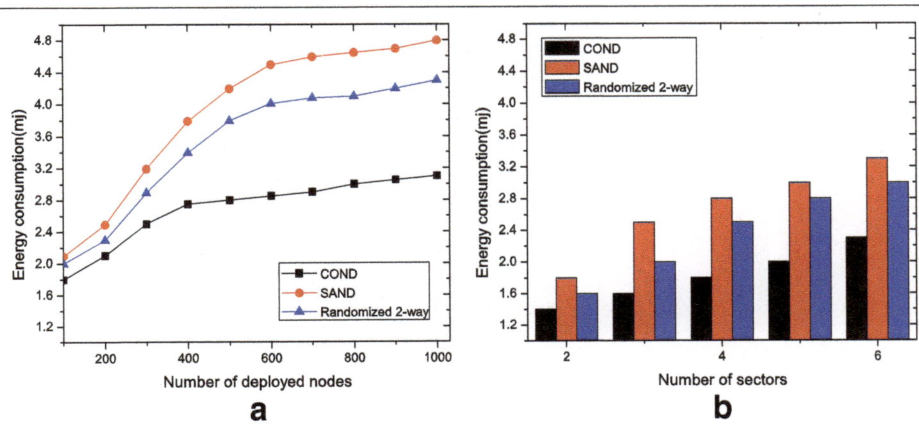

Fig. 9 Energy consumption performances of the studied algorithms. **a** Impact on varying number of sensors. **b** Impact on varying number of sectors

with the growing number of slots as the discovery ratio increases, as shown in Fig. 8d. It increases slightly when the number of time slots crosses 30 since the possibility of using unnecessary slots is increased.

6.3.4 Average energy cost for neighbor discovery

In the last set of simulation experiments, we measure the energy consumption for varying number of nodes and sectors. The graphs of Fig. 9 show that the average energy cost per neighbor discovery increases with the growing number of nodes and sectors in all the studied systems. As the neighbor discovery rate decreases with the growing number of nodes, the energy cost increases. However, the energy cost in our proposed COND mechanism is significantly lower than those in the other two systems. This result is achieved because of efficient neighbor discovery approach used in the proposed COND.

7 Conclusions

In this paper, we investigated distributed solutions to the problem of neighbor discovery in DSNs and introduced indirect neighbor discovery through collaboration among the already discovered nodes. We also introduced dynamic polling period in each sector following the number of already discovered nodes in it. The joint employment of the above strategies helped the proposed COND system to achieve excellent performances in neighbor discovery. Our in-depth look into the simulation trace files depicts that the collaboration-based indirect neighbor discovery helped to reduce the discovery latency significantly while the utilization of time slots in each frame was greatly enhanced due to the use of dynamic polling periods. The results from simulation experiments, carried out in NS-3, showed the achievements of as high as 80 and 75% performance gains in neighbor discovery latency and discovery ratio, respectively, compared to SAND, a leading work in the literature.

In the future, we shall concentrate on developing a theoretical model for analyzing delay tuning parameter and study the effects of node mobility on performances of the proposed neighbor discovery mechanism.

Acknowledgements
This work was financially supported by the King Saud University through Vice Deanship of Research Chairs.

Competing interests
The authors declare that they have no competing interests.

Author details
[1]Green Networking Research (GNR) Group, Department of Computer Science and Engineering, University of Dhaka, Dhaka 1000, Bangladesh. [2]Institute of Information Technology, University of Dhaka, Dhaka 1000, Bangladesh. [3]Chair of Pervasive and Mobile Computing, College of Computer and Information Sciences, King Saud University, Riyadh 11543, Saudi Arabia.

References

1. W Sun, Z Yang, X Zhang, Y Liu, Energy-efficient neighbor discovery in mobile ad hoc and wireless sensor networks: a survey. Commun. Surv. Tutorials IEEE. **16**(3), 1448–1459 (2014)
2. T Voigt, L Mottola, K Hewage, in *European Conference on Wireless Sensor Networks, Lecture Notes in Computer Science, vol 7772*. Understanding link dynamics in wireless sensor networks with dynamically steerable directional antennas (Springer, Berlin, 2013), pp. 115–130
3. F Babich, M Comisso, Throughput and delay analysis of 802.11-based wireless networks using smart and directional antennas. Commun. IEEE Trans. **57**(5), 1413–1423 (2009)
4. FN Nur, S Sharmin, MA Razzaque, MS Islam, MM Hassan, A low duty cycle MAC protocol for directional wireless sensor networks. Wirel. Pers. Commun., 1–25 (2016). doi:10.1007/s11277-016-3728-4
5. S Sharmin, FN Nur, MA Razzaque, MM Rahman, in *Networking Systems and Security (NSysS), 2015 International Conference On*. Network lifetime aware area coverage for clustered directional sensor networks, (Dhaka, 2015), pp. 1–9. doi:10.1109/NSysS.2015.7042949
6. M Islam, M Ahasanuzzaman, M Razzaque, MM Hassan, A Alamri, *et al*, in *Wireless and Mobile, 2014 IEEE Asia Pacific Conference On*. A distributed clustering algorithm for target coverage in directional sensor networks (IEEE, Bali, 2014), pp. 42–47
7. L Chen, Y Li, AV Vasilakos, in *IEEE INFOCOM 2016 - The 35th Annual IEEE International Conference on Computer Communications*. Oblivious neighbor discovery for wireless devices with directional antennas, (San Francisco, 2016), pp. 1–9. doi:10.1109/INFOCOM.2016.7524570
8. MJ McGlynn, SA Borbash, in *Proceedings of the 2nd ACM International Symposium on Mobile Ad Hoc Networking & Computing, (MobiHoc '01)*. Birthday protocols for low energy deployment and flexible neighbor discovery in ad hoc wireless networks (ACM, New York, 2001), pp. 137–145
9. FN Nur, S Sharmin, MA Razzaque, MS Islam, in *Networking Systems and Security (NSysS), 2015 International Conference On*. A duty cycle directional MAC protocol for wireless sensor networks, (Dhaka, 2015), pp. 1–9. doi:10.1109/NSysS.2015.7042950
10. G Jakllari, I Broustis, T Korakis, SV Krishnamurthy, L Tassiulas, in *Personal, Indoor and Mobile Radio Communications, 2005. PIMRC 2005. IEEE 16th International Symposium on*. Handling asymmetry in gain in directional antenna equipped ad hoc networks, vol. 2, (Berlin, 2005), pp. 1284–12882
11. Y Wang, B Liu, L Gui, in *Wireless Communications & Signal Processing (WCSP), 2013 International Conference On*. Adaptive scan-based asynchronous neighbor discovery in wireless networks using directional antennas (IEEE, Hangzhou, 2013), pp. 1–6
12. H Cai, T Wolf, in *Computer Communications (INFOCOM), 2015 IEEE Conference On*. On 2-way neighbor discovery in wireless networks with directional antennas (IEEE, Kowloon, 2015), pp. 702–710
13. R Murawski, E Felemban, E Ekici, S Park, S Yoo, K Lee, J Park, ZH Mir, Neighbor discovery in wireless networks with sectored antennas. Ad Hoc Netw. **10**(1), 1–18 (2012)
14. FN Nur, S Sharmin, MA Habib, MA Razzaque, MS Islam, in *2016 IEEE Region 10 Conference (TENCON)*. Collaborative neighbor discovery in directional wireless sensor networks, (Singapore, 2016), pp. 1097–1100. doi:10.1109/TENCON.2016.7848178
15. H Wu, S Nabar, R Poovendran, in *Proceedings of the 4th International ICST Conference on Simulation Tools and Techniques*. An energy framework for the network simulator 3 (ns-3) (ICST, 2011), pp. 222–230. Network simulator version 3 (ns-3). https://www.nsnam.org/. Accessed 17 Mar 2017
16. B Liu, B Rong, R Hu, Y Qian, Neighbor discovery algorithms in directional antenna based synchronous and asynchronous wireless ad hoc networks. Wireless Commun. IEEE. **20**(6), 106–112 (2013)
17. T Korakis, G Jakllari, L Tassiulas, CDR-MAC: a protocol for full exploitation of directional antennas in ad hoc wireless networks. Mobile Comput. IEEE Trans. **7**(2), 145–155 (2008)

18. E Felemban, S Vural, R Murawski, E Ekici, K Lee, Y Moon, S Park, SAMAC: A cross-layer communication protocol for sensor networks with sectored antennas. Mobile Comput. IEEE Trans. **9**(8), 1072–1088 (2010)
19. Q Wang, H-N Dai, Z Zheng, M Imran, AV Vasilakos, On connectivity of wireless sensor networks with directional antennas. Sensors. **17**(1), 134 (2017)
20. S Sharmin, FN Nur, MA Razzaque, MM Rahman, in *2016 IEEE Region 10 Conference (TENCON)*. Network lifetime aware coverage quality maximization for heterogeneous targets in dsns, (Singapore, 2016), pp. 3030–3033. doi:10.1109/TENCON.2016.7848603
21. MA Habib, S Saha, FN Nur, MA Razzaque, in *2016 IEEE Region 10 Conference (TENCON)*. Starfish routing for wireless sensor networks with a mobile sink, (Singapore, 2016), pp. 1093–1096. doi:10.1109/TENCON.2016.7848177
22. L Wei, B Zhou, X Ma, D Chen, J Zhang, J Peng, Q Luo, L Sun, D Li, L Chen, Lightning: a high-efficient neighbor discovery protocol for low duty cycle WSNs. IEEE Commun. Lett. **20**(5), 966–969 (2016)
23. AS Tehrani, AF Molisch, G Caire, in *Global Communications Conference (GLOBECOM)*. Directional zigzag: neighbor discovery with directional antennas (IEEE, San Diego, 2015), pp. 1–6
24. G Jakllari, W Luo, SV Krishnamurthy, An integrated neighbor discovery and mac protocol for ad hoc networks using directional antennas. Wireless Commun. IEEE Trans. **6**(3), 1114–1024 (2007)
25. J Du, E Kranakis, OM Ponce, S Rajsbaum, Neighbor discovery in a sensor network with directional antennae. Adhoc Sensor Wireless Netw. **30**(3/4), 261–286 (2016)
26. P Papadimitratos, M Poturalski, P Schaller, P Lafourcade, D Basin, S Capkun, J-P Hubaux, Secure neighborhood discovery: a fundamental element for mobile ad hoc networking. Commun. Mag. IEEE. **46** (2008). doi: 10.1109/MCOM.2008.4473095
27. H Cai, B Liu, L Gui, M-Y Wu, in *Communications (ICC), 2012 IEEE International Conference On*. Neighbor discovery algorithms in wireless networks using directional antennas (IEEE, Ottawa, 2012), pp. 767–772
28. W Zhang, L Peng, R Xu, L Zhang, J Zhu, in *2016 25th Wireless and Optical Communication Conference (WOCC)*. Neighbor discovery in three-dimensional mobile ad hoc networks with directional antennas, (2016), pp. 1–5. doi:10.1109/WOCC.2016.7506603
29. I Koutsopoulos, L Tassiulas, Fast neighbor positioning and medium access in wireless networks with directional antennas. Ad Hoc Netw. **11**(2), 614–624 (2013)
30. S Vasudevan, J Kurose, D Towsley, in *INFOCOM 2005. 24th Annual Joint Conference of the IEEE Computer and Communications Societies. Proceedings IEEE*. On neighbor discovery in wireless networks with directional antennas, vol. 4 (IEEE, 2005), pp. 2502–2512
31. Z Zhang, MF Iskander, Z Yun, A Host-Madsen, Hybrid smart antenna system using directional elements-performance analysis in flat rayleigh fading. IEEE Trans. Antennas Propag. **51**(10), 2926–2935 (2003)
32. B Zhang, F Yu, Z Zhang, in *2009 11th IEEE International Conference on High Performance Computing and Communications*. A high energy efficient localization algorithm for wireless sensor networks using directional antenna, (Seoul, 2009), pp. 230–236
33. W Ye, J Heidemann, D Estrin, in *Proceedings. Twenty-First Annual Joint Conference of the IEEE Computer and Communications Societies*. An energy-efficient mac protocol for wireless sensor networks, vol. 3, (New York, 2002), pp. 1567–15763. doi:10.1109/INFCOM.2002.1019408
34. T van Dam, K Langendoen, in *Proceedings of the 1st International Conference on Embedded Networked Sensor Systems*. SenSys '03. An adaptive energy-efficient mac protocol for wireless sensor networks (ACM, New York, 2003), pp. 171–180. doi:10.1145/958491.958512
35. J Polastre, J Hill, D Culler, in *Proceedings of the 2Nd International Conference on Embedded Networked Sensor Systems*. SenSys '04. Versatile low power media access for wireless sensor networks (ACM, New York, 2004), pp. 95–107. doi:10.1145/1031495.1031508
36. Y-B Ko, V Shankarkumar, NH Vaidya, in *Proceedings IEEE INFOCOM 2000. Conference on Computer Communications. Nineteenth Annual Joint Conference of the IEEE Computer and Communications Societies (Cat. No.00CH37064)*. Medium access control protocols using directional antennas in ad hoc networks, vol. 1, (Tel Aviv, 2000), pp. 13–211

2

Kernel PCA feature extraction and the SVM classification algorithm for multiple-status, through-wall, human being detection

Wei Wang*, Min Zhang, Dan Wang and Yu Jiang

Abstract

Ultra-wideband (UWB) radar with strong anti-jamming performance and high-range resolution can be used to separate multiple human targets in a complex environment. In recent years, through-wall human being detection with UWB radar has become relatively sophisticated. In this paper, the method of kernel principal component analysis (KPCA) feature extraction and the support vector machine (SVM) classification algorithm a~e applied to identify and classify the multiple statuses of through-wall human being detection. This method makes full use of the KPCA of powerful, nonlinear feature extraction and SVMs, which can solve the problem of multiple-status detection and nonlinear pattern recognition. The experimental data that come from KPCA feature extraction are used as input to the SVM classification algorithm, some of which are used to train the model and the others to test the model. Experimental results showed that KPCA feature extraction and the SVM classification algorithm effectively distinguished four statuses of through-wall human being detection and achieved the desired results.

Keywords: Kernel principal component analysis, Support vector machines, Feature extraction, Classification

1 Introduction

Ultra-wideband (UWB) radar can emit pulses of very short duration to penetrate walls, bulkheads, and other obstacles. The UWB radar pulse has the advantages of high accuracy, strong penetrating ability, high resolving power, good anti-stealth ability, and low power consumption. Hence, it has great potential in radar detection, imaging, precise positioning and target recognition. In the fight against terrorism, disaster relief, public security during rioting and urban street fighting and other fields, it has a significant application. It operates mainly through the acquisition and analysis of the echo signal to carry information to detect objects hidden in penetrating media, such as clay walls, dielectric plates, concrete, etc. The UWB signal can realize ranging accuracy in the order of centimeters, can identify and distinguish different target types, can overcome the absorbing effect of narrowband radar, can suppress clutter echo effects in complex backgrounds, and can provide non-intrusive detection. Therefore, the technology has broad and important application prospects as the best choice for all-round human being detection. The focus of this paper is on using P410 UWB radar for multiple-status, through-wall, human being detection [1].

UWB radar has good penetration ability and is not susceptible to weather, temperature, humidity or other physical factors. Therefore, in recent years, its application regarding through-wall detection has increasingly attracted the interest of scholars. In reference [2], this paper mainly introduces the normalized difference square matrix method, and the reference moving average method with the Discrete Fourier Transform (DFT) as the detection techniques for the periodic respiratory motion of the human target. The results of experiments to detect objects behind gypsum walls and concrete walls have been separately proved for human target detection. In reference [3], this paper mainly introduces the through-wall human being detection model based on UWB radar, deducing the wavelet packet transform of the target criterion, and designing the procedure for through-wall human being detection with statistical process control. The experimental data are collected for both stationary

* Correspondence: weiwang@mail.tjnu.edu.cn
Tianjin Key Laboratory of Wireless Mobile Communications and Power Transmission, Tianjin Normal University, Tianjin 300387, China

and moving status of human beings behind a brick wall. In reference [4], this paper mainly introduces an efficient method of TOA (time of arrival) estimation using UWB through-wall radar to detect and track moving targets behind a wall based on the TWRI (through-wall radar imaging) algorithm. The processed result of the experimental data obtained from the UWB through-wall radar shows its detection and tracking effects on moving targets. In reference [5], this paper mainly introduces residual subspace analysis, addressing the anomaly detection problem in large-scale data-mining applications, and suggests a framework using compressed sensing (CS) theory. The experimental results were based on the benchmark PETS 2007 and 83 GB of real footage from three public train stations. The results of this paper proposed a method that is scalable and, importantly, its performance is comparable to conventional methods for anomaly detection when the complete data are available. Reference [6] processes fuzzy pattern recognition and the genetic algorithm to identify the multiple-status human being behind a brick wall, and recognizes the principle of maximum degree of membership function to establish target prediction function by fuzzy pattern recognition. The results show fuzzy pattern recognition performance for the multi-status human being behind the wall. Reference [7], mainly describes a complete UWB signaling tomography system for high-contrast or large-object recognition, applied to breast cancer screening. This article focuses on the implementation of a two-degree, free-imaging set to deal with asymmetric objects with lifelike breast "ghosts." In reference [8], a radar sensor network (RSN) was used to arrive at a unified analytical framework that takes all factors into account and allows the uncertainty derivation of probability detection and location. The experimental results allow the system designer to have a clear understanding of the impact of each system parameter and the trade-off between performance and complexity. In reference [9], Yarovoy has examined the detection and positioning of human beings in complex environments with UWB radar and it has been shown that the range of a person varies by up to 0.6 cm because of their breathing movements, and novel motion or breathing detection has been presented on the basis of the measurements from radar return spatial variations. The experimental results of human being radar return have been analyzed in the frequency band of 1 to 2 GHz. In reference [10], this paper describes a complex process based on the M-sequence UWB radar estimation method for behind-wall moving-target tracking, and introduces the phase-task, solution signal processing method. Experimental results were based on the scenario of tracking a single moving object through a concrete wall, and the UWB radar signal is used to deal with the performance of the demonstration trajectory estimation method.

In recent years, multiple-status, through-wall human detection being has attracted wide attention and been applied to many fields such as national economy, space technology, and national defense. The kernel principal component analysis (KPCA) method can deal with the nonlinear relationship between variables as a multivariable statistical process monitoring effective algorithm using this method to establish a multiple-status fault detection model. The experimental results are based on three case studies: (1) a two-dimensional toy example, (2) a realistic simulation usually used as a benchmark example, known as the Tennessee Eastman Process, and (3) real data from a methanol plant [11]. Reference [12] shows that reducing the impact of channels and handsets on system performance is one of the major issues that improve the accuracy of the most state-of-the-art speaker recognition algorithms. By adapting the model between the different channel conditions to explore the SVM framework-specific techniques to obtain completely nonlinear channel compensation, the observed changes in the particular type of marker channel are less sensitive to the system. In [13], Solomonoff proposed the SVM method to adjust the model to improve the accuracy of the speaker recognition algorithm and reduce the impact of the channel and the mobile phone on the system performance, and discussed SVM framework-specific technology to obtain complete nonlinear channel compensation. The experimental results are based on systems that are less sensitive to the particular type of marker channel changes observed during training. In [14], this paper mainly discusses the road detection algorithm of the front-view single camera using the road probability distribution model (RPDM) and the online learning method. The combination of the dynamic RPDM and the fuzzy support vector machine (FSVM) makes the algorithm self-supervising and optimizes learning from the previous results.

In this paper, we proposed an algorithm for through-wall human being detection based on KPCA feature extraction and the SVM classification algorithm. We also used linear PCA feature extraction and the SVM classification algorithm compared with the method of this paper. The remainder of the paper is organized as follows: in Section 1, we introduce the merit of UWB radar in the field of target recognition and elaborated the main research contents of this paper. In Section 2, we introduce the theory of KPCA and SVMs. The algorithm of construction and implementation is in Section 3. Experimental results for multiple-status human being detection are shown in Section 4. In Section 5 are the Discussion and our Conclusions.

2 Theory

2.1 KPCA nonlinear feature extraction theory [15, 16]

Principal component analysis (PCA) is a linear dimensionality reduction and feature extraction method for high-dimensional data. It maps the input data from the original high-dimensional space to the characteristic subspace, extracts the main feature vector of the input data, and achieves the purpose of analyzing the original data with the main component. In general, PCA can only be effectively performed on a set of observations that are best described by second-order correlations and vary linearly, or the observations are generated by a Gaussian distribution. It is well known, however, that the variations of the actual data are nonlinear and highly non-Gaussian, and that the majority of the data cannot be described by second-order correlations. Therefore, when PCA is employed, the performance is very poor. In this paper, KPCA is a nonlinear PCA method based on kernel functions which intrinsically constructs a nonlinear mapping from the input space to the feature space F by a nonlinear transformation Φ, and performs linear PCA in F. Between two input examples, x and y, in the original space, one can avoid performing the nonlinear mappings and computing both the dot products in the feature space by using a kernel function of form:

$$k(x,y) = \Phi(x)\cdot\Phi(y).$$

The conceptual framework of the KPCA method is shown schematically in Fig. 1.

There are many forms of kernel functions. According to Mercer's theorem of functional analysis, if the kernel function is a continuous kernel of a positive integral operator, there exists a map, Φ, into a dot product space, F, such that the formula holds. As long as the requirement on the kernel function satisfies Mercer's theorem and selects the fitting kernel function, it can achieve a good

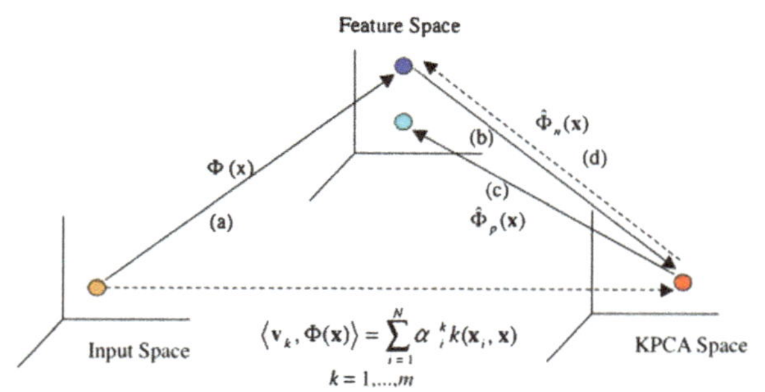

Fig. 1 The conceptual framework of the kernel principal component analysis (KPCA) method

dimensionality reduction effect. Below are some typical kernel functions, such as:

Polynomial kernel $k(x,y) = \langle x,y\rangle^d$,

Sigmoid kernel $k(x,y) = \tanh(\beta_0\langle x,y\rangle + \beta_1)$,

Radial basis kernel $k(x,y) = \exp\left(\frac{\|x-y\|^2}{c}\right)$,

where d, β_0, β_1, and c are specified a priori by the user. The polynominal kernel and radial basis kernel always satisfy Mercer's theorem, whereas the sigmoid kernel satisfies it only for certain values of β_0 and β_1. Due to the good performance of the radial basis function, in practical application the radial basis function is generally chosen as the kernel function of KPCA; so, in this paper, we use the radial basis kernel as the KPCA kernel function.

Giving a set of input data (with zero mean) X ($x_1, \cdots, x_N \in R^m$ that N is the number of samples, m is the dimension of the measurement variables) through PCA and KPCA algorithm calculating their covariance matrix, the covariance of PCA can be expressed by formula (1) and the covariance of KPCA can be expressed by formula (2) in a linear feature space F instead of the nonlinear input space:

$$C = \frac{1}{N}x_i x_i^T = \frac{1}{N}XX^T \text{and} \tag{1}$$

$$C^F = \frac{1}{N}\sum_{j=1}^{N}\Phi_j(x)\Phi_j(x)^T, \tag{2}$$

where it is assumed that $\sum_{k=1}^{N}\Phi(x_k) = 0$, and the $\Phi(\cdot)$ is a nonlinear mapping function that projects the input vectors from the input space to F. Note that the dimensionality of the feature space can be arbitrarily large or possibly infinite. To calculate the covariance matrix, one has to solve the eigenvalue problem in the feature space:

$$\lambda v = C^F v, \tag{3}$$

where eigenvalues $\lambda \geq 0$ and eigenvector $v \in F$, the eigenvector v for any $\lambda \neq 0$ can be linearly represented by $\Phi(x_i)$:

$$v = \sum_{i=1}^{N} a(i)\Phi(x_i) \tag{4}$$

Eq. (2) can be converted into the kernel eigenvalue problem:

$$N\lambda a = Ka, \tag{5}$$

where an $N * N$ matrix K is a kernel matrix, $K = k_{ij} = (\Phi(x_i)\cdot\Phi(x_j)) = k(x_i, x_j)$, and α is the feature vector of the kernel matrix. When reconstruct input data from feature space, we use the fellow Eq. (5):

$$y_k = \langle v_k, \Phi(x) \rangle = \sum_{i=1}^{N} a_i^k \langle \Phi(x_i), \Phi(x) \rangle \quad (6)$$

Before applying the KPCA algorithm, we assume that the data have been standardized, but in the actual application, without knowing the specific form of Φ, the data are not standardized. This can be done by substituting the kernel matrix K with:

$$\widehat{K} = K - 1_N K - K 1_N + 1_N K 1_N, \quad (7)$$

where $1_N = \left(\frac{1}{N}\right)_{N\times N}$

2.2 SVM classification algorithm theory [17–19]

Support vector machines (SVMs) are a supervised, machine learning method for classification, pattern recognition, regression analysis, and other learning tasks based on statistical learning theory (SLT). In this method, support vectors are used to represent decision boundaries; then, one maps the linearly indivisible data of the low-dimensional input space into a high-dimensional feature space to make it linearly separable; this it is based on the structural risk minimization theory and then constructs an optimal separating hyperplane in this space. The basic idea for SVMs is show in Fig. 2.

Given a training set of N data points (y_j, x_j) , j = 1 , 2 , 3 , ... , N where $x_j \in R^n$ is the input pattern and $y_j \in \{1, -1\}$ is the output pattern. The SVM method approach aims at constructing a classifier of the form:

$$y(x) = \text{sign}\left[\sum_{j=1}^{N} a_j y_j \Psi(x, x_j) + b\right], \quad (8)$$

where sign (·) is a symbolic function used to classify, α_j represents positive real constants and b is a real constant. For $\Psi(\cdot, \cdot)$ one typically has the following choices: $\Psi(x, x_k) = x_k^T x$ (linear SVM) and $\Psi(x, x_k)$ is a kernel function similar to KPCA. Here, we also select the radial basis kernel as the SVM kernel function and use the nonlinear SVM in this paper.

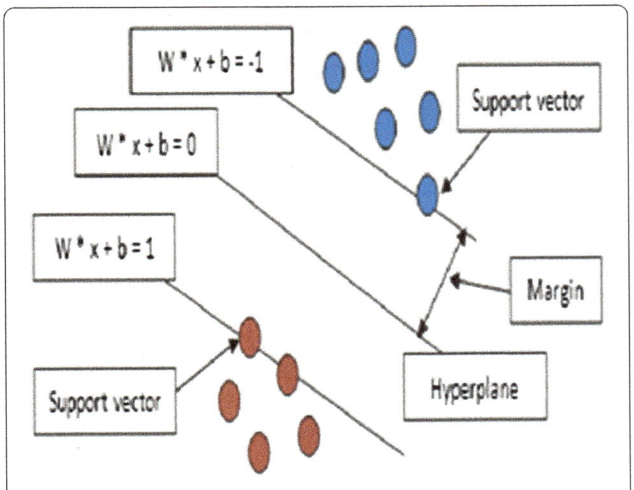

Fig. 2 Basic idea for support vector machines (SVMs)

The classifier is constructed as follows. One assumes that:

$$\omega^T \phi(x_i) + b \geq 1, \text{if } y_i = +1, \quad (9)$$

$$\omega^T \phi(x_i) + b \leq -1, \text{if } y_i = -1, \quad (10)$$

which is equivalent to:

$$y_j\left[\omega^T \phi(x_j) + b\right] \geq 1, j = 1, 2, 3, ..., N, \quad (11)$$

where $\phi(\cdot)$ is a nonlinear function which maps the input space into a higher-dimensional feature space. But, in order to have the possibility to violate formula (10), the function is not clearly constructed. We introduce the variables ξ_j to solve the following primal optimization problem and to obtain the separating hyperplane in the higher-dimensional space:

$$\min_{\omega,\xi_j} \frac{1}{2}\omega^T\omega + C\sum_{j=i}^{N}\xi_j, \quad (12)$$

subject to

$$y_j\left(\omega^T \phi(x_i) + b\right) \geq 1 - \xi_j \quad (13)$$

$$\xi \geq 0, j = 1, 2, ..., N$$

3 Algorithm and implementation

In order to better understand the feature extraction algorithm of this paper, we give the training steps of the network as follows:

Algorithm 1 Kernel Principal Component Analysis algorithm

Input: training set data, testing set data

Output: feature exaction data

1: standardize the input sample data to generate samples matrix X

2: select the radial basis function, select the appropriate c value and calculate the matrix K ,according to equation (6) to obtain the matrix K

3: find the eigenvector and eigenvalues of K according to equation (4), and select the eigenvectors corresponding to the largest eigenvalues.

4: calculate the projection of selected sample data according to equation (5), as input data for supporting vector machine recognition, classification for multiple statuses detection of through wall human being.

In this paper, we used the ROC (receiver operating characteristic) curve to estimate the performance of the classification algorithm more comprehensively. The ROC curve has the advantages of high credibility, accurate description of the target, and of not being affected by the data environment. It is based on a series of two different classification methods (cutoff or decision threshold). In the ROC curve, the FPR (false positive rate) is used as the abscissa and the

TPR (true positive rate) is used as the ordinate. Compared with the traditional evaluation methods, the performance of the ROC curve analysis method is better and there are no excessive restrictions. Therefore, the ROC curve analysis method is applicable to a wider range. The evaluation index of the ROC is shown in the following [20] (Table 1):

TP is the positive sample which is predicted to be true by the model, FN is the positive sample which is predicted to be false by the model, FP is the false sample which is predicted to be true by the model, and TN is the false sample which is predicted to be false by the model. The calculation formulas of FPR and TPR are as follows:

$$FPR = \frac{FP}{TN + FP}, \; TPR = \frac{TP}{TP + FN}. \tag{14}$$

4 Experimental process and results

In this experimental, the UWB radar module is the P410 from the Time Domain Company. The P410 equipment development board is composed of a FPGA, a DSP, a network port, a UWB transceiver, a fan and many other components and is connected to the computer by net mouth using the UDP protocol to set the signal pulse length, frequency, and other parameters. The center frequency of the P410 UWB radar is 4.3 GHz. The wall material is a 23.5-cm thick brick. The P410 UWB radar is placed at a distance of 20 cm from the brick wall and the distance between the UWB radar and the ground is half of the brick wall. In this experimental environment, there are four statuses for through-wall human being detection. The first is nobody behind the brick wall, the second is two persons walking 1 m away from the brick wall, the third is three persons breathing normally 2 m away from the brick wall, and the last is two persons breathing rapidly 1 m away from the brick wall. The data in the four statuses consist of 500 groups of pulses and pulse sampling points set at 1000.

Based on the computer of P42.5GHZ,512M RAM and matlab R2009a simulation experiment, we selected the 300 groups of data from the four kinds of states through the feature extraction as the training set, the rest for the testing set. In order to compare the effects of feature extraction, the feature extraction method of this paper is compared with the PCA feature extraction method regarding the classification result. Additionally, the experiments were performed on the same training and testing data using the two feature extraction algorithms combined with SVMs. Through the theoretical knowledge in Section 2, we know that the KPCA method has the same mathematical and statistical properties as the linear PCA in the feature space, such as lack of correlation of the principal components; the principal component can represent the maximum variance of the sample data; the sample is reconstructed with the least squares error; in addition, it can extract more sample information than the linear PCA. Under the premise of the same classification performance, KPCA requires fewer principal elements than PCA, and it does not need to solve the nonlinear optimization problem, only the eigenvalue decomposition of the matrix, compared with other nonlinear feature extraction methods.

In the simulation results, the eigenvalues by descending order of PCA and KPCA are shown in the following: Fig. 3 shows the matrix eigenvalues of PCA and Fig. 4 the matrix eigenvalues of KPCA. In this picture, the horizontal axis represents the number of sample eigenvalues, and the vertical axis represents the eigenvalues. Compared with the two graphs, we can see that the KPCA tends to zero at the 300th eigenvalue, but the PCA starts to tend to zero at the 500th eigenvalue. If the variance contribution rate that the sum of the selected eigenvalues divided by the sum of all the eigenvalues reaches 90%, the PCA needs the first 100 eigenvalue representations, that is the 1000-dimensional column vector can be compressed into a vector of only 100-dimension. However, the KPCA only needs the first 70 eigenvalues to make the contribution rate of 90%. In other words, the 1000-dimensional column vector can be compressed into a vector of only 70-dimension. Under the same classification performance, the KPCA requires fewer principal elements and lower mean square error, and the effect of reducing the dimension is more obvious.

In this paper, the data by feature extraction of the training set is input into the SVMs for the training model. After the model training is finished, the test set data are input to the trained model for testing, and the experimental results are showed in Table 2. In Table 2, nine indicators are listed, and the feature extraction performance of the two algorithms is compared by each indicator. The accuracy of the KPCA algorithm is significantly higher than that of the PCA algorithm. The kappa statistics are an important indicator of the degree

Table 1 The evaluation index of receiver operating characteristic (ROC)

		Predicted		
		1	0	Sum
Actual	1	True positive (TP)	False negative (FN)	Actual positive (TP + FN)
	0	False positive (FP)	True negative (TN)	Actual negative (FP + TN)
Sum		Predicted positive (TP + FP)	Predicted negative (FN + TN)	TP + FP + FN + TN

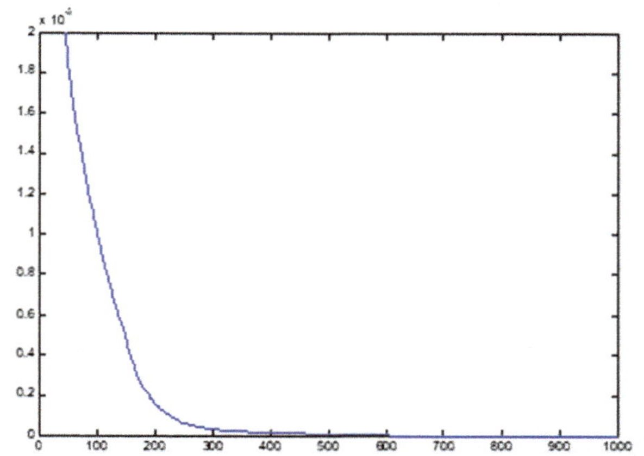

Fig. 3 Matrix eigenvalues of principal component analysis (PCA)

of consistency in the evaluation. If the kappa value is greater than 0.75, the consistency is good, indicating consistency between 0.75 and 0.4; less than 0.4 indicates poor consistency. From the table, we can see that the kappa value of the KPCA algorithm is greater than 0.75, the consistency is better, and the kappa value of PCA algorithm is less than 0.75 consistency. The mean absolute error and root mean squared error are used to measure the difference between the predicted value of the classifier and the actual result and the smaller the value the better the classification performance; so, it is apparent that the KPCA is superior to PCA. When the absolute error cannot reflect the error of the true size, the relative absolute error and root relative squared error can reflect the size of the error by the proportion of the error to the true value. Additionally, the smaller the relative absolute error is, the better the classification performance of the feature extraction algorithm is. To sum up, the KPCA and SVM combination algorithms are better than those of PCA and SVMs in comparison with each index of the two algorithms.

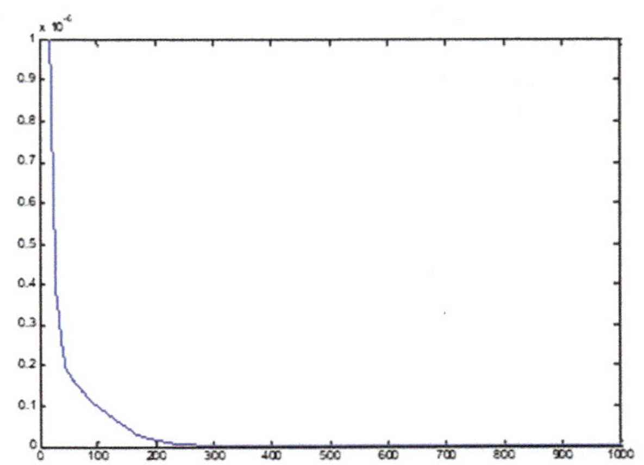

Fig. 4 Matrix eigenvalues of kernel principal component analysis (KPCA)

Table 2 Experimental results compared with PCA + SVM, KPCA + SVM

Classification algorithm parameters	PCA + SVM		KPCA + SVM	
Correctly classified instances	633	79.125%	654	81.75%
Incorrectly classified instances	167	20.875	146	18.25%
Kappa statistic	0.7217		0.7567	
Mean absolute error	0.2827		0.2794	
Root mean squared error	0.3604		0.3558	
Relative absolute error	75.3889%		74.5%	
Root relative squared error	83.2388%		82.1584%	
Total number of instances	800		800	

KPCA kernel principal component analysis, *PCA* principal component analysis, *SVM* support vector machine

This paper also used the ROC curve to analyze the experimental results. The ROC curve is closer to the upper left corner, it represents the algorithm having higher accuracy. As shown in Fig. 5, the ROC curve of through-wall human being detection by the two algorithms, using

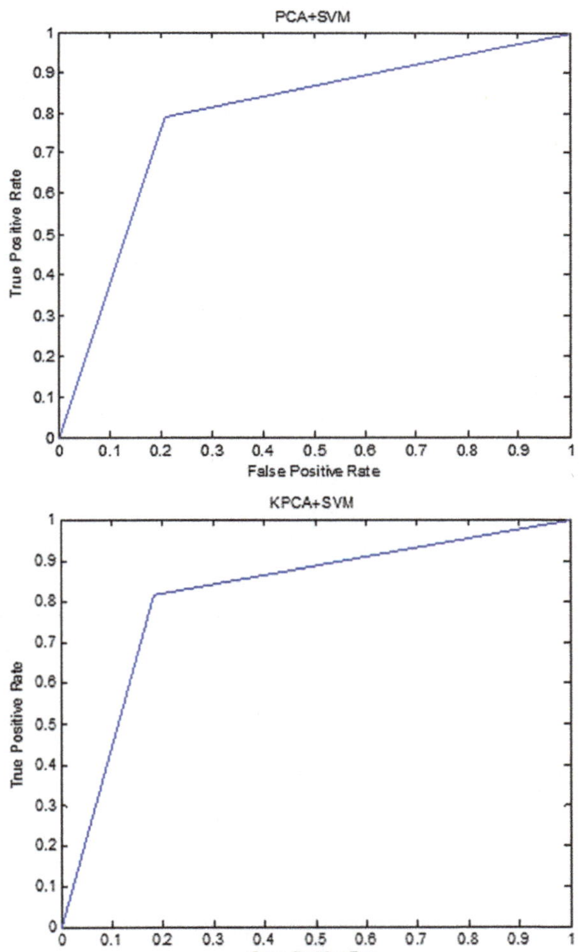

Fig. 5 Receiver operating characteristic (ROC) of through-wall human being detection

the KPCA algorithm, is significantly better than that of the PCA algorithm. The ROC curve of through-wall human being detection is shown as follows:

5 Conclusion

We have presented a framework for through-wall human being detection under four statuses with P410 UWB radar using the SVM classification algorithm. In the aforementioned classification algorithm, we needed to extract the feature of the data to reduce the dimension and the time complexity. Compared to other feature extraction methods, KPCA has the following main advantages: (1) nonlinear data is effectively dealt with, (2) no nonlinear optimization is involved, (3) the calculations in KPCA are as simple as in standard PCA, and (4) the number of PCs need not be specified prior to modeling. So, in this paper, the combination of KPCA and SVMs with this method combines the advantages of KPCA component analysis and SVMs in the application of pattern recognition, so the practical application can be better than the performance of the two methods alone. Experimental results show that the algorithm can effectively distinguish between there being no person behind the brick wall, two persons walking 1 m away from the brick wall, three persons breathing normally 2 m away from the brick wall, and two persons breathing rapidly 1 m away from the brick wall which has important theoretical significance and practical application value. Compared with the traditional PCA method, the simulation results show that the proposed method has better stability and reliability, can improve the recognition rate effectively, and can effectively optimize the extraction of the radar target's principal feature, accelerate the identification of its speed, improve the recognition performance of the target, and it also has good promoting ability. Further research work is being focused on a small sample of multiple-status, human being detection and aims to improve recognition rate with different kernel functions.

Acknowledgements

This paper is supported by Natural Youth Science Foundation of China (61501326,61401310), and Natural Science Foundation of China (61271411). It also supported by Tianjin Research Program of Application Foundation and Advanced Technology (15JCZDJC31500), and Tianjin Science Foundation (16JCYBJC16500).

Authors' contributions

WW after the wall of human detection research, to participate in the idea of the article. MZ conducts human detection test data to measure and draft manuscripts. DW participates in data simulation. YJ participated in the design of the study and conducted a statistical analysis. All authors read and approved the final manuscript.

Competing interests

The authors declare that they have no competing interests.

References

1. B. Zhang, W. Wang, Through-wall detection of human being with compressed UWB radar data. EURASIP J. Wireless Comm. & Networking. **2013**(1), 1–7 (**2013**)
2. A. Kumar, Q. Liang, Z. Li, B. Zhang, Experimental study of through-wall human being detection using ultra-wideband (UWB) radar. Globecom Workshops, 1455–1459 (2012)
3. W. Wang, B. Zhang, J. Mu, Through wall detection of human being based on SPC and wavelet packet transform by UWB radar, proceeding of IEEE International Conference on Communications, pp. 869–878 (2013)
4. S. WU, G. Fang, Detection and tracking of moving target wall using UWB through-wall radar, proceeding of IEEE on intelligent system design and engineering application (ISDEA), Changsha,605–608 (2010)
5. P. Duc-Son, V. Svetha, M. Lazarescu, S. Budhaditya. Anomaly detection in large-scale data stream networks. J. Data Mining and Knowledge Discovery. doi:10.1007/s10618-012-0297-3,2014:145-189
6. W. Wang, D. Wang, Through-wall multi-status target identification in smart and autonomous systems with UWB radar. IEEE nternet of Things Journal.. 10.1109/JIOT.2017.2695644,2017:86-89
7. M. Guardiola, S. Capdevila, S. Blanch, J. Romeu, UWB high-contrast robust tomographic imaging for medical applications, Electromag. in Adv. Appl. :560–563 (2009)
8. S. Bartoletti, A. Conti, A. Giorgetti, Analysis of UWB radar sensor networks. IEEE International Conference on Communications, 1–6 (2010)
9. A.G. Yarovoy, L.P. Ligthart, J. Matuzas, B. Levitas, UWB radar for human being detection. IEEE Aerosp. Electron. Syst. Mag. **21**(1'), 22–26 (2006)
10. D. Kocur, J. Rovnakova, M. Svecova, Through wall tracking of moving targets by M-sequence UWB radar. Stud. Comput. Intell. **243**, 349–364 (2009)
11. M. Maestri, M. Cassanello, G. Horowitz, *Kernel PCA Performance in Processes with Multiple Operation Modes, Special Issue-Wcce8*, pp. 56–60 (2009)
12. K.I. Kim, K. Jung, H.J. Kim, *Face Recognition Using Kernel Principal Component Analysis. IEEE Signal Processing Letters* **9**(2), 251–263 (2002)
13. A. Solomonoff, C. Quillen, W. Campbell, Channel compensation for SVM speaker recognition, MIT Lincoln Laboratory Lexington, MA, USA, 248:250–560 (2002)
14. S. Zhou, K. Iagnemma, *Self-Supervised Learning Method for Unstructured Road Detection Using Fuzzy Support Vector Machines, Supported by the National Natural Science Foundation of China and the Robotic Mobility Group in Massachusetts Institute of Technology, at Boston, MA, USA*, pp. 134–152 (2010)
15. L. Jong-Min, Y. ChangKyoo, S.W. Choi, P.A. Vanrolleghem, I.-B. Lee, Nonlinear process monitoring using kernel principal component analysis. Chem. Eng. Sci. **1**, 223–234 (2004)
16. B. Scholkopf, S. Alexander, M. Klaus-Robert, Nonlinear component analysis as a kernel eigenvalue problem. Neural Comput. 10 (5) **1**, 1299–1399 (1998)
17. W. Wang, Z. Chen, Throat polyp detection based on compressed big data of voice with support vector machine algorithm. EURASIP J. on Adv. in Sig. Proc. **2014**, 1 (**2014**)
18. J.A.K. Suykens, J. Vandewalle, Least squares support vector machine classifiers. Neural Proc. Lett. **9**, 293–300 (1999)
19. C. Chang, C. Lin, *LIBSVM: A Library for Support Vector Machines. National Taiwan University*.. doi:10.1145/1961189.1961199,2011:127-130
20. M. O'Connell, J.L. Szalma, ROC-Estimator Software and ROC Analysis, Proceedings of the Human Factors and Ergonomics Society 57th Annual Meeting, pp. 123–134 (2013)

Battery aging-aware energy management of green small cells powered by the smart grid

Mouhcine Mendil[1,2*], Antonio De Domenico[1], Vincent Heiries[1], Raphael Caire[2] and Nouredine Hadjsaid[2]

Abstract

Mobile operators are deploying energy-harvesting heterogeneous networks due to their foreseen advantages such as self-sustainable capability and reduced operating expenditure, which cannot be offered by conventional grid powered communications. However, the used energy storage is subject to irreversible aging mechanisms, requiring intelligent management that considers both the energy cost and battery life cycle. In this paper, we propose a cognitive energy management strategy for small cell base stations powered by local renewable energy, a battery, and the smart grid to simultaneously minimize electricity expenditures of the mobile operators and enhance the life span of the storage device. Non-linear battery models and aging processes are considered to formulate the energy cost optimization problem. Simulation results in different configurations show that a degradation-aware policy significantly improves the battery lifetime, while achieving considerable cost savings.

Keywords: Cognitive energy management, Green communication, Energy harvesting, Small cell, Battery, Smart grid

1 Introduction

In the last decades, mobile user density and data traffic volume have exponentially increased all over the world. To respond to this trend, the mobile network operators (MNOs) have deployed small cell base stations (SBSs) to enhance their network service capabilities [1]. According to [2], this solution results in a significant energy demand augmentation essentially generated by the base station power consumption that represents 75 to 80% of the entire mobile network. Based on this, the deployment of heterogeneous cellular networks (HetNets) requires an efficient energy management to ensure their economic and environmental sustainability.

A multitude of concepts have recently been proposed to improve the energy efficiency in wireless communications, addressing network planning, protocols, and equipments [3]. In addition, renewable energy (RE) usage in cellular networks has drawn attention for its numerous benefits such as decreasing carbon emissions, enabling long-term cost savings thanks to reduced operating expenditure (opex) [4], and feeding off-grid base stations where the connection to the power grid is expensive or impossible [5]. In this context, cognitive techniques have been explored to improve energy-harvesting communications [6]. More specifically, the cognitive radio offers the ability to sense the conditions of the wireless communication networks and interact with the environment to adjust some parameters such as the transmission power, frequency band, and modulation mode [7].

The current energy-harvesting technologies require local energy storage to absorb the production fluctuation and ensure a continuous equilibrium between energy offer and demand. However, the typical used energy storage, i.e., electric battery, generates expensive investment cost and is subject to irreversible degradations. Such phenomenon, called battery aging, has been intensively studied and been classified into two categories: cycle aging, which is due to the energy exchanges with the battery, and calendar aging that appears when the battery is on rest [8]. In this context, Barre et al. [9] have presented a comprehensive review of techniques, models, and algorithms used for Li-ion battery aging estimation.

The presence of energy storage requires intelligent energy management policies to optimize the energy cost of grid-connected base stations (BSs) with energy harvesting. Some studies in the literature have addressed this problem by using either offline or online optimization

*Correspondence: mouhcine.mendil@gmail.com
[1]CEA, LETI, MINATEC, F-38054, Grenoble, France
[2]University Grenoble Alpes, G2Elab, F-38054, Grenoble, France

approaches. The first category assumes perfect knowledge of the stochastic energy variables [10–12]. In [11] for example, the power consumption of a HetNet powered by RE and equipped with an infinite-capacity battery has been optimized by supervising the BS transmit power and the battery usage. In particular, non-linear model predictive control theory has been used to manage the stored energy considering the average electricity price and power production. Zhang et al. [12] have proposed an energy-aware traffic offloading for a HetNet with multiple SBSs. They have used queuing theory to model the energy production and consumption and deduced an efficient power control according to the statistical information of energy arrival and traffic load. Online policies have also been proposed by assuming a casual knowledge of the environment. In this context, stochastic optimization has been implemented by assuming that the statistics of the energy processes are known and that past observations can correct the energy forecasts [13–15]. The authors in [13] have investigated an online stochastic approach based on multi-period recourse. Mao et al. [15] have considered a hybrid energy supply for the HetNet and formulated the energy cost minimization problem as a discrete Markov decision process. The monotony properties of the optimal policy have been inferred to simplify and solve the optimization problem using the backward induction algorithm. In [16], we have proposed an energy controller that uses reinforcement learning techniques to elaborate an optimal energy flow policy without prior knowledge of the environment stochastic behavior.

These works have focused on maximizing the energy saving, and none has integrated both calendar and cycle battery agings in the energy management framework. As a matter of fact, the maximal use of the battery flexibility enables large cost savings but can lead to rapid battery life loss. However, the battery is an expensive investment of the system, and enhancing its life span is vital for an efficient return on investment. Consequently, there is a trade-off between pure cost-efficient and battery aging-aware strategies that has not been evaluated so far. This motivated us to investigate the design of a cognitive energy controller that optimizes the operating energy cost while using the battery in the most effective way to avoid accelerated cycle and calendar agings. The current study extends our work [17] by introducing the battery aging models and formulating the energy cost problem such that the battery degradation factors are reduced while the optimal cost saving is still achieved. Additionally, the battery life evolution is studied to show the impact of the proposed energy strategies on the aging process.

1.1 Contributions and organization

The contributions of this paper are summarized as follows:

- Different from existing works, we propose a cognitive energy flow management framework for grid-connected energy-harvesting SBSs to jointly optimize the energy opex and the battery life cycle. The cognitive decision architecture is centered around the battery and uses realistic models to capture the non-linear battery behavior and aging mechanisms. Without loss of generality, the present framework is implemented in the offline case, in which we consider non-casual information about the environment variables. System simulations show that the proposed controller achieves considerable cost reduction compared to simpler strategies.
- The trade-off between pure energy cost optimization and battery aging-aware policies is explored. The energy cost and battery aging are evaluated in different configurations. Simulation results show that the proposed energy management allows considerable battery life extension such that the battery lasts five times longer compared to a pure energy cost optimization strategy. In exchange, the opex is slightly increased but this additional expense remains negligible with respect to the current battery replacement cost.

The paper is structured as follows. Section 2 introduces our proposed architecture and the associated system models. Section 3 provides a formulation of the joint cost minimization and battery life preservation problem for a green small cell. Section 4 presents the simulation results. The paper is concluded and perspectives are discussed in Section 5.

2 System architecture

The proposed architecture, illustrated in Fig. 1, is composed of:

- The SBS: wireless communication station covering a small area (10 m to 1 km), used to offer high data rate services to mobile users.
- The photo-voltaic (PV) system: equipment harvesting solar energy to produce electricity. It is one of the two energy sources of the system.
- The two-way link to the smart grid (SG): the SG is the second energy source of the system. The two-way energy connection enables to buy or sell electricity to the power grid.
- The battery: storage device that offers flexibility in the energy utilization. It can store the electricity coming from the SG and the PV system to feed the SBS or sell the energy back to the SG.
- The cognitive energy supervision system (ESS): controller that schedules the energy flows to reduce the electricity bill and improve the battery life span.

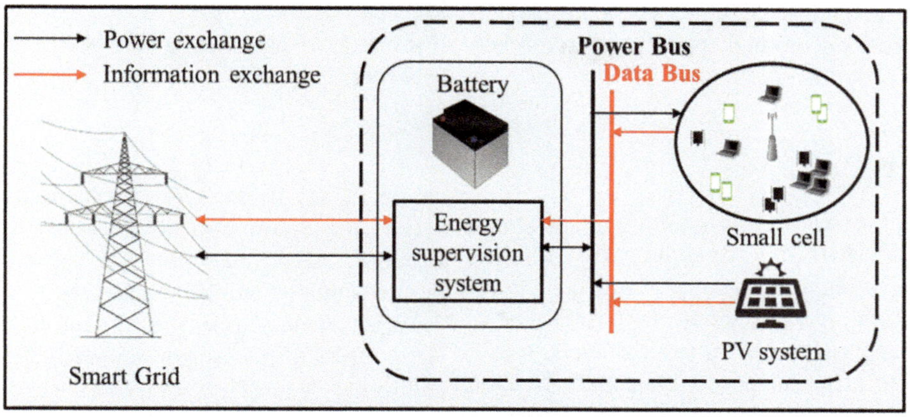

Fig. 1 System architecture of the proposed cognitive energy supervisor for small cells

In the following, we present the chosen model for each component of the system.

2.1 Small cell base station power consumption model

The SBS load ρ is modeled as a non-homogeneous Poisson process, whose intensity varies during time. We suppose that, depending on the load, the SBS can operate in two modes: active when $\rho > 0$ and sleep when $\rho = 0$. The relation between the SBS power consumption P_{BS} [W] and the traffic load is given by the following equation [18]:

$$P_{BS}(t) = \begin{cases} P_0 + \Delta_p \cdot \rho(t) \cdot P_{max}, & \text{if } 0 < \rho(t) \le 1 \\ P_{sleep}, & \text{if } \rho(t) = 0 \end{cases}, \tag{1}$$

where P_0 is the power consumption at the minimum non-zero output power, Δ_p is the slope of the input-output power consumption, P_{max} is the maximum output power, and P_{sleep} is the power consumed in sleep mode.

2.2 Energy storage model

We choose a Lithium-ion battery as the power storage device in our architecture for its several advantages such as high energy density and low self-discharge. The battery can store the electricity provided by the PV system and the SG and discharge the stored energy to feed the SBS and sell it back to the SG. In this paper, the battery state is jointly described by its state of charge (SOC) and its state of health (SOH), which both depend on the (dis)charge power. The SOC is an expression of the battery momentary storage level as a percentage of its nominal capacity C_N [Ah], which corresponds to the battery capacity at the beginning of life, and the SOH is a metric that reflects the general condition of a battery and its ability to deliver the specified performance compared with a new battery.

2.2.1 State of charge model

The SOC variation is generally calculated using current integration. The rate at which the battery is charged or discharged, noted C_{rate} [s^{-1}], corresponds to the charge or discharge current intensity $i(t)$ [A] relative to the battery nominal capacity:

$$C_{rate}(t) = \frac{i(t)}{3600 \cdot C_N}. \tag{2}$$

Periodically, for a given C_{rate}, we use the Ampere-Hour integral model to estimate the SOC variation [19]:

$$z(t + \Delta t) = z(t) + \eta \int_t^{t+\Delta t} C_{rate}(u)du, \tag{3}$$

where $z(t)$ is the SOC at time t, Δt represents the time step between two SOC estimations, and η is the battery Coulombic efficiency, equals to η_{dis} when discharging and η_{chg} when charging.

2.2.2 Battery power model

The battery is an electrochemical system composed of several modules. Each module is composed of cells organized in series and parallel. Without loss of generality, we suppose that the battery contains n_s modules connected in series, where each cell module comprises one cell. In this configuration, the relation between the (dis)charge current $i(t)$ and the voltage of the kth cell V_k [V] is the following [20]:

$$V_k(t) = \text{OCV}(z(t)) + R_k \cdot i(t), \tag{4}$$

where OCV [V] is the open circuit voltage and R_k [Ω] is the kth cell internal resistance, which depends on several parameters such as the SOC, the current intensity, the temperature, and the SOH [21]. The OCV-SOC dependency can be constructed experimentally by disconnecting the battery from any load for a long duration until reaching equilibrium and then measuring its voltage, for different SOC values [22]. The obtained data can be used to elaborate an analytical OCV model. In this paper, we consider an n-order polynomial approximation model such that [23]:

$$\mathrm{OCV}(z(t)) = \sum_{j=0}^{n} a_j \cdot z^j(t), \tag{5}$$

where n is a natural number and $(a_j)_{j=1..n}$ are the polynomial coefficients calculated from the experimental OCV-SOC dependency function.

As a sign convention, we assume that the charge (resp. discharge) current and power have a positive (resp. negative) sign. Consequently, the power P_{batt} $[W]$ of the battery can be calculated using the sum of all cell power:

$$P_{\text{batt}}(t) = \sum_{k=1}^{n_s} i(t) \cdot V_k(t). \tag{6}$$

By combining Eqs. (2) to (6), the battery power can be expressed as a function of two consecutive SOC values:

$$P_{\text{batt}}(z(t), z(t+\Delta t)) = \sum_{k=1}^{n_s} \sum_{j=0}^{n} A_{j,k} z^j(t) z(t+\Delta t) - B_{j,k} z^{j+1}(t) + \alpha^2 \cdot R_k \cdot z^2(t+\Delta t), \tag{7}$$

such that

$$A_{j,k} = \alpha \cdot \left(a_j - 2\alpha \cdot R_k \cdot \delta_{1,j}\right),$$
$$B_{j,k} = \alpha \cdot \left(a_j - \alpha \cdot R_k \cdot \delta_{1,j}\right),$$
$$\alpha = \frac{3600 \cdot C_N}{\eta \cdot \Delta t},$$

and $\delta_{1,j}$ is the Kronecker symbol, equals to 1 when $j = 1$ or 0 otherwise.

2.2.3 State of health models

The SOH degradation is inevitable in a battery life cycle. It is manifested as a loss of available capacity (energy loss) and/or an increase in impedance (power loss). In this paper, we assume that the SOH reflects the capacity evolution:

$$SOH(t) = \frac{C_{\text{ref}}(t)}{C_N}, \tag{8}$$

where $C_{\text{ref}}(t)$ is the reference capacity defined as the battery maximum storage capacity at time t. The degradation of the battery reference capacity can be caused by two aging situations: during use (cycle aging) and on storage (calendar aging) [8]. In the following, these two aging mechanisms are considered independent and thus additive.

Cycle aging: is modeled as reference capacity losses, which depends linearly on the battery SOC variations [24]. At each time step, the new SOH is obtained by Eq. (9):

$$\mathrm{SOH}(t+\Delta t) = \mathrm{SOH}(t) - Z \cdot |\mathrm{SOC}(t) - \mathrm{SOC}(t+\Delta t)|. \tag{9}$$

The experimental results of [25] determine the linear aging coefficient Z for different battery technologies. However, because the cycle aging is amplified outside the recommended operating range $\Delta_{\text{soc}} = [20\%, 90\%]$ of the battery SOC (see Fig. 2 [26]), the aging coefficient Z is defined as:

$$Z = \begin{cases} 85 \times 10^{-6}, & \text{if } 20\% \le \mathrm{SOC}(t) \le 90\% \\ \chi \times 85 \times 10^{-6}, & \text{otherwise} \end{cases} \quad \forall t, \tag{10}$$

where χ is a scalar strictly greater than one.

Calendar aging: The battery temperature is an important parameter to model calendar aging, especially for Li-ion technology. The thermal model is used to estimate the cell temperature in response to the current, voltage,

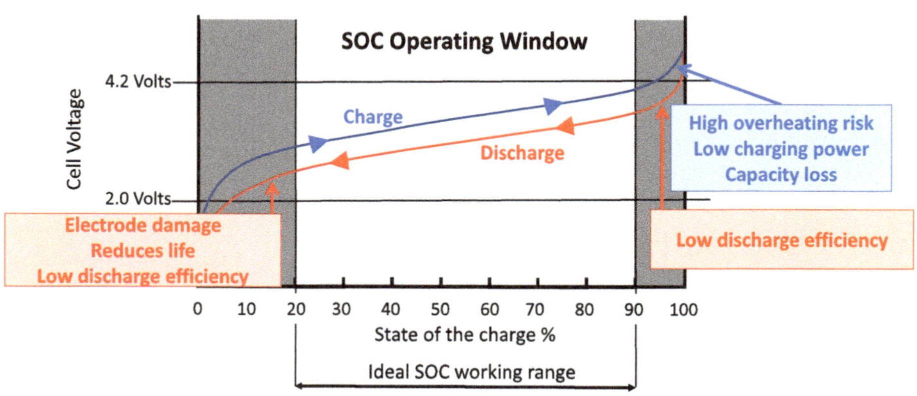

Fig. 2 Recommendations for the operating range of SOC for lithium ion battery [26]

and ambient temperature. According to the energy conservation law, the temperature change for a single-cell battery is given by [27]:

$$m_c \cdot c_h \frac{dT}{dt} = Q_g - Q_r, \tag{11}$$

where m_c [g] is the cell mass, c_h $[J/(g.K)]$ is the specific heat capacity at constant pressure, T [°C] is the temperature within the cell, Q_g $[W]$ is the rate of heat generated by the single cell, and Q_r $[W]$ is the rate of heat removed from the cell by the cooling. The heat generation for a battery cell k can be approximated by Joule heating law:

$$Q_g = R_k \cdot i^2. \tag{12}$$

For simplicity, the heat generation and temperature within the battery are assumed to be uniformly distributed. The single-cell thermal model is thus supposed to represent the overall internal battery temperature.

Ecker et al. [28] have proposed a calendar lifetime prediction model describing the degradation of the battery C_{ref} over time. The model shows exponential dependency with the battery voltage V and temperature T, and square root dependency with the time of rest. The degradation of SOH after a time rest Δt (expressed in weeks) is defined as follows:

$$\frac{\mathrm{SOH}(t+\Delta t)}{\mathrm{SOH}(t)} = 1 + c_a \cdot c_V^{\frac{V-V_0}{\Delta V}} \cdot c_T^{\frac{T-T_0}{\Delta T}} \cdot \sqrt{\Delta t}, \tag{13}$$

where T_0 and V_0 are reference temperature and voltage, ΔT and ΔV are reference temperature and voltage variation, and c_a, c_V, and c_T are fitting parameters based on accelerated calendar aging test data. Given this model, we can conclude that high voltages, and therefore high SOCs (Eq. (4)), contribute to an accelerated battery degradation during rest. Also, the calendar aging grows exponentially with the temperature. Knowing the relation between the current intensity and the heat generated within the battery (Eq. (12)), it is clear that a high current rate increases the internal temperature and therefore leads to faster calendar aging.

2.2.4 *Battery aging constraints*

The aforementioned aging models suggest that Li-ion batteries must be used within a safe operating area restricted by temperature, current, and SOC windows. Not respecting these restrictions leads to a rapid attenuation of the battery performance (capacity loss and decrease of charge and discharge efficiencies). Also, avoiding a long battery rest duration can considerably lower the calendar aging. In this paper, we consider three aging constraints to preserve the battery from rapid degradations:

- We restrict the battery usage on the specific range of the SOC $\Delta_{SOC} = [20\%, 90\%]$. As discussed earlier, operating the battery outside this range accelerates the cycle aging by factor χ. In addition, the calendar aging is amplified when the battery voltage is high, which corresponds to a high SOC.
- We avoid using high (dis)charge currents that cause accelerated cycle aging (due to deep cycling) and calendar aging (due to heat generation). The current restriction can be reformulated as a limitation of the SOC variation in each decision period:

$$\forall t, \Delta\mathrm{SOC}_{min} \le z(t+\Delta t) - z(t) \le \Delta\mathrm{SOC}_{max} \tag{14}$$

where $\Delta\mathrm{SOC}_{max} \ge 0$ (resp. $\Delta\mathrm{SOC}_{min} \le 0$) is the maximum variation of the SOC during charge (resp. discharge).

- We prevent the battery from long resting to lower the calendar aging impact. It is possible to completely avoid rest periods and force the battery into permanent cycling. However, according to some researches [29, 30], providing batteries with a rest period after (dis)charging might be essential for relaxation of gradients generated due to the passage of current and could enable capacity recovery. Such phenomenon is not included in our models, but we can take it into consideration by allowing at most one time step rest between charges and discharges. Equation 15 expresses this constraint by imposing a minimum variation of the SOC (be it positive or negative) over any two consecutive time steps:

$$\forall t, [z(t+2\Delta t) - z(t+\Delta t)]^2 + [(z(t+\Delta t) - z(t)]^2 \ge \epsilon, \tag{15}$$

where ϵ is strictly positive.

2.3 Harvested energy model

Our architecture uses solar panels to capture solar energy and convert it into electricity via the photo-voltaic (PV) effect. The solar radiation I_g $[\mathrm{W}/m^2]$ depends on several factors including geographical location and time of the day.

Let I_t and T_t be the random variables corresponding to the solar radiation and the ambient temperature at hour t, respectively. Given the correlation between the solar radiation and the temperature, we suppose that the combined daily radiation-temperature vector $(I_1, \ldots, I_{24}, T_1, \ldots, T_{24})$ follows a multivariate Gaussian distribution $GP([\boldsymbol{\mu}_{\text{irrad}}, \boldsymbol{\mu}_{\text{temp}}], \boldsymbol{\Sigma}_{\text{irrad-temp}})$, where $\boldsymbol{\mu}_{\text{irrad}}$ (resp. $\boldsymbol{\mu}_{\text{temp}}$) is a vector of size 1×24 composed of the hourly average radiations (resp. temperatures) of the day, and $\boldsymbol{\Sigma}_{\text{irrad-temp}}$ is the covariance matrix 48×48. We compute $\boldsymbol{\mu}_{\text{irrad}}$, $\boldsymbol{\mu}_{\text{temp}}$, and $\boldsymbol{\Sigma}_{\text{irrad-temp}}$ as the means and the covariance of successive realizations related to historical measures of solar radiations and associated temperatures during 5 years [31]. It is noteworthy that by using real historical data, all the phenomena that influence the temperature and the solar radiation are captured in the obtained stochastic process. Then, the hourly photovoltaic output power P_{PV} [W] is given by the following relation [32]:

$$P_{PV}(t) = \eta_{PV} \cdot S \cdot I_g(t), \tag{16}$$

where η_{PV} is the energy conversion efficiency of the solar panel and S [m^2] is the panel surface.

2.4 Energy price model

In the SG, reducing the peak to average consumption ratio is one of the main keys to maintain a smooth balance between the power consumption and production. To this purpose, dynamic pricing can be adopted to adapt consumption profiles to the energy availability.

In this paper, we consider a stochastic dynamic energy price. Let $p(t)$ [\$/kWh] be the random variable corresponding to the buying price (i.e., the cost of energy from the SG) at hour t. The vector $(p(1), \ldots, p(24))$ of daily energy buying price is supposed to follow a multivariate Gaussian distribution $GP(\boldsymbol{\mu}_{\text{price}}, \boldsymbol{\Sigma}_{\text{price}})$, where $\boldsymbol{\mu}_{\text{price}}$ is a vector of size 1×24 composed of the hourly average buying price of the day and $\boldsymbol{\Sigma}_{\text{price}}$ is the covariance matrix 24×24. We compute $\boldsymbol{\mu}_{\text{price}}$ and $\boldsymbol{\Sigma}_{\text{price}}$ as the mean and the covariance of successive realizations related to historical data of electricity pricing for residential customers during 5 years [33]. Moreover, we consider that the price of energy sold to the SG is proportional to the buying price such that $p_{\text{sell}} = \kappa \cdot p$, where κ is the price factor.

3 Cognitive energy supervision system

We aim at minimizing the energy expenditures of the energy-harvesting SBS while reducing the battery aging. This is achieved by scheduling the power flow between the energy sinks and sources over a time horizon discretized into N decision periods. During each time step, we consider that the SBS load, the PV power, and the energy price remain constant. The ESS, in charge of the energy management, is composed of two layers:

1. The high level controller (HLC) schedules the consecutive battery SOCs during the optimization horizon. The obtained energy strategy minimizes the energy cost and reduces the battery aging.
2. The low level controller (LLC) implements the HLC's energy strategy by controlling the power flow between each subsystem in real time such that the energy balance is respected.

Figure 3 represents the cognitive cycle on which the ESS is based. First, the HLC uses the statistical models of the energy variables to solve the long-term (which can be days, hours, or minutes) aging-aware energy cost optimization problem. Specifically, the policy planned by the HLC consists of a succession of the battery SOCs during the optimization period (the SOC variation means that the battery is being charged or discharged, see Section 2.2). Next, according to the SOC strategy, the LLC manages in the short term (minutes to milliseconds) the energy exchange with the SG and between each subsystem of the green small cell. Given the energy conservation law, the LLC senses the realizations of each energy variable and adjusts the power flows to meet the targeted SOC values while respecting the equilibrium between the power supply and demand. In other words, if the energy provided by the PV system and/or the battery is not sufficient to power the SBS and/or the battery, the LLC purchases the missing quantity from the SG. Similarly, the LCC sells the energy surplus when the provided energy exceeds the consumption.

This study focuses on the HLC. We suppose that the LLC is available and operates in real time. The constrained cost optimization problem $\mathbf{P_1}$ at the HLC level aims to find the optimal SOC strategy $\mathbf{z}^* = (z^*(1), \ldots, z^*(N+1))$ and is defined as follows:

$$\mathbf{P}_1 : \quad \underset{(z(1),\ldots,z(N+1))\in[0,1]^{N+1}}{\operatorname{argmin}} \sum_{t=1}^{N} p(t) \cdot [E_b(t) + \kappa E_s(t)],$$

subject to

$$E_b(t) = \max\left(0, P_{BS}(t) + P_{Batt}(z(t), z(t+1)) - P_{PV}(t)\right), \tag{17}$$

$$E_s(t) = \min\left(0, P_{BS}(t) + P_{Batt}(z(t), z(t+1)) - P_{PV}(t)\right), \tag{18}$$

$$z(t) \in \Delta_{SOC}, t = 1, \ldots, N+1, \tag{19}$$

$$\Delta SOC_{\min} \le z(t+1) - z(t) \le \Delta SOC_{\max}, t = 1, \ldots, N, \tag{20}$$

$$[z(t+2) - z(t+1)]^2 + [(z(t+1) - z(t)]^2 \ge \epsilon, t = 1, \ldots, N-1, \tag{21}$$

where $(z(1), \ldots, z(N+1))$ is the decision vector that represents the battery SOCs over the optimization horizon and $E_b \ge 0$ (resp. $E_s \le 0$) is the amount of energy bought from (resp. sold to) the SG. The objective function of $\mathbf{P_1}$ corresponds to the long-term cost due to power transactions with the electrical grid. At all time steps, the balance between the power supply and demand is illustrated by the constraints (Eqs. (17) and (18)). Note that the battery can be either an energy source (when $P_{Batt}(z(t), z(t+1)) \le 0$) or a load (when $P_{Batt}(z(t), z(t+1)) \ge 0$). When the

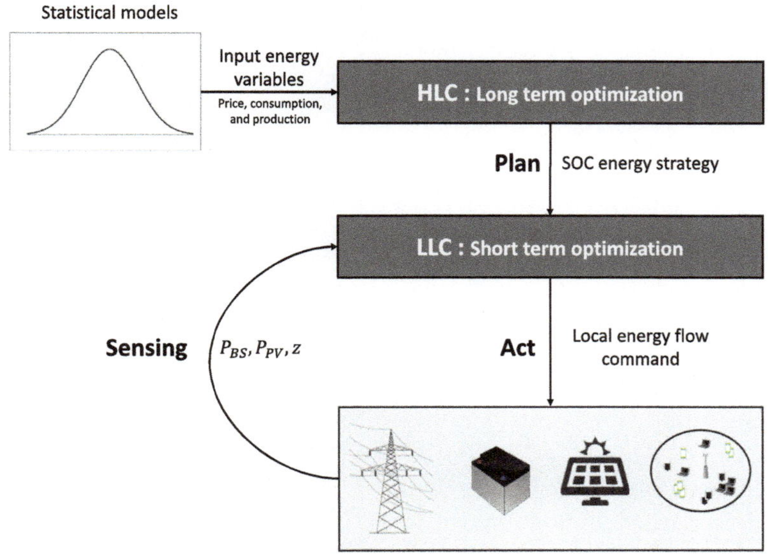

Fig. 3 Proposed cognitive energy supervision infrastructure

energy consumed is greater than the energy provided by the PV/battery system (i.e., $P_{\rm BS}(t) + P_{\rm Batt}(z(t), z(t+1)) - P_{\rm PV} \geq 0$), the controller perceives a cost $p(t) \cdot E_b(t) \geq 0$ corresponding to the energy bought from the SG. In contrast, when the energy available is superior to the energy consumption (i.e., $P_{\rm BS}(t) + P_{\rm Batt}(z(t), z(t+1)) - P_{\rm PV} \leq 0$), the controller receives a negative cost $\kappa p(t) \cdot E_s(t) \leq 0$ (that can be seen as a reward) associated with the energy sold to the SG. The aim of the HLC is to jointly minimize the cumulative positive costs and maximize the cumulative rewards, which corresponds to minimize the negative costs. In addition, during all the decision periods, Eqs. (19), (20), and (21) represent the constraints on the SOC that have to be respected to improve the battery life span (see Section. 2.2.4).

4 Results and discussion

The simulation has been accomplished for finite horizons of 24 h, that is $N = 24$ and $\Delta_t = 1$ h. The stochastic variables are generated each hour of the day according to their respective models. The profiles illustrated in Fig. 4 describe the hourly average SBS load, solar radiation, ambient temperature, and energy buying price used in our simulations to model the average SBS load ρ, the average solar radiation $\mu_{\rm irrad}$, the average ambient temperature $\mu_{\rm temp}$, and the average energy buying price $\mu_{\rm price}$, respectively. Regarding the SBS, the general trend is a progressive increase of the traffic load such that the peak is reached around 21:00–22:00. Notice that the load between 3:00 and 10:00 is absent since it is completely handled by the under-layer macro base station. Concerning the solar

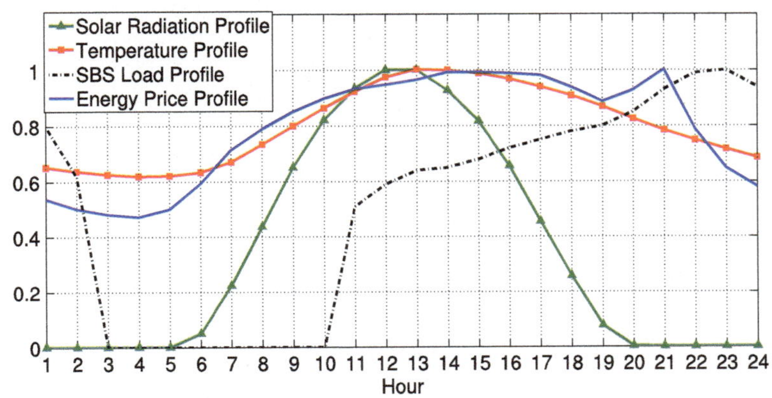

Fig. 4 Normalized profiles of the average solar radiation, average ambient temperature [31], SBS load (based on [36]), and average energy price [33]

Table 1 Simulation parameters

	Parameter	Value	Parameter	Value
SBS	P_0	13.6 W	Δ_p	4
	P_{max}	0.13 W	P_{sleep}	8.6 W
Battery	n_s	5	C_N	7 Ah
	$\forall k\ R_k$	50 mΩ	η_{chg}	96%
	η_{dis}	100%	z_0	30%
	ΔSOC_{min}	−30%	ΔSOC_{max}	30%
	ϵ	10^{-4}		
Solar panel	η_{PV}	14%	S	0.25 m^2
Energy price	κ	100%		

Table 2 Battery thermal and aging simulation parameters [28, 35]

Parameter	Value	Parameter	Value
c_h	1900 J/(kg K)	V_0	3.5 V
c_a	−0.0064	T_0	25 °C
c_V	1.1484	ΔV	0.1 V
c_T	1.5479	ΔT	10 °C
χ	5	m_c	40g

radiation and ambient temperature, the profiles are both bell-shaped and the peak is reached around midday. In particular, the solar radiation is available only at daytime. Finally, the SG electricity price is characterized by two separated peak hours (between 14:00–16:00 and at 21:00) and off-peak intervals during night and evening.

We also consider our experimental data for Li-ion batteries to model the OCV (Eq. (5)) as a second order polynomial such that $OCV(z(t)) = 2.9 + 0.13 \cdot z(t) - 0.008 \cdot z^2(t)$. Other simulation settings for each component of the system are summarized in Tables 1 and 2. Without loss of generality, we assume that the battery parameters (nominal capacity, cell resistance, and charge/discharge efficiency) are independent of the current intensity and temperature. Also, it is noteworthy that the usage of the hour as a time step does not lead to loss of generality. In reality, we can consider another time scale (e.g., minutes) for the power control and the dynamics of the energy variables without any further changes in the energy management framework.

In this paper, we adopt an offline approach to solve the non-linear constrained problem $\mathbf{P_1}$. We suppose that the ESS operates in the *ideal* case where the realizations of the SBS load, solar radiation, temperature, and energy price over all the optimization period are considered known. By doing so, we can assess the ESS maximum performance, which can serve as an upper bound in the realistic case where the stochastic variables cannot be perfectly forecast. In this configuration, the simulation has been performed for 1825 different days (5 years), such that $\mathbf{P_1}$ is solved each 24 h by running the interior-point algorithm implemented in MATLAB with several initializations to reduce the probability of local minima [34]. Furthermore, we evaluate the trade-off between the SBS energy opex savings and the battery life span preservation by solving $\mathbf{P_1}$ considering three constraint sets:

1. $\mathcal{C}_1 = \{$Eqs. (17) and (18)$\}$. The decision-making does not take into consideration the battery life span preservation.
2. $\mathcal{C}_2 = \{$Eqs. (17) to (20)$\}$. The power flow strategy includes as constraints the recommended battery operating SOC interval and maximum (dis)charge rate.
3. $\mathcal{C}_3 = \{$Eqs. (17) to (21)$\}$. In addition to the recommended battery operating SOC interval and maximum (dis)charge rate, the power flow strategy includes the battery rest time limitation.

The results obtained for each 24 h are averaged and presented in the following subsections.

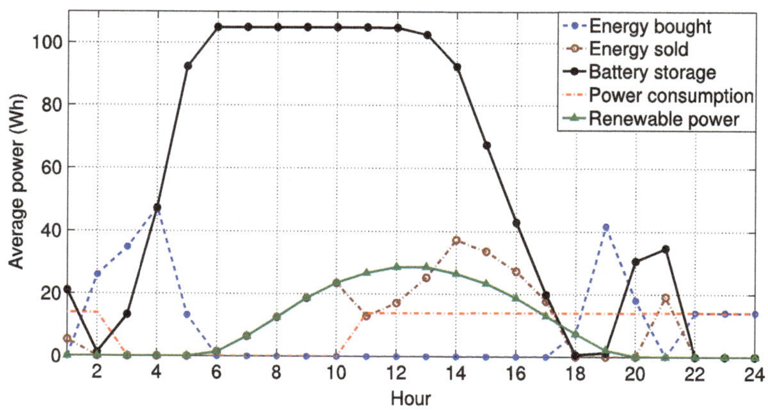

Fig. 5 Energy transaction, consumption, production, and storage with the *ideal* strategy under $\mathcal{C}_1$, averaged over 5 years

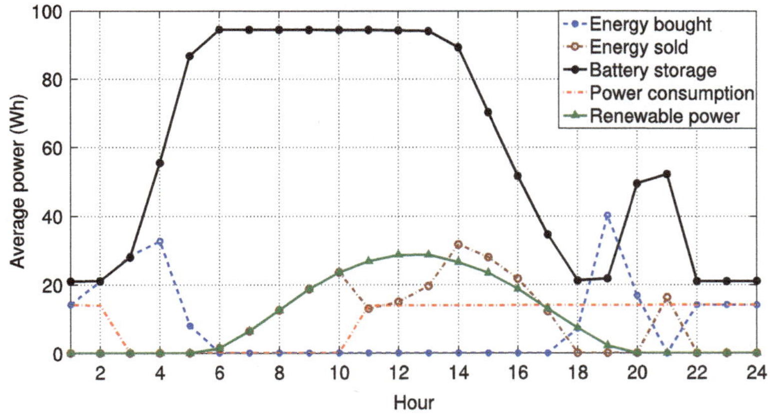

Fig. 6 Energy transaction, consumption, production, and storage with the *ideal* strategy under $\mathcal{C}_2$, averaged over 5 years

4.1 Power flow management

The energy flows obtained under the constraint sets $\mathcal{C}_1$, $\mathcal{C}_2$, and $\mathcal{C}_3$ are respectively presented in Figs. 5, 6, and 7. In these figures, we have represented the average SBS energy consumption, RE production, energy transactions with the SG, and energy stored in the battery. We observe that, in average, the cognitive ESS buys electricity at night, when the PV system cannot produce any energy, to power the SBS and/or store it into the battery. Additionally, we can note that the amount of energy purchased from the SG depends closely on the energy price (the lower the price, the higher the amount of energy bought). Once the PV production becomes available or when the price is high, the cognitive ESS prioritizes the use of the energy produced by the PV panel and the energy already stored in the battery to feed the SBS and sells a quantity of the surplus to the SG. Note that the global behavior of the three policies is similar. However, there are some differences between these strategies, generated by the battery constraints imposed in each case. In $\mathcal{C}_1$ (Fig. 5), the absence of any battery operational limitations enables the ESS to charge and discharge the battery at full (from 0 Wh to 105 Wh), using high (dis)charge current intensities (for example, the SOC variation between 4:00 and 5:00 is equal to 45%$\geq$ ΔSOC_{max}). Some limitations are introduced in $\mathcal{C}_2$ (Fig. 6) such that the SOC values are restricted in Δ_{SOC} (corresponding to the battery power level interval [20, 95] Wh), and all the SOC variations are limited between ΔSOC_{min} and ΔSOC_{max}. Finally, $\mathcal{C}_3$ (Fig. 7) includes $\mathcal{C}_2$ constraints in addition to the calendar aging restriction that avoid a long battery inactivity. For example, and contrarily to $\mathcal{C}_2$, slight battery (dis)charges have been introduced between 7:00 and 13:00 to generate a battery activity and prevent large capacity losses due to the calendar aging process.

Figure 8 compares the *ideal* strategy under each constraint sets. The battery power flow under the constraint set $\mathcal{C}_1$ is characterized by high battery currents and extreme SOCs. The amounts of energy exchanged with the SG are large and the battery state varies rapidly. In contrast, the strategies under $\mathcal{C}_2$ and $\mathcal{C}_3$ involve progressive battery (dis)charge such that the aging constraints are

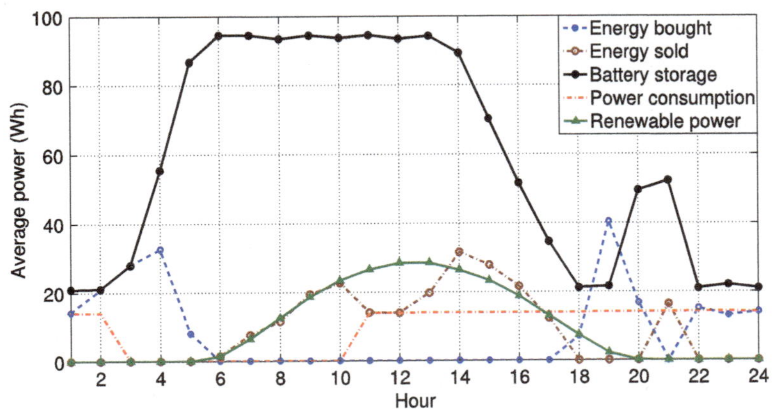

Fig. 7 Energy transaction, consumption, production, and storage with the *ideal* strategy under $\mathcal{C}_3$, averaged over 5 years

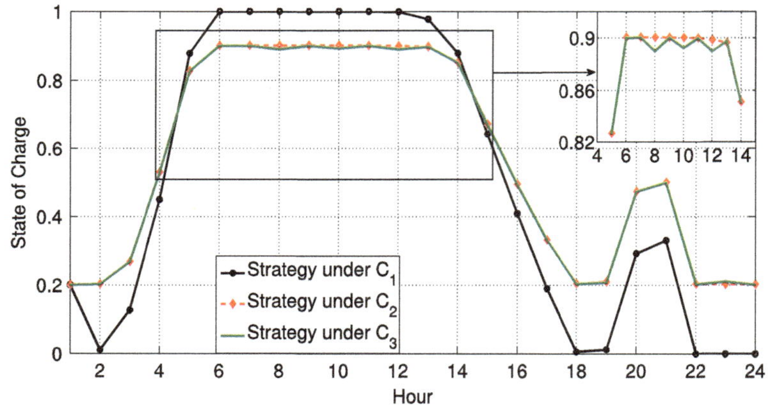

Fig. 8 Average SOC of the *ideal* strategy under the constraint sets $\mathcal{C}_1$, $\mathcal{C}_2$, and $\mathcal{C}_3$

respected. In reality, these two strategies are very similar such that the only differences are noticed between 7:00–13:00 and 22:00–0:00 regarding the battery storage level and the energy bought from the grid. This difference is due to Eq. (21), which restricts the battery rest during two consecutive time steps and generates a sawtooth pattern with respect to the SOC.

4.2 Battery aging

In this section, we investigate the impact of each power scheduling policy on the battery aging. As mentioned before, such aging phenomenon can be dissociated into two parts: the cycle aging and the calendar one.

Figure 9 shows the average time evolution of the SOH due to cycle aging for the ideal strategy in the three constraint configurations. Each time the battery is charged or discharged, the cycle aging contributes to decrease its life span whereas no degradation occurs during rest. The battery degradation observed under $\mathcal{C}_1$ (0.08% SOH loss per day) are substantially high compared to the strategy under $\mathcal{C}_2$ and $\mathcal{C}_3$ (0.02% SOH loss per day). The accelerated aging is explained by the profound depth of (dis)charge cycles and the battery operating in dangerous SOC areas (below 20% and above 90%). Besides, the cycle aging in $\mathcal{C}_3$ is slightly higher than $\mathcal{C}_2$ because of the additional cycling that prevent long battery rests.

Similarly, Fig. 10 illustrates the battery calendar aging in the three constraint configurations. The calendar aging corresponds to the SOH loss when the battery is not used and depends on the rest duration, momentary voltage, and temperature. In the unconstrained case $\mathcal{C}_1$, the internal battery temperature peaks illustrated in Fig. 11 are caused by deep battery (dis)charges. The corresponding current intensities are such high that the generated heat by Joule effect increases the battery internal temperature to attain large values (30 °C). In this situation, the accumulated calendar aging can also be considerable (0.12% SOH loss in 1 day). The SOC constraints $\mathcal{C}_2$ contribute to slightly reduce the impact of the calendar aging process thanks to current and voltage limitations (0.09% per day). However, it is only by restricting the rest time, such as we do by introducing $\mathcal{C}_3$, that calendar aging can be considerably avoided (0.02% per day).

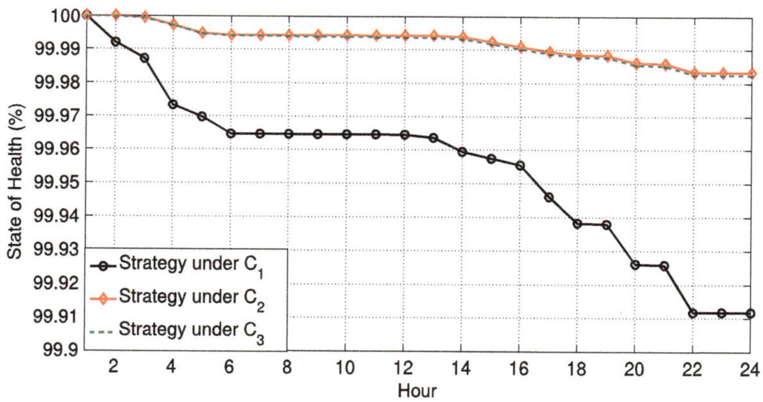

Fig. 9 Average battery cycle aging of the *ideal* strategy under the constraint sets $\mathcal{C}_1$, $\mathcal{C}_2$, and $\mathcal{C}_3$

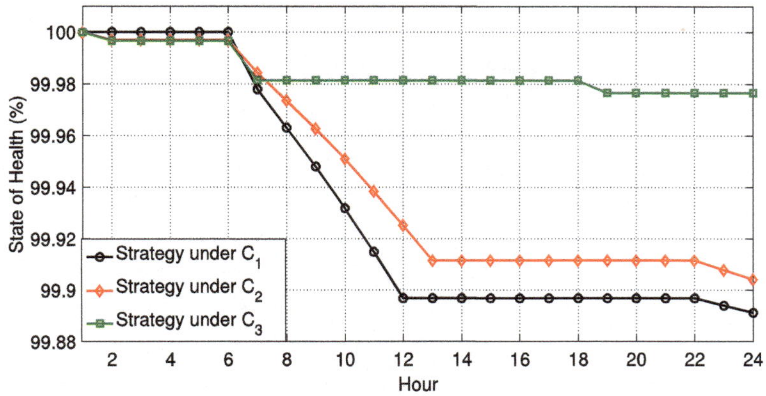

Fig. 10 Average battery calendar aging of the *ideal* strategy under the constraint sets $\mathcal{C}_1$, $\mathcal{C}_2$, and $\mathcal{C}_3$

Finally, by summing the cycle and calendar aging effects, we conclude that the respect of the SOC constraints enables considerable reduction of the battery degradation rate. This allows in average 51% (resp. 30%) of the battery SOH preservation per year when operating under $\mathcal{C}_3$ (resp. $\mathcal{C}_2$) compared to the unconstrained case $\mathcal{C}_1$.

4.3 Economic performance

In this part, we assess the economic performance of the proposed energy management by comparing the energy cost of the ideal strategy with two other schemes:

1. The *reference strategy* that systematically buys all the energy needed to feed the SBS from the SG. The battery and the PV system are not used.
2. The *naive strategy* is a greedy strategy that aims to decrease the immediate energy cost, regardless of any long-term cost saving opportunity. At each decision period, the PV production is first dedicated to cover the SBS consumption. In case the production exceeds the consumption, the surplus is sold to the SG. Otherwise, the missing energy is purchased from the SG. Consequently, the battery is never used.

For this analysis, we assume a battery investment cost of \$0.4/Wh that includes the purchase and installation fees. Figure 12 represents the normalized average hourly cost of the ideal (under the three constraint sets), naive and, reference strategies. In the reference strategy, the cost is exclusively generated by the SBS consumption. Once the PV production is considered, it appears that the naive strategy overlaps with the reference strategy during night, while energy production during day is sufficient to feed the SBS and sell energy back to the SG (negative cost). Concerning the ideal strategy, we first notice the energy purchase used to charge the battery during time intervals that match with off-peak price periods. The energy stored is then discharged when the electricity price is high to achieve a higher benefit compared to the aforementioned strategies. In particular, the ideal strategy under $\mathcal{C}_1$ spends and earns more than those under $\mathcal{C}_2$ and $\mathcal{C}_3$ given the additional allowed battery flexibility.

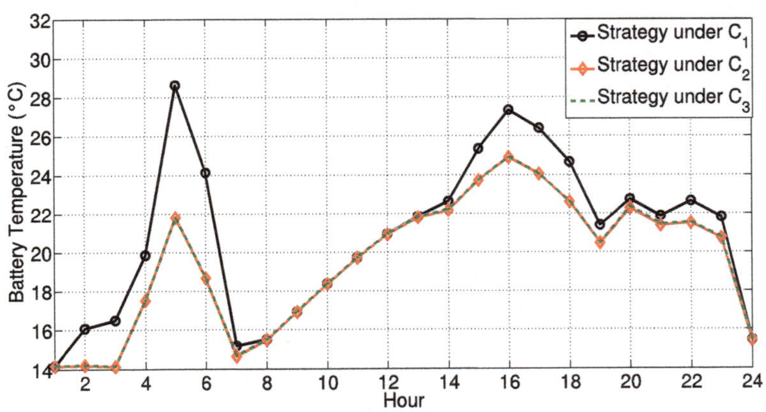

Fig. 11 Average battery temperature of the *ideal* strategy under the constraint sets $\mathcal{C}_1$, $\mathcal{C}_2$, and $\mathcal{C}_3$

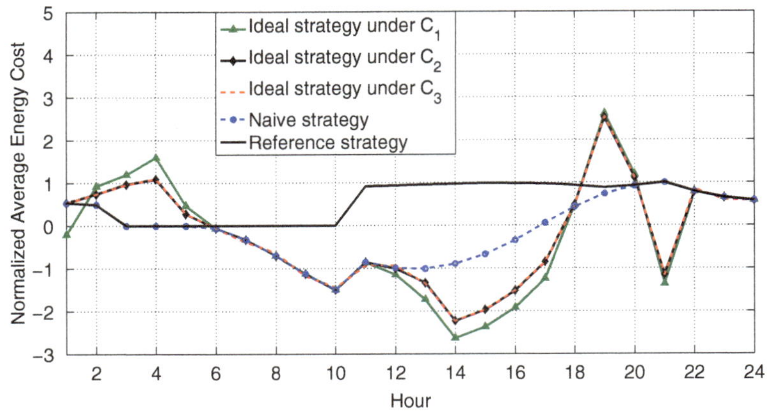

Fig. 12 Normalized daily energy cost of the *ideal* (under $\mathcal{C}_1$, $\mathcal{C}_2$, and $\mathcal{C}_3$), the *naive*, and the *reference* strategies, averaged over 5 years

Table 3 shows the accumulated energy costs averaged over 5 years and normalized with respect to the reference policy. Negative costs mean that not only cost saving is achieved but also the mobile operator is making profits. The largest cost saving of 145% compared to the reference is naturally achieved in the ideal case under $\mathcal{C}_1$. With the constraints $\mathcal{C}_2$ and $\mathcal{C}_3$, the proposed cognitive energy management framework performs approximately 132% cost reduction. On the contrary, the naive strategy achieves only one quarter of the ideal strategy cost savings.

It is clear that an efficient use of the battery enables more flexibility in the energy trading. In fact, the ESS can exploit the price variation to buy energy at low cost, not only to match an imminent energy demand but also to store it in prevision of future consumption that can generate heavy expenditures from the SG. Also, the RE energy can be saved in the battery until interesting selling prices are offered by the SG. However, there is a trade-off between how much the battery can be used to realize cost savings and the aging issues. In fact, when the battery aging constraints ($\mathcal{C}_2$ and $\mathcal{C}_3$) are respected, the cost saving is reduced by about ten points, which corresponds to a loss of \$1.6 in 1 year. At the same time, 30% (resp. 51%) of the battery initial SOH is preserved per year, equivalent to \$13 (resp. \$20) cost saving each year under $\mathcal{C}_2$ (resp. $\mathcal{C}_3$), which is by far more profitable given the current battery cost. In other words, it means that the implementation of the ideal energy strategy requires the battery replacement after 1.5 years under $\mathcal{C}_1$, 3 years under $\mathcal{C}_2$, and 7 years under $\mathcal{C}_3$.

Table 3 Normalized average opex for the *ideal* (with and without SOC constraints), *naive*, and *reference* strategies

Strategy	Ideal under $\mathcal{C}_1$	Ideal under $\mathcal{C}_2$	Ideal under $\mathcal{C}_3$	Naive	Reference
Cost	−0.45	−0.332	−0.329	−0.11	1

5 Conclusions

In this paper, we have presented a cognitive energy controller for a small cell base station connected to the smart grid and equipped with a battery and renewable production. This architecture had for purpose to jointly optimize the energy cost and reduce the battery aging effects. Obtained simulation results have shown that the energy supervision system achieves very large cost reduction compared to basic strategies while enhancing the storage life span. In particular, the battery aging constraints allows to considerably reduce the calendar and cycle degradation (up to 51% in average of the initial state of health preserved per year). Furthermore, the respect of these constraints resulted in only ten points decrease of the average opex cost saving, which is negligible considering current battery costs. As future work, we plan to study the proposed cognitive energy supervision framework considering casual information about the environment energy variable. We also aim to realize a demonstrator of the proposed solution to assess its performances in real-life conditions.

Funding

The research leading to these results is funded by the French Agence Nationale de la Recherche in the framework of the SOGREEN project (ANR-14-CE29-0025-01).

Authors' contributions

MM, ADeD, VH, RC, and NH substantially contributed to the conception or design of the work or the acquisition, analysis, or interpretation of data for the work. MM, ADeD, and VH drafted the work and revised it critically for important intellectual content. MM, ADeD, VH, RC, and NH agreed to be accountable for all aspects of the work in ensuring that questions related to the accuracy or integrity of any part of the work are appropriately investigated and resolved. MM, ADeD, VH, RC, and NH contributed to the final approval of the version to be published.

Competing interests

The authors declare that they have no competing interests.

References

1. Small Cells Deployment Market Status 2015 Report. http://scf.io. Accessed 20 Feb 2016
2. HAH Hassan, L Nuaymi, A Pelov, in *Online Conference on Green Communications (GreenCom), 2013 IEEE*. Renewable energy in cellular networks: a survey (IEEE, 2013), pp. 1–7
3. A De Domenico, E Calvanese Strinati, A Capone, Enabling green cellular networks: a survey and outlook. Comput. Commun. **37**, 5–24 (2014)
4. G Piro, M Miozzo, G Forte, N Baldo, LA Grieco, G Boggia, P Dini, HetNets powered by renewable energy sources: sustainable next-generation cellular networks. IEEE. Internet Comput. **17**(1), 32–39 (2013)
5. H Wang, H Li, C Tang, L Ye, X Chen, H Tang, S Ci, Modeling, metrics, and optimal design for solar energy-powered base station system. EURASIP J. Wirel. Commun. Netw. **2015**(1), 39 (2015)
6. J Palicot, in *Proceedings of the 2009 International Conference on Wireless Communications and Mobile Computing: Connecting the World Wirelessly*. Cognitive radio: an enabling technology for the green radio communications concept (ACM, 2009), pp. 489–494
7. S Jin, X Ma, W Yue, Energy-saving strategy for green cognitive radio networks with an LTE-advanced structure. J. Commun. Netw. **18**(4), 610–618 (2016)
8. M Broussely, P Biensan, F Bonhomme, P Blanchard, S Herreyre, K Nechev, RJ Staniewicz, Main aging mechanisms in li-ion batteries. J. Power Sources. **146**(1-2), 90–96 (2005)
9. A Barré, B Deguilhem, S Grolleau, M Gérard, F Suard, D Riu, A review on lithium-ion battery ageing mechanisms and estimations for automotive applications. J. Power Sources. **241**, 680–689 (2013)
10. N Michelusi, L Badia, R Carli, L Corradini, M Zorzi, Energy management policies for harvesting-based wireless sensor devices with battery degradation. IEEE Trans. Commun. **61**, 4934–4947 (2013)
11. C Liu, B Natarajan, Power management in heterogeneous networks with energy harvesting base stations. Phys. Commun. **16**(C), 14–24 (2015)
12. S Zhang, N Zhang, S Zhou, J Gong, Z Niu, X Shen, Energy-aware traffic offloading for green heterogeneous networks. IEEE J. Selected Areas Commun. **34**(5), 1116–1129 (2016)
13. R Kaewpuang, D Niyato, P Wang, Decomposition of stochastic power management for wireless base station in smart grid. IEEE Wirel. Commun. Lett. **1**(2), 97–100 (2012)
14. J Leithon, TJ Lim, S Sun, in *Smart Grid Communications (SmartGridComm), 2013 IEEE International Conference on*. Online energy management strategies for base stations powered by the smart grid (IEEE, 2013), pp. 199–204
15. Y Mao, J Zhang, KB Letaief, Grid energy consumption and QoS tradeoff in hybrid energy supply wireless networks. IEEE Trans. Wirel. Commun. **15**(5), 3573–3586 (2016)
16. M Mendil, A De Domenico, V Heiries, R Caire, N Hadj-said, in *Personal, Indoor, and Mobile Radio Communications (PIMRC), 2016 IEEE 27th Annual International Symposium on*. Fuzzy Q-learning based energy management of small cells powered by the smart grid (IEEE, 2016), pp. 1–6
17. M Mendil, A De Domenico, V Heiries, R Caire, N Hadj-said, in *International Conference on Cognitive Radio Oriented Wireless Networks*. Energy management of green small cells powered by the smart grid (Springer, 2016), pp. 642–653
18. G Auer, V Giannini, C Desset, EA Godor, How much energy is needed to run a wireless network? IEEE Wirel. Commun. **18**(5), 40–49 (2011)
19. L Lu, X Han, J Li, J Hua, M Ouyang, A review on the key issues for lithium-ion battery management in electric vehicles. J. Power Sources. **226**, 272–288 (2013)
20. R Restaino, W Zamboni, in *IECON 2012-38th Annual Conference on IEEE Industrial Electronics Society*. Comparing particle filter and extended kalman filter for battery state-of-charge estimation (IEEE, 2012), pp. 4018–4023
21. A Barre, F Suard, M Gérard, M Montaru, D Riu, Statistical analysis for understanding and predicting battery degradations in real-life electric vehicle use. J. Power Sources. **245**, 846–856 (2014)
22. W-Y Chang, The state of charge estimating methods for battery: a review. ISRN Appl. Math. **2013**, 1–7 (2013)
23. C Weng, J Sun, H Peng, in *ASME 2013 dynamic systems and control conference*. An open-circuit-voltage model of lithium-ion batteries for effective incremental capacity analysis (American Society of Mechanical Engineers, 2013), pp. V001T05A002–V001T05A002
24. Y Riffonneau, S Bacha, F Barruel, S Ploix, Optimal power flow management for grid connected PV systems with batteries. IEEE Trans. Sustainable Energy. **2**(3), 309–320 (2011)
25. E Lemaire-Potteau, F Mattera, A Delaille, P Malbranche, in *23rd European Photovoltaic Solar Energy Conference (Valencia, Spain, 2008)*. Assessment of storage ageing in different types of PV systems: technical and economical aspects, (Valencia, 2008), pp. 2765–2769
26. Lithium battery failures. http://www.mpoweruk.com/lithium_failures.htm. Accessed 20 Feb 2016
27. A Cordoba-Arenas, S Onori, G Rizzoni, A control-oriented lithium-ion battery pack model for plug-in hybrid electric vehicle cycle-life studies and system design with consideration of health management. J. Power Sources. **279**, 791–808 (2015)
28. M Ecker, JB Gerschler, J Vogel, S Käbitz, F Hust, P Dechent, DU Sauer, Development of a lifetime prediction model for lithium-ion batteries based on extended accelerated aging test data. J. Power Sources. **215**, 248–257 (2012)
29. A Eddahech, O Briat, J-M Vinassa, Lithium-ion battery performance improvement based on capacity recovery exploitation. Electrochimica Acta. **114**, 750–757 (2013)
30. M Rashid, A Gupta, Effect of relaxation periods over cycling performance of a li-ion battery. J. Electrochem. Soc. **162**(2), 3145–3153 (2015)
31. G SOLAR, Solar radiation time series. http://geomodelsolar.eu/data/full-time-series. Accessed 20 Feb 2016
32. G Rami, T Tran-Quoc, N Hadjsaid, JL Mertz, in *Power Engineering Society General Meeting, 2004. IEEE*. Energy supply for remote base transceiver stations of telecommunication (IEEE, 2004), pp. 1916–1921
33. Ameren: Ameren price database. https://www2.ameren.com/RetailEnergy/rtpDownload. Accessed 20 Feb 2016
34. A Forsgren, PE Gill, MH Wright, Interior methods for nonlinear optimization. SIAM Rev. **44**(4), 525–597 (2002)
35. K Uddin, A Picarelli, C Lyness, N Taylor, J Marco, An acausal li-ion battery pack model for automotive applications. Energies. **7**(9), 5675–5700 (2014)
36. G Auer, O Blume, V Giannini, I Godor, M Imran, Y Jading, E Katranaras, M Olsson, D Sabella, P Skillermark, *et al*, D2. 3: Energy efficiency analysis of the reference systems, areas of improvements and target breakdown. EARTH. **20**(10) (2010)

Performance of fractional delay estimation in joint estimation algorithm dedicated to digital Tx leakage compensation in FDD transceivers

Robin Gerzaguet[1]*, Laurent Ros[2], Fabrice Belvèze[3] and Jean-Marc Brossier[2]

Abstract

This paper deals with the performance of the fractional delay estimator in the joint complex amplitude/delay estimation algorithm dedicated to digital Tx leakage compensation in FDD transceivers. Such transceivers suffer from transmitter-receiver signal leakage. Combined with non linearity of components in the receiver path, the baseband received signal is impaired by a baseband polluting signal. This baseband polluting term depends on the equivalent Tx leakage channel which models leakages of the receiver path. In (Gerzaguet et al., Digit. Signal Proc 51:35–46, 2016), we have proposed a Tx leakage interference cancellation algorithm based on joint estimation of the complex gain and the fractional delay of the equivalent Tx leakage channel and we have derived the asymptotic performance of the complex gain estimator, that showed the necessity of the fractional delay estimation. In this paper, we propose a comprehensive study of the fractional delay estimation algorithm and its analytic performance. The study is based on the analysis of the S-curve and loop noise variance of the timing error detector, from which an approximation of the asymptotic performance of the joint estimation algorithm is derived.

Keywords: Tx leakage, FDD transceiver, Digital compensation, Least-mean-square algorithm, Joint estimation, S-curve

1 Introduction

Cognitive radios offer the possibility to improve the spectrum use and to adapt the transmission scheme to optimize the sharing of the available bandwidth, which leads to important constraints on the hardware components, located on the physical layer [1]. These constraints thus lead to performance limitations [2], and on this paper we focus on a hardware impairment that may occur in compact wireless transceivers when the radio follows a Frequency Division Duplexing (FDD) scheme. In FDD framework, transmission and reception are done simultaneously, using two different carrier frequencies [3].

A surface acoustic wave (SAW) duplexer is often used to connect the received (Rx) and the transmitted (Tx) path to a common antenna [4] (see Fig. 1). As the duplexer does not provide infinite attenuation between the Rx path and the Tx path, the transmitted signal can leak into the Rx path [5] leading to the so-called Tx Leakage (TxL) phenomenon. As the uplink and the downlink bands are spectrally separated, the received signal will not be directly impaired by the leakage of the transmitted signal which is filtered by the Rx Low Pass Filter (LPF) after the demodulation.

However, due to the non linearity and imperfections of components in the analog Rx stage, especially the low noise amplifier (LNA) [6] and the demodulator [7], intermodulation products can lead to a pollution in baseband. This polluting term corresponds to the square component of the Tx leakage signal, shifted to baseband [8]. As a consequence, this polluting signal will impair the received signal and can severely degrade the performance. The pollution is potentially detrimental in the cell edge context

*Correspondence: robin.gerzaguet@irisa.fr
This article is an extended version of the conference paper (Gerzaguet et al., Proc. International Conference on Cognitive Radio Oriented Wireless Networks (Crowncom), 2016) presented at the international conference Conference on Cognitive Radio Oriented Wireless Networks (Crowncom) 2016
[1]University of Rennes 1, IRISA, 38000 Grenoble, France
Full list of author information is available at the end of the article

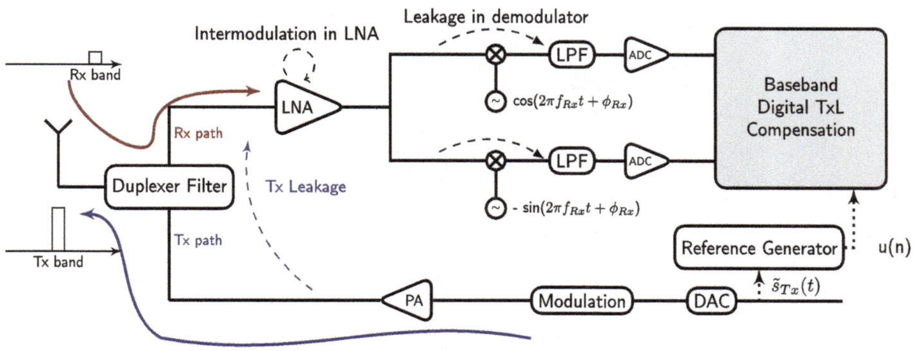

Fig. 1 Transceiver scheme. Classical frequency division duplexing chain in a RF transceiver with Tx leakage baseband pollution. The PA denotes the power amplifier, the ADC is the analog to digital converter, and the DAC denotes the digital to analog converter

(i.e., when the receiver is far from the base station), where the power of the received signal is low, and the power of the transmitted signal is strong [9] leading to a strong polluting signal and thus a low signal to interference ratio.

To avoid this pollution, passive methods based on analog filtering can be implemented. Such mitigation methods consist in adding a band pass filter to attenuate the leaked transmitted signal in the Rx stage before or after the LNA [10]. As radio frequency (RF) transceivers contain more and more digital parts, and as signal-processing techniques are becoming an area of interest for RF impairments problematic [11], several digital compensation methods have been investigated in the past few years for the Tx leakage (TxL) compensation [5, 9, 12].

In this paper, we complete the performance analysis of our previously proposed joint estimation (JE) algorithm dedicated to Tx leakage compensation [12, 13]. This algorithm estimates the equivalent TxL channel (approximated by a time varying complex gain and a fractional delay (FD)). It works with two reference-based least-mean-square (LMS) algorithms. As [13] and [12] mainly focus on the performance of the complex gain estimator (with and without a priori knowledge of the FD), we focus in this paper on the FD estimation part. The two main objectives, in addition to propose a comprehensive study, are (i) to prove that the FD estimation algorithm can lock around the desired FD value (ii) and to derive an approximation of the asymptotic performance of the JE algorithm.

The FD estimation algorithm has similitudes with data-aided algorithm that were designed for phase and timing synchronization as well as for automatic gain control or channel complex gain estimation [14]. However in the proposed estimation process, the received signal is considered as noise, and an image of the Tx signal is considered as a pilot sequence (used as reference). In this paper, as we focus on the FD estimator, we assume a constant complex channel gain and we derive analytic formulae of the S-curve of the FD detector [14] (defined as the expectation of the so-called error signal that updates the FD estimation). A linear approximation of the estimation error variance of the FD stage is established as a function of the slope of the S-curve, which represents the gain of the equivalent timing-error detector. This result is finally used to derive an approximation of the asymptotic performance of the JE algorithm, in terms of signal to interference ratio (SIR), using the primary results obtained in [13].

This paper is organized as follows. We give the baseband polluting model in Section 2. We recall the joint estimation of the complex gain and the fractional delay in Section 3. We derive the performance of the fractional delay estimation algorithm in Section 4. Section 5 validates our method and theoretical results through simulations.

2 Baseband model and issues

The discrete time observation model sampled at $T_{\text{Rx}} = 1/F_{\text{Rx}}$ is expressed as

$$d(n) = \underbrace{x(n) + b(n)}_{x_b(n)} + s_{TxL}(n) \tag{1}$$

where n is the time index, $x(n)$ is the data-bearing signal, which is assumed to be white and zero-mean with variance σ_x^2, $b(n)$ is the white additive Gaussian noise, of variance σ_b^2, $s_{TxL}(n)$ is the TxL polluting signal and $x_b(n) = x(n) + b(n)$ is then the desired (or expected) noisy received signal without interference, with variance $\sigma_{xb}^2 = \sigma_x^2 + \sigma_b^2$.

The baseband polluting model of the TxL phenomenon is due to the cascade of several impairments, and more precisely to the combination of duplexer finite isolation, low-noise amplifier non linearity and coupling on oscillator [5, 6]. As a consequence, the baseband polluting term can be expressed as [13]

$$s_{TxL}(n) = \left[h_{\mathrm{Rx}}(t) \star |h_{\mathrm{TxL}}(t) \star \tilde{s}_{Tx}(t)|^2\right]_{t=nT_{Rx}} \quad (2)$$

where h_{TxL} and h_{Rx} denote, respectively, the duplexer equivalent channel and the Rx chain equivalent channel. They respectively model the impact of the Duplexer and the Rx chain. $\tilde{s}_{Tx}(t)$ corresponds to the complex baseband transmitted signal and $\star$ denotes the convolution. We suppose that the duplexer is frequency flat as it is customarily done in the literature [5, 15, 16] and modeled as a complex gain introduced by the coupling at the oscillator. We assume that the Rx filter is known (or estimated). Due to digital and analog blocks, the polluting signal has a delay Δ, which is composed of an integer part D and a fractional part δ, $\Delta = (D + \delta)T_{\mathrm{Rx}}$. From (2), the polluting term can be expressed as:

$$U(t) = h_{\mathrm{Rx}}(t) \star |\tilde{s}_{Tx}(t)|^2 \quad (3)$$

$$s_{TxL}(n;\Delta) = \beta_{\mathrm{TxL}}(n) \times [U(t-\Delta)]_{t=nT_{Rx}} \quad (4)$$

where $\beta_{\mathrm{TxL}}(n)$ models the global TxL complex channel gain at time index n (modeling the impact of both the duplexer and the demodulator leakage) assumed to be slowly varying w.r.t the sampling time T_{Rx}. We have proposed in [13] a TxL interference compensation based on two reference-based LMS algorithms that jointly estimates the complex gain and fractional delay of the TxL channel. The proposed algorithm can estimate time-varying parameters estimation, as presented in [12] with the tracking of the complex gain. We have shown that an optimal step-size for the complex gain estimation stage can be obtained. This optimal step-size depends on the time varying complex gain model. Since we focus in this paper on the static performance of the FD estimator, we assume a constant TxL channel for the theoretical analysis:

$$\beta_{\mathrm{TxL}}(n) = \beta_{\mathrm{TxL}}\ ,\ \forall\ n$$

We also assume that the integer part D of the delay is known, as in practice it can be easily estimated with a correlation process or with a parallel structure, as described by [5], and without loss of generality, we take $D = 0$. In [12], we have shown that the fractional delay cannot be neglected and we have proposed a joint estimation algorithm that is recalled in the next part. This structure is driven by a reference signal $u(n)$, synthesized in the reference generator (see Fig. 1 and (5)) from the baseband Tx samples, with the known or estimated Rx filter h_{Rx}:

$$u(n) = [U(t)]_{t=nT_{Rx}} = \left[h_{\mathrm{Rx}}(t) \star |\tilde{s}_{Tx}(t)|^2\right]_{t=nT_{Rx}} \quad (5)$$

One can see that the reference signal $u(n)$ corresponds to the polluting signal with $\delta = 0$ and $\beta_{\mathrm{TxL}} = 1$. It should also be noted that, in practice, additional processing such as an upsampler, a LPF, and a resampler may be required for the reference generation.

3 TxL compensation based on joint estimation algorithm

The compensation algorithm is based on the reference $u(n)$ from (5) synthesized from the Tx stage, using the knowledge of the transmitted samples $\tilde{s}_{Tx}(t)$, and the model of the Rx filter chain. From this noise reference $u(n)$, the compensation algorithm acts as a noise cancellation algorithm that first produces joint estimations of the fractional delay $\hat{\delta}_n$ and the complex gain $\hat{\beta}_{TxL}(n)$ in order to delay and scale the reference $u(n)$ before it can be properly subtracted to the observed signal $d(n)$ (see (1)). The structure of the joint estimation algorithm is described on Fig. 2. It is composed of two blocks, one dedicated to the complex gain estimation, with a classic one tap LMS approach and the other one with a fractional delay estimation with another LMS. At each iteration, the reference signal $u(n)$ is delayed by the estimated FD $\hat{\delta}_n$ with an interpolation structure that can be for example a Farrow structure [17] that produces:

$$u_{\hat{\delta}_n}(n) = \sum_{j=0}^{L} u(n-j)\left(\prod_{\substack{i=0\\ i\neq j}}^{L} \frac{\hat{\delta}_n - i}{j-i}\right) \quad (6)$$

The fractional delay estimation algorithm is based on the minimisation of the instantaneous square error $|e(n)|^2$. Finally, the joint estimation algorithm can be expressed as [12]:

$$e(n) = d(n) - \hat{\beta}_{\mathrm{TxL}}(n)u_{\hat{\delta}_n}(n) \quad (7a)$$

$$\hat{\beta}_{\mathrm{TxL}}(n+1) = \hat{\beta}_{\mathrm{TxL}}(n) + \mu u_{\hat{\delta}_n}(n)e(n) \quad (7b)$$

$$\Delta_u(n) = u(n-1) - u(n) \quad (7c)$$

$$\delta_{n+1} = \hat{\delta}_n + \nu\Re\left\{\left[\hat{\beta}_{TxL}(n)\Delta_u(n) - u_{\hat{\delta}_n}(n)L(n)\right]e^*(n)\right\} \quad (7d)$$

$$L(n+1) = \left(1 - \mu u_{\hat{\delta}_n}(n)^2\right)L(n) + \mu d(n)\Delta_u(n) + 2\mu\hat{\beta}_{\mathrm{TxL}}(n)u_{\hat{\delta}_n}(n)\Delta_u(n)\ , \quad (7e)$$

where $e(n)$ is the compensated output, $\hat{\beta}_{\mathrm{TxL}}(n)$ the complex gain estimation, μ the step-size of the complex gain estimator, $L(n) = \partial\hat{\beta}_{\mathrm{TxL}}(n)/\partial\delta$ and ν, the constant step size of the fractional delay estimator. It can be seen that this algorithm is recursive, online (as it provides a compensated output $e(n)$ at each iteration), and with low complexity.

4 Performance of the FD estimation

In this section, we focus on the FD estimation algorithm. To perform the analytical study of this algorithm, we assume a perfectly known and constant channel ($\hat{\beta}_{\mathrm{TxL}}(n) = \beta_{\mathrm{TxL}}$). This algorithm can be considered as

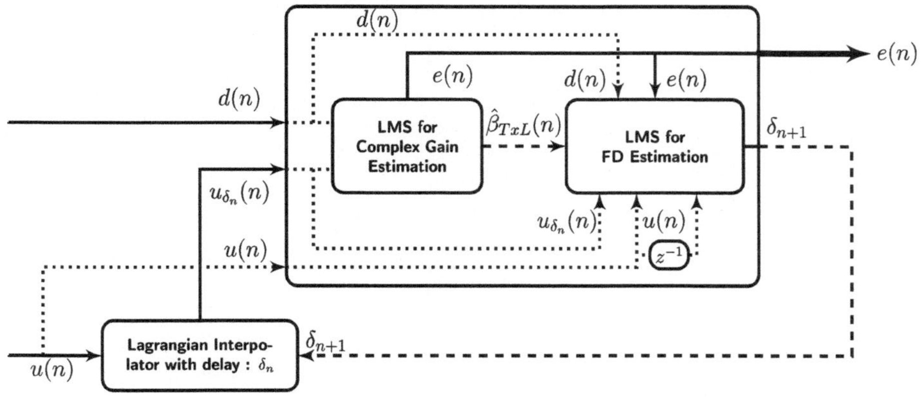

Fig. 2 Joint estimation framework for the TxL compensation

a delay-locked loop synchronization algorithm and its performance can be studied with the same approach as described in [14, 18]. Thus, the FD estimation algorithm is equivalent to a first order tracking loop algorithm (see Fig. 3) piloted by its delay error signal denoted $e_{\hat{\delta}_n}$. The stability and the performance of the proposed FD estimation algorithm can be studied using classical error detector open-loop analysis tools.

The equivalent delay error detector provides a delay error signal $e_{\hat{\delta}_n}$ function of the delay δ and of the estimate $\hat{\delta}_n$. Assuming fixed delay δ and fixed estimate $\hat{\delta}_n$, the delay error signal can be decomposed into the sum of the conditional expectation $E\left[e_{\hat{\delta}_n}|\hat{\delta}_n\right]$ and an additional random zero mean term. The first term S_c is denoted S-curve and is function of the FD estimation error $\epsilon_{\hat{\delta}_n} = \delta - \hat{\delta}_n$ and the second term N_T is called loop noise and is denoted $N_T(n)$.

$$
\begin{aligned}
S_c(\epsilon_{\hat{\delta}_n}) &= E\left[e_{\hat{\delta}_n}|\hat{\delta}_n\right] \\
N_T(n) &= e_{\hat{\delta}_n} - S_c(\epsilon_{\hat{\delta}_n})
\end{aligned}
$$

According to (7c)-(7d), the S-curve can be expressed as

$$
\begin{aligned}
S_c(\delta, \hat{\delta}_n) = |\beta_{\mathrm{TxL}}|^2 \Big[&- \mathrm{sinc}(\delta) + \mathrm{sinc}(1-\delta) \\
&+ \mathrm{sinc}(\hat{\delta}_n) - \mathrm{sinc}(1-\hat{\delta}_n) \Big] \sigma_{Tx}^4
\end{aligned} \tag{8}
$$

where $|\beta_{\mathrm{TxL}}|^2$ corresponds to the power of the TxL channel, that is assumed to be known. If the delay error is small, it is possible to linearize the S-curve around its stable equilibrium point [19]:

$$
S_c(\epsilon_{\hat{\delta}_n}) \approx \mathrm{K_D} \times \epsilon_{\hat{\delta}_n} \tag{9}
$$

where $\mathrm{K_D} = |\beta_{\mathrm{TxL}}|^2 \times \sigma_{Tx}^4$ (with σ_{Tx}^2 the variance of the transmitted samples) is the slope of the S-curve at the stable equilibrium point. As $\mathrm{K_D} > 0$, the FD estimator is stable around δ (i.e., the loop will lock within the range of FD). Besides, as $S_c(0) = 0$, the FD estimator is unbiased. Assuming a white loop noise for small FD error, the variance of the FD estimation stage can be derived as [14]:

$$
\sigma_{\epsilon_\delta}^2 = \frac{\nu}{\mathrm{K_D}\,(2 - \nu \mathrm{K_D})} \Gamma_{N_T}[0] \tag{10}
$$

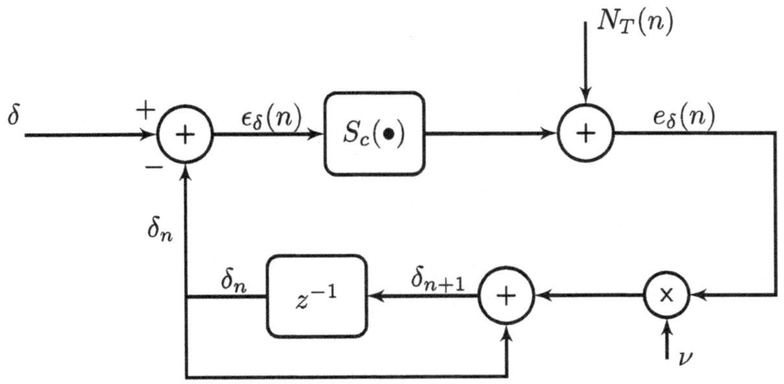

Fig. 3 Equivalent scheme of the FD estimation algorithm as a first-order tracking loop

with $\Gamma_{N_T}[0]$ the loop noise auto-correlation at zero delay, that can be expressed as

$$\Gamma_{N_T}[0] = \sigma_{\text{Tx}}^2 |\beta_{\text{TxL}}|^2 \sigma_{xb}^2 , \tag{11}$$

From (9), (10), and (11), and under the aforementioned assumptions, the variance of the first order FD estimation can finally be approximated as a function of the step-size ν by:

$$\sigma_{\epsilon_\delta}^2 \approx \frac{\nu \sigma_{xb}^2}{4\left[1 - \nu \sigma_{\text{Tx}}^4 |\beta_{\text{TxL}}|^2\right]} \tag{12}$$

Some interpretations and remarks can be done from (12):

- Since the fractional delay is constant, the variance of the fractional delay estimation converges to zero when the step-size ν tends toward to zero. In practice, decreasing the step-size would decrease the estimation variance but would also lead to a longer transient state before convergence.
- The variance is proportional to the power of the noisy received signal $\sigma_{xb}^2 = \sigma_x^2 + \sigma_b^2$. This has to be taken into account when using the algorithm in practical case. In the simulation part (see Section 5), the step-size ν is thus normalized by the power of the noisy signal. The presence of the noisy signal acts as a perturbation for the fractional delay estimation. If the entry signal would only the Tx polluting term, the fractional delay estimation would be straightforward.
- A stability condition can be deduced from (12) $\nu < 1/\left(\sigma_{\text{Tx}}^4 |\beta_{\text{TxL}}|^2\right)$.

In [13], we have derived the performance of the complex gain estimator (assuming a constant TxL channel), in presence of a non compensated FD. The performance have been assessed in terms of signal to interference ratio (SIR) of the complex gain estimator. Based on these results, we can now express the performance of the JE algorithm, using the error signal defined in (7a) (i.e., the output of the compensation structure) and the variance of the FD estimator expressed in (12) :

$$\begin{aligned} \text{SIR}_{\text{comp}} &= 10\log_{10}\left(\frac{\sigma_{xb}^2}{||e - x_b||^2}\right) \\ &\approx -10\log_{10}\left[\frac{\mu\sigma_u^2}{2-\mu\sigma_u^2} + |\beta_{\text{TxL}}|^2 \frac{\sigma_{\epsilon_\delta}^2}{\sigma_{x_b}^2} + \frac{\mu\sigma_u^2\left(|\beta_{\text{TxL}}|^2\right)\sigma_{\epsilon_\delta}^2}{\sigma_{x_b}^2\left(2-\mu\sigma_u^2\right)}\right] \end{aligned} \tag{13}$$

The argument in the $-10\log_{10}$ function in (13) is composed by the addition of three terms that limit the performance:

- The first one is related to the complex gain estimation stage. If the step-size μ of the complex gain algorithm is small, it can be approximated as $\mu\sigma_u^2/2$.
- The second and third term are due to the fractional delay estimation stage. The second term is due to the residual fractional delay due to the fractional delay estimation error (see the estimation error expression in (54) [12]).
  The third term models the impact of the fractional delay when fed back in the complex gain estimator (see the misalignment expression in (56) [12]). Both terms show the impact of the residual fractional delay on the performance and the overall performance depends on the algorithm step-sizes μ, ν and on the pollution statistics $\left(\text{channel amplitude} |\beta_{\text{TxL}}|^2 \text{ and noisy signal variance} \sigma_{x_b}^2\right)$.

5 Simulations

The performance of the proposed algorithm, and more precisely, the performance of the FD estimation algorithm are further analyzed by simulations.

- On the Fig. 4, we first plot the theoretical, linearized theoretical, and simulated S-curve versus the estimation error of the FD estimator in open loop. It shows that the theory is corroborated and that the S-curve can be linearized when the estimation error is low (between −10 and 10 percent of the sampling time). It also demonstrates that the FD estimation loop will lock and is unbiased as $\mathcal{S}_c(0) = 0$.
- On Fig. 5, we represent the theoretical and simulated loop noise zero-delay auto-correlation $\Gamma_{N_T}[0]$ versus the FD estimation error. When the FD estimation error is low, it can be seen that the assumption of a white loop noise with variance expressed in (11) leads to accurate results.
- On Fig. 6, the simulated delay error variance $\sigma_{\epsilon_\delta}^2$ is compared to the theory derived in (12). The curves are obtained for different fractional delay values and different normalized fractional delay estimator step-size values. One can see that the delay error variance is independent from the initial FD value and linear with respect to the FD step-size (for small step-size ν); which is aligned with the theoretical equation derived in (12).
- On Fig. 7, we finally consider the performance of the JE algorithm considering the impact of the complex gain step-size. We consider that the data bearing signal is a white Gaussian signal polluted by a TxL signal that follows (4). The polluting SIR (i.e., the ratio between the power of the TxL polluting signal and $\sigma_{x_b}^2$) is set to 0 dB and the power of the desired noisy signal is $\sigma_x^2 + \sigma_b^2 = 10^{-8}$. The noisy data signal has thus a low power whereas the transmitted signal (of unitary variance, $\sigma_{\text{Tx}}^2 = 1$) is strong (cell edge context). At the output of the demodulator, the polluting signal has the same power as the desired

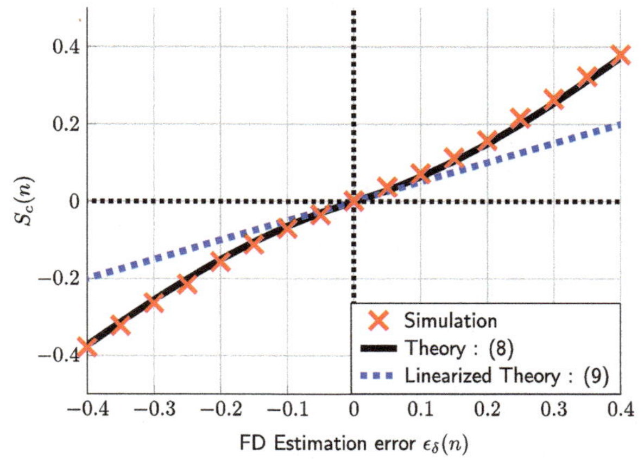

Fig. 4 Theoretical, simulated and linearized S-curve versus the FD estimation error

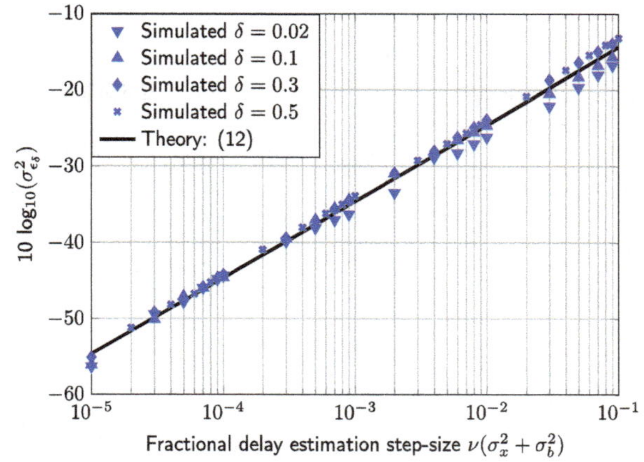

Fig. 6 Theoretical and simulated error delay variance versus the fractional delay estimator step-size

noisy signal (initial SIR of 0dB) due to the attenuation induced by the pollution. The complex gain algorithm is driven by a non interpolated reference $u(n)$ defined with (5) and the JE algorithm uses a Farrow structure to apply the estimated FD to $u(n)$. We also assume a constant TxL channel, and we represent the performance of the complex gain estimator and the JE algorithm for several values of FD. The FD estimation step-size is set to $\nu = 10^{-6}/\sigma_{xb}^2$.
The main objectives of this Figure are (i) show the impact of a non compensated fractional delay (ii) show the potential benefits of the FD estimation (iii) demonstrates that the theoretical performance (13) corroborates with the simulated results. We indeed show that the performance can be dramatically reduced if the FD estimation algorithm is not enabled. Besides, the JE algorithm greatly improves the asymptotic performance. If there is no FD and the FD estimation stage is disabled, the performance

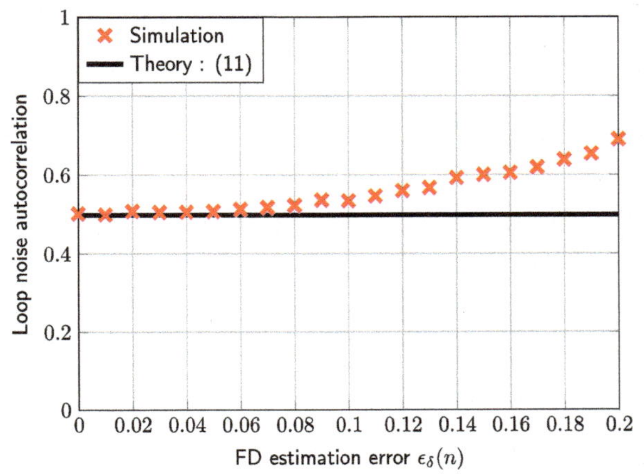

Fig. 5 Theoretical and simulated zero-delay autocorrelation of loop noise versus estimation error of the FD estimator

becomes linear w.r.t the complex gain step-size μ (for small μ) as, in this case, (13) becomes $-10\log\left(\mu\sigma_u^2/(2-\mu\sigma_u^2)\right)$.
If there is no FD but the JE scheme is enabled, one can see a performance floor at around 60 dB when the complex gain estimation step-size tends toward to zero. This is due to the residual error introduced with the FD estimation stage. From (12) and (13), with $\mu = 0$, the asymptotic SIR for the JE becomes

$$\mathrm{SIR_{comp}} = -10\log\left(\frac{\nu|\beta_{\mathrm{TxL}}|^2}{4\left(1-\nu\sigma_{\mathrm{Tx}}^4|\beta_{\mathrm{TxL}}|^2\right)}\right).$$

We can finally conclude that the asymptotic performance of the structure can be well approximated and interpreted with (13); leading to a direct link between a desired asymptotic level and a step-size value for the FD estimator ν.

- To better stress the impact of the FD estimation stage on the complex gain stage, we finally represent the performance of the complex gain algorithm and the joint estimation algorithm with a constant gain step-size and a varying FD step-size on Fig. 8. The same polluting model as in Fig. 7 have been used. The step-size of the complex gain algorithm is set to $\mu = 10^{-5}$ have been used. The red curves (which represents the gain estimator algorithm without taking into account the presence of a fractional delay) are constant (as the FD stage is deactivated, and as the TxL channel is constant). The higher the FD, the poorer the performance as a non compensated FD severally degrades the performance (see (17) of [13] for the theoretical expression of the performance loss). It is thus shown that the performance can be dramatically reduced if the FD estimation algorithm is not enabled. The blue curves represent the

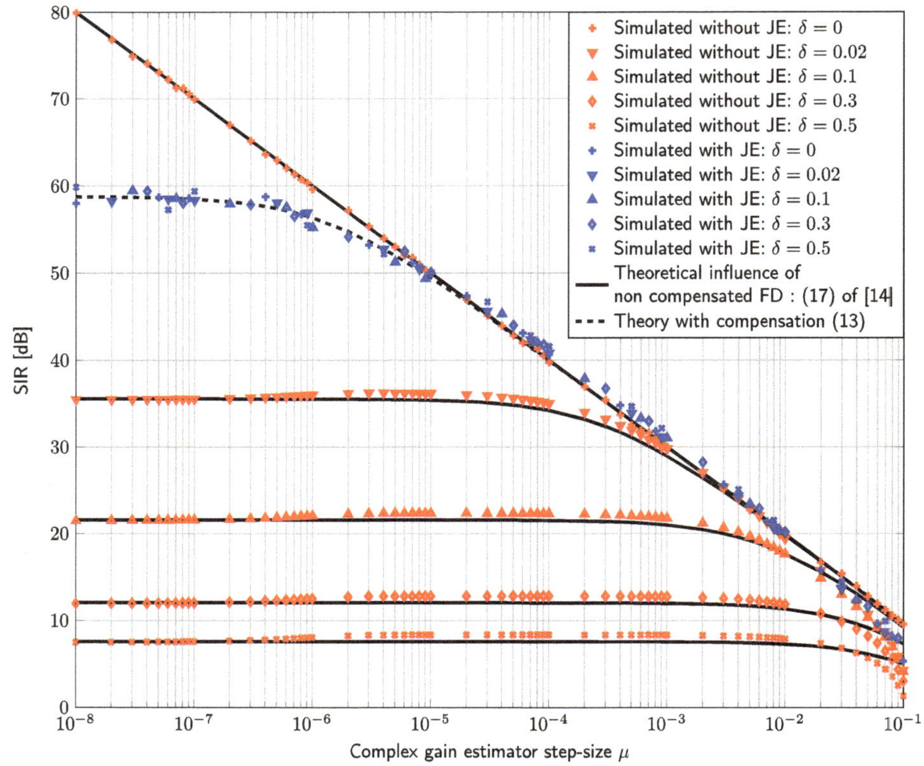

Fig. 7 JE asymptotic performance. Asymptotic performance of the compensation algorithm for several values of the complex gain step-size and for different values of fractional delays for a fixed fractional delay step-size $\nu = 10^{-6}/\sigma_{xb}^2$

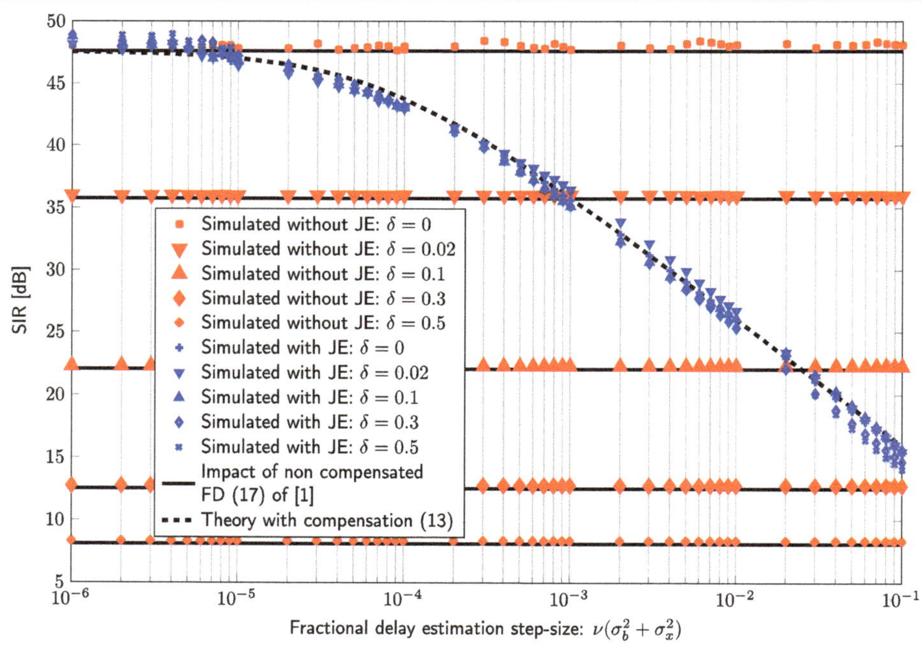

Fig. 8 JE asymptotic performance. Asymptotic performance of the compensation algorithm for several values of the fractional delay step-size and for different values of fractional delays for a fixed fractional delay step-size $\mu = 10^{-5}$

performance when the FD stage is activated, and the corresponding theoretical expression derived in (13) is also represented. Is is shown that the asymptotic performance of the structure can be approximated with good accuracy with (13) leading to a direct link between a desired asymptotic performance level and step-size value for the FD estimator ν. If the step-size of the FD stage is high, the performance can be low as the variance of the stage (12) is proportional to $\nu\sigma_{xb}^2$. In this case, the FD stage can be seen as a disruptive element for the complex gain stage. With a low step-size ν, the performance is greatly enhanced (at the price of a longer transient state) and the performance converges to the performance of the complex gain without FD (first term of the theoretical expression derived in (13) as $\mathrm{SIR_{comp}} \approx -10\log_{10}\left[\mu\sigma_u^2/(2-\mu\sigma_u^2)\right]$).

6 Conclusions

This paper deals with the performance of a joint estimation algorithm dedicated to the compensation of the digital Tx leakage in RF transceivers. The strong constraints that apply to cognitive radios lead to a potential detrimental loss of performance due to hardware impairments and we focus here on a pollution that occurs in FDD transceivers, related to the pollution of an image of the transmitter stage on the receiver stage. Based on the initial algorithm proposed in [13], in this paper we focus on the fractional delay estimation algorithm that can be considered as a first-order tracking loop synchronisation algorithm piloted by an interpolated reference signal. We have firstly derived the analytical S-curve, defined as the conditional expectation of the error signal that controls the loop, and we have deduced an approximation of the asymptotic estimation error variance of the FD estimation algorithm. As the proposed approximated theoretical performance formula of the whole JE algorithm shows good accordance with simulations, this result can be used to properly tune the delay estimation algorithm step-size .

Funding
The authors declare that they have no funding sources.

Contributions
The two main objectives of the proposed paper, in addition to propose a comprehensive study on the Tx Leakage digital compensation, are (i) to prove that the initially proposed FD estimation algorithm can lock around the desired FD value (ii) and to derive an approximation of the asymptotic performance of the JE algorithm. This leads to a direct link between a desired asymptotic performance level for the compensation structure and a step-size value for the FD estimator

Competing interests
The authors declare that they have no competing interests.

Author details
[1]University of Rennes 1, IRISA, 38000 Grenoble, France. [2]GIPSA-Lab, Image and signal department, 38400 Saint Martin d'Heres, France. [3]ST-Microelectronics, 38000 Grenoble, France.

References

1. D Cabric, RW Brodersen, in *Proc. IEEE 16th International Symposium on Personal, Indoor and Mobile Radio Communications (PIMRC)*. Physical layer design issues unique to cognitive radio systems, vol. 2, (2005), pp. 759–7632
2. H Tang, in *First IEEE International Symposium on New Frontiers in Dynamic Spectrum Access Networks (DySPAN)*. Some physical layer issues of wide-band cognitive radio systems, (2005), pp. 151–159
3. H Martikainen, in *Proc. European Wireless Conference (EW)*. Analysis of duplexing modes in the IEEE 802.16 wireless system, (2010)
4. ME Knox, in *Proc. IEEE 13th Annual Wireless and Microwave Technology Conference (WAMICON)*. Single antenna full duplex communications using a common carrier, (2012), pp. 1–6
5. A Frotzscher, G Fettweis, in *Proc. IEEE Vehicular Technology Conference (VTC)*. A stochastic gradient LMS algorithm for digital compensation of Tx leakage in Zero-IF-Receivers, (2008), pp. 1067–1071
6. B Razavi, Design considerations for direct-conversion receivers. IEEE Trans. Circ. Syst. (Analog Digit Signal Process). **44**, 428–435 (1997)
7. D Manstretta, M Brandolini, F Svelto, Second-order intermodulation mechanisms in CMOS downconverters. IEEE J. Solid-State Circ. **38**, 394–406 (2003)
8. C Lederer, M Huemer, in *Proc. IEEE International Conference on Wireless Information Technology and Systems (ICWITS)*. The influence of DC offsets on the digital cancellation of second-order Tx intermodulation distortions in homodyne receivers, (2012), pp. 1–4
9. A Kiayani, L Anttila, M Valkama, in *Proc IEEE 56th International Midwest Symposium on Circuits and Systems (MWSCAS)*. Modeling and dynamic cancellation of Tx-Rx leakage in FDD transceivers, (2013), pp. 1089–1094
10. K Dufrene, R Weigel, in *The European Conference on Wireless Technology*. Highly linear IQ downconverter for reconfigurable wireless receivers, (2005), pp. 19–22
11. G Fettweis, M Lohning, D Petrovic, M Windisch, P Zillmann, W Rave, in *Proc. IEEE 16th International Symposium on Personal, Indoor and Mobile Radio Communications (PIMRC)*. Dirty RF: a new paradigm, (2005), pp. 2347–23554
12. R Gerzaguet, L Ros, F Belveze, J-M Brossier, Performance of a digital transmitter leakage LMS-based cancellation algorithm for multi-standard radio-frequency transceivers. Digit Signal Process. **51**, 35–46 (2016)
13. R Gerzaguet, L Ros, F Belveze, J-M Brossier, in *Proc. 21st International Conference on Electronics Circuits and Systems (ICECS)*. Joint Estimation of Complex Gain and Fractional Delay for Tx Leakage Compensation in FDD Transceivers, (2014)
14. U Mengali, AN D'Andrea, *Synchronization Techniques for Digital Receivers. Applications of Communications Theory* (Springer, 1997)
15. M Faulkner, DC offset and IM2 removal in direct conversion receivers. IEE Proceedings on Communications. **149**(3), 179–184 (2002)
16. C Lederer, M Huemer, in *Proc. IEEE Radio and Wireless Symposium (RWS)*. LMS based digital cancellation of second-order Tx intermodulation products in homodyne receivers, (2011), pp. 207–210
17. CW Farrow, in *IEEE International Symposium on Circuits and Systems*. A continuously variable digital delay element, (1988), pp. 2641–26453
18. E Simon, K Raoof, L Ros, in *Proc. IEEE International Conference on Acoustics, Speech, and Signal Processing (ICASSP)*. Optimization of symbol timing recovery for multi-user ds-cdma receivers, vol. 4, (2003), pp. 604–74
19. H Meyr, M Moeneclaey, F S.A, *Digital Communication Receivers* (Proakis, 1998)

5

A novel physical layer scheme based on superposition codes

Pietro Savazzi* and Anna Vizziello

Abstract

The recently proposed superposition codes (SCs) have been mathematically proved to be decoded at any rate below the capacity, for additive white Gaussian noise (AWGN) channels. The main objective of this paper is to study the feasibility of a novel SC approach as an alternative to the traditional way of designing modern physical (PHY) layer schemes. Indeed, currently, PHY solutions are based on the decomposition into two separate problems of modulation shaping and coding over finite alphabets. Since superposition codes are defined over real numbers, modulation and coding can be jointly realized. Moreover, a fast decoding method is developed and tested by comparing the obtained results with both the uncoded system performance and two approximate message passing (AMP) algorithms. Finally, possible perspective to fifth generation (5G) applications exploiting SC solutions are outlined in the paper, and some interesting relations with sparse signal recovery are analyzed for further future research lines.

Keywords: Superposition code, Physical layer

1 Introduction

Some recent works have proven that sparse superposition codes (SCs), defined over real numbers, can be decodable at any rate less than the Shannon capacity with an exponentially small probability of error [1, 2]. In more details, a maximum likelihood decoding allows reliable communications for all rates up to the Shannon capacity [1], while a fast decoding algorithm and its soft decoding version are proposed in [2] and [3] respectively.

According to the traditional information theory, the maximum transmission rate over an additive white Gaussian noise (AWGN) channel is [4]:

$$C = \frac{1}{2}\log_2\left(1+\frac{P}{\sigma^2}\right) \tag{1}$$

where P is the codeword power and $\sigma^2 = \frac{N_0}{2}$ is the Gaussian noise variance.

After the fundamental work of Shannon, the communication model for standard physical (PHY) layer design is usually split into two separate blocks: the channel coder and the modulation shaping [5]. Further, the best channel codes that may be found in the literature are based on elements defined over binary or q-ary Galois finite fields [6].

Although the original Shannon proof of the noisy channel coding theorem is based on considering large and random codes defined over the real numbers set [7], the numerical instability and the power constraints of real number error-correcting codes have limited their use in practical communication schemes. On the other hand, they have been mainly studied for tackling fault tolerance problems in digital computers operation [8].

Moreover, some important and interesting research lines have been outlined [9, 10] by approaching the channel coding problem with codeword elements defined over the real or complex field. In the last decade, complex-field coding has been mainly used for counteracting multipath fading in wireless channels [11].

In [12], a low complexity coding scheme for RFID transmissions has been proposed, considering a simple implementation of a similar superposition coding idea. Despite its simple implementation, the codec scheme can provide a significant coding gain.

Sparse superposition codes can be thought as a joint modulation-coding scheme, since the main idea is to add redundancy directly to a spread spectrum signal, by assigning different orthogonal spreading sequences to the same link.

*Correspondence: pietro.savazzi@unipv.it
Department of Electrical, Computer and Biomedical Engineering, University of Pavia, via Ferrata 5, Pavia, Italy

This aspect could be exploited for the fifth generation (5G) technologies [13], especially from the point of view related to the interference rejection potential embedded in the SC scheme, which may be used for both full-duplex communications and non-orthogonal multi-access protocols that are considered in 5G systems.

The main contributions of this paper are as follows:

- The investigation—from an application-oriented point of view—of the novel SC concept, in order to test its feasibility and possible performance, gains with respect to the state-of-the-art PHY design. While [1–3] analyze SC from a theoretical perspective, we develop a simulation study case, which is still missing, useful to understand the benefits of SC application.
- A fast and simple decoding algorithm, which is less parameter-dependent compared to [2] and with a less computational burden with respect to the technique presented in [14].
- An overview about some interesting perspectives for future 5G research and development exploiting SC concept.

The proposed decoding scheme represents the most relevant novelty of this work, and it may be interpreted as an orthogonal matching pursuit (OMP) sparse reconstruction algorithm, while its limited number of iterations may be thought as a simple stopping rule [15].

This algorithm is similar to those shown in [2], although the two approaches differ for the following point: at each iteration, we look for the maximum values of the vector $Z^{(i)}$, see Eq. 7 and step 4 of the proposed algorithm in Section 3, while in [2], the elements of the same vector are compared to a specific threshold parameter τ, see sections I.B and II of the cited reference.

Moreover, in Section II of [2], a combined statistics of the residuals $Y^{(i)}$, see Eq. 11, and a set of pacing parameters are used to theoretically compute the performance limit of the decoder. On the contrary, we have considered performance results obtained by simulations of the complete concatenated scheme.

Differently, in [1], the authors consider the performance limits of the optimal maximum likelihood detector, while in [3] a soft-decision decoder is taken under consideration with Bayes optimal statistics at each iteration. Finally, the works [14, 16–18] take into account approximate message passing decoding algorithms.

The paper is organized as follows: Section 2 is devoted to defining the superposition code and describing the system model. Section 3 details the proposed decoding scheme. Section 4 shows some interesting simulation results. Section 5 outlines relevant future 5G developments exploiting SCs and their relations with other approaches. Finally, some concluding remarks wrap up and close the paper in Section 6.

2 Sparse superposition codes

2.1 The coding algorithm

We here briefly outline the main concepts of SCs. The main insight of [1, 2] is to use superposition codes as a forward error correction (FEC) technique. In particular, we limit our analysis to binary transmission over an AWGN channel, although an extension to a non-binary-signed finite alphabet is possible to develop more spectral efficient schemes.

Specifically, let $u = (u_1, u_2, \ldots, u_k)$ be a string of k bits to be coded. The main idea is to map u to a vector c of length N, whose elements are real numbers. Thus, let us define the dictionary matrix $X = [X_1, X_2, \ldots, X_N] \in \mathbb{R}^{n\times N}$, whose entries are $X_{i,j} \sim \mathcal{N}(0,1)$, where $\mathcal{N}(0,1)$ represents a normal process with unit variance and zero mean. The codeword length equals n, so the code rate is $R = k/n$.

In order to obtain the codeword, a sparse vector $\beta \in \mathbb{R}^N$ is built as shown in Fig. 1. The matrix X is partitioned into $L = N/M$ sections, being L and M positive integers. The codewords are computed as $c = X\beta^T$, with β belonging to the set:

$$\mathcal{B} = \left(\beta : \beta = \sqrt{\tfrac{P}{L}} \text{ in only one of the } M \text{ elements of each section}\right) \tag{2}$$

Although [2] shows that rates closer to capacity at higher SNRs need an exponential power allocation profile, we use a constant power allocation. Indeed, in this way, we achieve even error probability along the codeword and a decoder with less parameters to be set than the scheme in [2], as explained in Section 3.

The detailed coding process is done as in the following:

- An input bit string u of length $k = L\log_2(M)$ is split into L substrings of size $\log_2(M)$
- Each substring of u is to be intended as the position index of the non-zero element of β in each of the L subsections
- The rate of the superposition code, i.e., the number of input bits per channel use, can be then computed as:

$$R = \frac{L\log_2 M}{n} \tag{3}$$

2.2 System model

The received sequence at the output of the AWGN channel is:

$$Y = c + \eta = X\beta^T + \eta \tag{4}$$

where η is the noise vector whose elements belong to $\mathcal{N}(0,\sigma^2)$.

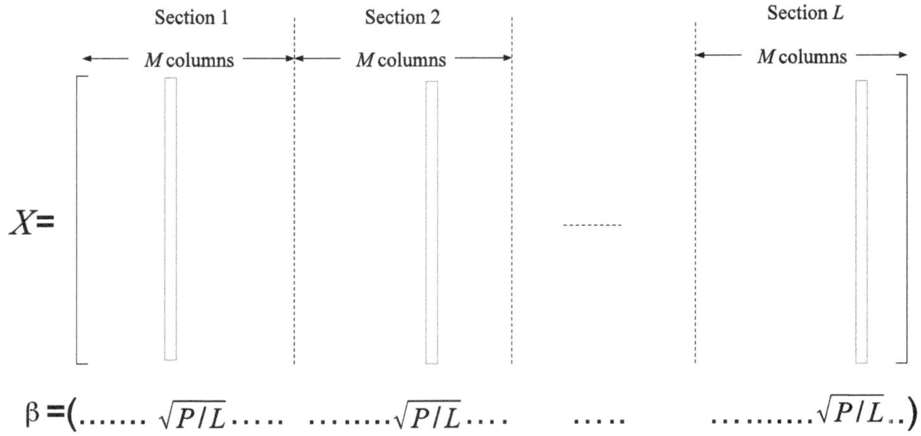

Fig. 1 Construction of the sparse vector β as reported in [1]

According to the previous definitions, the codeword power has to be P, since:

$$||\beta||^2 = P \Rightarrow \frac{E\left(||X\beta^T||^2\right)}{n} = P \tag{5}$$

Being T the bit duration, the energy per bit to noise power spectral density ratio can be defined as:

$$\frac{E_b}{N_0} = \frac{RPT}{\sigma^2} = \frac{RP}{N_0/2} \tag{6}$$

3 The decoding algorithm

We here propose a novel fast and simple decoding scheme for superposition codes. Similarly to the algorithm in [2], we propose an iterative decoding algorithm with a maximum number of iterations equal to L. However, the proposed solution is able to decode anyway the received codeword, without considering the possibility to mark erasures at the channel output as in [2]. In this way, the decoder always tries to output the more reliable decoded codeword.

This reasoning is consistent with the case study of FEC implementation, for which a decoder output may be always needed.

1. *Initialize*: $X^{(1)} = X$, $Y^{(1)} = Y$, $\hat{\beta} = 0_N$,

   $$J^{(1)} = [J_1, J_2, \ldots, J_L],$$

   where $J_t = [(t-1)M + 1, (t-1)M + 2, \ldots, tM]$ are the indexes of each of the L sections of length M
2. *For* $i = 1$*to* L
3. *Compute* the vector $Z^{(i)}$ as in the following:

   $$Z^{(i)} = \frac{X^{(i)T} Y^{(i)}}{||Y^{(i)}||} \tag{7}$$

4. *Find* the index p of the maximum positive value of the vector $Z^{(i)}$. Since we only look for the maximum value of the vector $Z^{(i)}$, the normalization by $||Y^{(i)}||$ may be neglected and Eq. (7) may replaced by the simpler $Z^{(i)} = X^{(i)T} Y^{(i)}$
5. *Identify* the section $J^{(i)}_{t_p}$ containing the position p and *update* the partially decoded word:

   $$\hat{\beta}(p) = \sqrt{(P/L)} \tag{8}$$

6. *if* $i < L$, *update*:

   $$X^{(i+1)} = \left\{X^{(i)}_{l \in J_i \setminus J^{(i)}_{t_p}}\right\} \tag{9}$$

   $$J_{i+1} = J_i \setminus J^{(i)}_{t_p} \tag{10}$$

   $$Y^{(i+1)} = Y^{(i)} - X^{(i)} \beta^T \tag{11}$$

7. *Return* to step (2).

After step (7), the vector β contains the sparse representation of the decoded word.

Since at each iteration step we choose one of the L groups corresponding to $\log_2 M$ decoded bits, according to the position of the maximum of $Z^{(i)}$, after L iterations the algorithm outputs the complete decoded word.

Once the decoding procedure of the joint demodulation-decoding scheme is completed, a further *concatenated code* can be used to correct any remaining residual errors, such as a Reed Solomon (RS) code [6] defined over $GF(M)$ [1, 2], in order to improve the error correction performance. The use of inner and outer codes is a common practice in modern communication standards, since they allow different levels of error protection depending on transmitted data requirements or channel conditions.

Matching the Galois field size with the section dimension M allows viewing each integer number belonging to the interval $[1, M]$ as a symbol in $GF(M)$.

This way of matching the Galois field size with a particular dimension of the inner concatenated scheme is similar in nature with the approach proposed in [19, 20].

Since $L \leq M$ [1], a shortened Reed Solomon [6] $(L, L-2t)$ code can be obtained from an (n_{RS}, k_{RS}) one, so that the total code rate of the concatenated scheme is:

$$R = \frac{k(L-2t)}{nL}, \quad (12)$$

where $n_{RS} - k_{RS} = 2t$ is even.

4 Simulation results

Simulation results have been obtained by using the set of parameters listed in Table 1. We have chosen a standard-de-facto 255,239 Reed Solomon shortened code, as the outer scheme. This choice influenced the other parameter settings as explained in the following.

- The number of bits $\log_2 M$ in each of the L subsections has been set equal to 8, corresponding to $M = 256$, in order to match the symbol dimension of the outer Reed Solomon code, as explained at the end of the previous section.
- Since $k = L \log_2 M$, we have chosen the number of subsections $L = 128$ in order to have a moderate codeword length equal to 2048 for a rate of $R = 0.5$.
- This value of L also satisfies the restricted isometry property (RIP) for random sensing, see Eq. (13) of [21], for which $L \leq \frac{n}{C \log M}$. For instance, that corresponds to $L < 369$ for $n = 2048$, and the constant $C = 1$.
- The constant power allocation is determined by $P = \frac{E_b}{LT}$, where E_b is the average signal energy per bit and T is the bit duration.

As shown in Figs. 2 and 3, the performance of the proposed scheme looks promising, especially for the SC code rate equal to $\frac{1}{3}$, shown in Fig. 2. The SC gain for the code rate $R = \frac{1}{2}$ could be improved by optimizing the code parameters, following for instance the procedure described in [2]. In particular, the code length could be chosen larger in order to improve the decoder performance, since the parameter L could be increased up to 256, keeping satisfied the RIP condition for $R = 0.5$.

Table 1 Simulation parameter setting

Simulation parameters	
L	128
M	256
Code rate SC code	1/3, 1/2
(n_{RS}, k_{RS})	(255, 239)
RS t	8
Code rate of the concatenated scheme	0.29167, 0.4375
Message length k	1024
Codeword length n	3072, 2048

In Fig. 4, the frame error rate (FER) of the proposed scheme is compared to those obtained by the approximate message-passing decoding algorithms presented in [14, 18]. The rate of all the schemes is 0.5, while the code parameters for the two AMP decoders are the same shown in Fig. 2 of [16], according to the code A case.

Although the proposed decoder performance is a little worse with respect the one obtained by the algorithm presented in [14] by Rush et al., the presented results are anyway interesting since they needed an extremely lower computational burdening. This may be appreciated by considering the computational complexity analysis in Table 2, where we compare the number of algebraic operations required by the proposed scheme and the AMP shown in [14] and analyzed in Table II of [16].

For the computational comparison analysis, we have only considered the AMP algorithm of [14] since it is more similar to the presented scheme, considering both implementation and performance. Anyway, it is possible to note that the actual complexity of the proposed algorithm is comparable to a simple correlation receiver for code division multiple access (CDMA) signals; nevertheless, the obtained coding gain is comparable to those attainable by more complex techniques.

5 Overview on 5G applications

The main drivers of the current research for 5G are Internet of Things (IoT), Gigabit wireless connectivity, and Tactile Internet among others [22], which call for several communication requirements. The underlying design principles, i.e., synchronism and orthogonality, of the PHY layer of the current Long-Term Evolution Advanced (LTE-A) radio access network constitute the major drawbacks for the conceived service architecture. On this purpose, as stated in [22], new waveforms are envisioned as possible solutions. In this section, we outline possible applications of this work on SCs to different aspects of wireless communications, with particular emphasis to 5G applications.

5.1 Compressed sensing

Sporadic traffic-generating devices (e.g., machine-type communication (MTC) devices in 5G IoT) awake now and then, thus they should transmit their data immediately and only coarsely synchronized, conversely to the fine synchronization procedure of LTE-A PHY layer random access [22]. On this purpose, sparse signal processing and new waveforms can be combined to efficiently deal with the sporadic traffic and control signaling in the uplink random access channel (RACH) [22]. The sparse representation of superposition codes, and sparse signal processing methodology in general, are effective for this application. Indeed, the detection and demodulation of MTC traffic in the PHY layer RACH benefits from the "bursty" nature

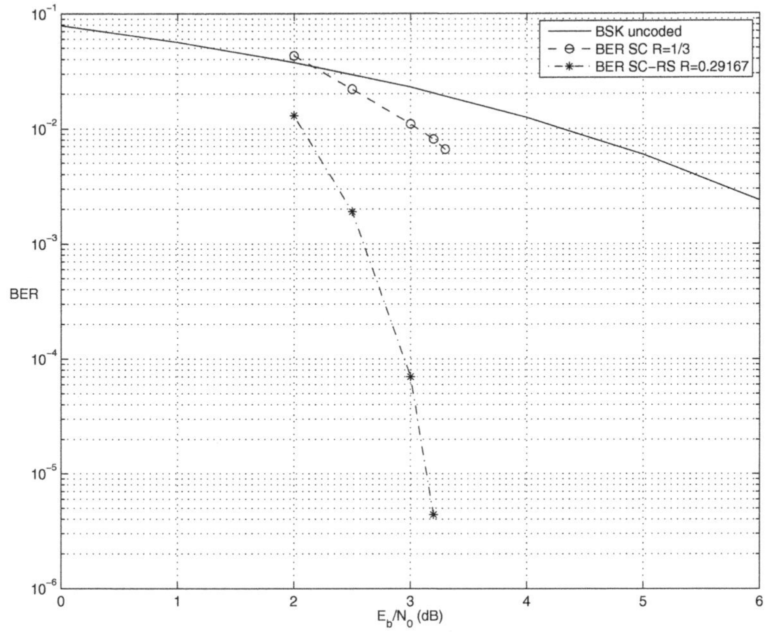

Fig. 2 BER versus $\frac{E_b}{N_0}$ for the uncoded, SC, and SC-RS architectures. The code rate is 0.29167

of signals ("sparse" in mathematical sense) and they can be described by a small set of parameters within a larger set of observables [22–24]. A possible solution is to make sparse signal processing usable for 5G RACH, exploiting joint sparsity of messages, mobile channels, and user activity [22].

5.2 Extension to more spectrally efficient schemes

Given the increasing amount of traffic data for 5G services, spectrally efficient solutions become mandatory. Although we will limit our analysis on SC to a preliminary binary case, it would be possible to increase the SC spectral efficiency by extending the presented results to the

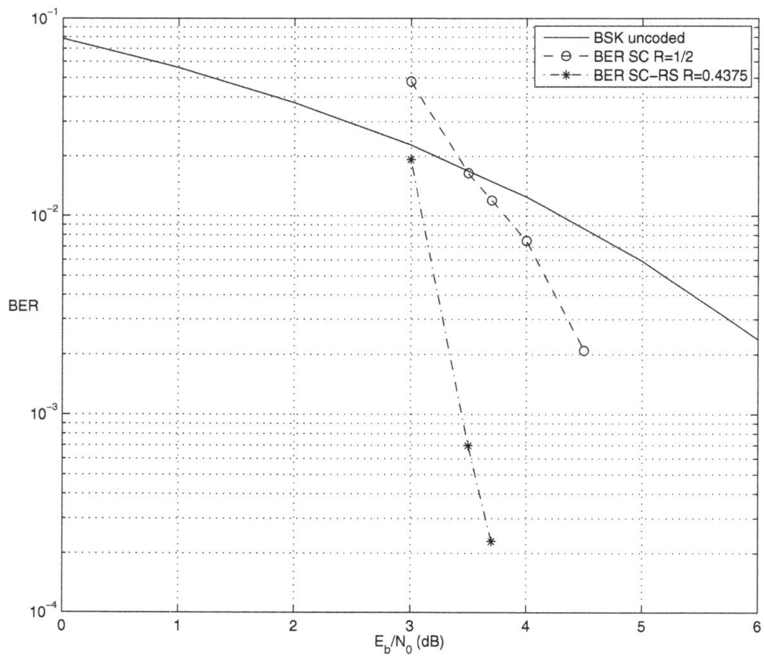

Fig. 3 BER versus $\frac{E_b}{N_0}$ for the uncoded, SC, and SC-RS architectures. The code rate is 0.4375

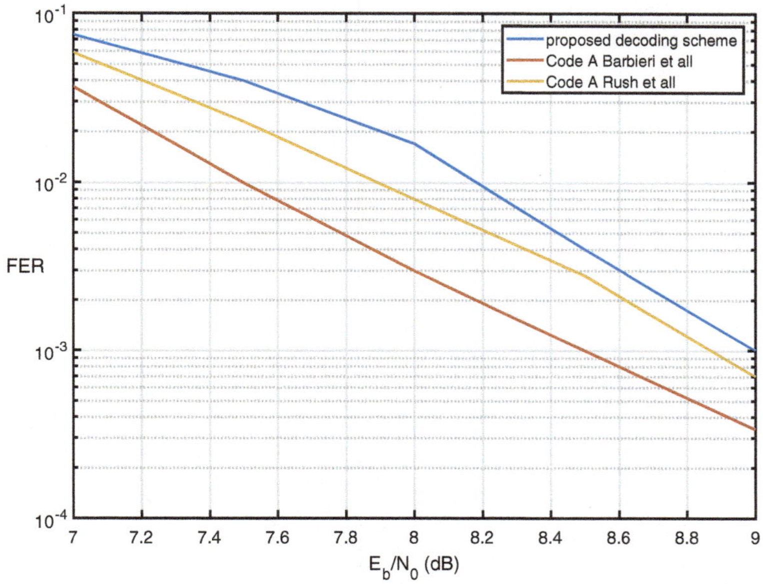

Fig. 4 FER versus $\frac{E_b}{N_0}$ comparison between the proposed scheme and two examples of SC decoded by the AMP algorithm. The code rate is 0.5 for all the three schemes

non-binary case. For instance, in [25] and [26], a signed superposition coding scheme is proposed, and in [27], an algorithm is developed to determine both the locations and amplitudes of the non-zero elements in the sparse vector.

5.3 Multiple access scheme

Superposition codes have been firstly introduced in [28] for multi-user channels, in order to compute the capacity region of the Gaussian broadcast channel [1]. In this sense, a similar scheme applies where the multiple user information is both superimposed and coded simultaneously for downlink multiple access channels.

Furthermore, this novel PHY technique could be suitable for non-orthogonal multiple access schemes [29, 30], since SC schemes can be interpreted as signal despreading with error correction capabilities, which could make non-orthogonal multiple access receivers more robust to user interferences. This would be useful also for 5G solutions, in particular for full-duplex links where the self-interference level is very severe.

Table 2 Computational complexity comparison

Number of operations per each iteration	Proposed decoder	AMP, Rush et al. [14]
Sum	$2ML(n-1)+n$	$2ML(n+1)-n-L$
Multiplication	$n(2ML+1)$	$n(2ML+1)+3ML$
Division	–	$2ML+n+1$
Exponential	–	ML

5.4 Synchronization

Synchronism is the underlying design principles of the PHY layer of today's LTE-A radio access network [22]; thus, novel synchronization solutions for the future 5G service architecture will be really useful. On this purpose, thinking about SC codes as a novel method for both modulation and coding, at least for the inner code, an interesting perspective is to develop joint algorithm for decoding and channel parameter synchronization to be applied to schemes like the ones in [31, 32].

In more details, the code structure, i.e., the sensing matrix X which is known at the receiver, may be used for both frame synchronization and fine clock recovery, instead of using pilot symbols. In this way, the spectral efficiency may be improved by decreasing the number of required pilots.

6 Conclusions

In this work, we have highlighted the feasibility of real number superposition coding. This field, closely related to compressed sensing algorithms, seems to be very promising because the simple decoding algorithm could be efficiently applied to very high data-rate signals, by using a mixed digital-analog system. This can relax the analog-to-digital conversion requirements since the algorithm acts on real numbers. On this purpose, further deep investigations will be essential.

Ongoing research works are devoted to extend the proposed scheme to non-binary signaling techniques in order to improve the final spectral efficiency. Furthermore, a real implementation in a software-defined-radio (SDR) platform will enable a more precise characterization of the true system performance, by carefully considering the synchronization related issues.

An important objective will be to look for performance as close as possible to channel capacity with dictionaries of moderately small sizes, allowing small signal spaces. From a communication theory point of view, starting from [3], new probabilistic algorithms capable of achieving performances closer to channel capacity may be sought. We believe that concatenation with a robust outer code, including Reed Solomon and q-ary LDPC, shall be necessary in order to obtain good performance.

Authors' contributions

Both the authors have conceived the proposed algorithm and developed the MATLAB code used for producing the presented results. The final manuscript has been written and revised by the two authors. Both authors read and approved the final manuscript.

Competing interests

The authors declare that they have no competing interests.

References

1. A Joseph, AR Barron, Least squares superposition codes of moderate dictionary size are reliable at rates up to capacity. IEEE Trans. Inf. Theory. **58**(5), 2541–2557 (2012)
2. A Joseph, AR Barron, Fast sparse superposition codes have near exponential error probability for R < C. IEEE Trans. Inf. Theory. **60**(2), 919–942 (2014)
3. AR Barron, S Cho, in *Proc. IEEE Int. Symp. on Information Theory (ISIT)*. High-rate sparse superposition codes with Iteratively Optimal Estimates (IEEE, Cambridge, 2012)
4. CE Shannon, A mathematical theory of communication. Bell Syst. Tech. J. **27**(5), 379–423623656 (1948)
5. GD Forney Jr., G Ungerboeck, Modulation and coding for linear Gaussian channels. IEEE Trans. Inf. Theory. **44**(6), 2384–2415 (1998)
6. S Lin, DJ Costello, *Error Control Coding, Second Edition*. (Prentice-Hall, Inc, Upper Saddle River, 2004)
7. CE Shannon, Communication in the presence of noise. Proc. IRE. **37**(1), 10–21 (1949)
8. Z Chen, in *Proceedings of the Conference on High Performance Computing Networking, Storage and Analysis. SC '09*. Optimal real number codes for fault tolerant matrix operations (ACM, New York, 2009), pp. 29–12910
9. TG Marshall Jr., Coding of real-number sequences for error correction: a digital signal processing problem. IEEE J Sel. Areas Commun. **2**(2), 381–392 (1984)
10. TH Liew, LL Yang, L Hanzo, Systematic redundant residue number system codes: analytical upper bound and iterative decoding performance over AWGN and Rayleigh channels. IEEE Trans. Commun. **54**(6), 1006–1016 (2006)
11. Z Wang, GB Giannakis, Complex-field coding for OFDM over fading wireless channels. IEEE Trans. Inf. Theory. **49**(3), 707–720 (2003)
12. E Goldoni, P Savazzi, A Vizziello, A novel channel coding scheme for rfid generation-2 systems. Electronics. **6**(1), 1–9 (2017). http://www.mdpi.com/2079-9292/6/1/4, doi:10.3390/electronics6010004
13. IF Akyildiz, S Nie, S-C Lin, M Chandrasekaran, 5g roadmap: 10 key enabling technologies. Comput. Netw. **106**, 17–48 (2016)
14. C Rush, A Greig, R Venkataramanan, Capacity-achieving sparse superposition codes via approximate message passing decoding. IEEE Trans. Inf. Theory. **63**(3), 1476–1500 (2017)
15. TT Cai, L Wang, Orthogonal matching pursuit for sparse signal recovery with noise. IEEE Trans. Inf. Theory. **57**(7), 4680–4688 (2011)
16. C Condo, WJ Gross, in *2015 IEEE Workshop on Signal Processing Systems (SiPS)*. Sparse superposition codes: a practical approach (IEEE, Hangzhou, 2015), pp. 1–6
17. C Condo, WJ Gross, Implementation of sparse superposition codes. IEEE Trans. Signal Proc. **65**(9), 2421–2427 (2017)
18. J Barbier, F Krzakala, in *2014 IEEE International Symposium on Information Theory*. Replica analysis and approximate message passing decoder for superposition codes (IEEE, Honolulu, 2014), pp. 1494–1498
19. A Marinoni, P Savazzi, P Gamba, Efficient detection and decoding of q-ary ldpc coded signals over partial response channels. EURASIP J. Wirel. Commun. Netw. **2013**(1), 18 (2013)
20. A Marinoni, P Savazzi, in *Proc. IEEE International Conference on Communications (ICC)*. Ecient receivers for q-ary LDPC coded signals over partial response channels (IEEE, Cape Town, 2010)
21. EJ Candes, MB Wakin, An introduction to compressive sampling. IEEE Signal Proc. Mag. **25**(2), 21–30 (2008)
22. G Wunder, P Jung, M Kasparick, T Wild, F Schaich, Y Chen, ST Brink, I Gaspar, N Michailow, A Festag, L Mendes, N Cassiau, D Ktenas, M Dryjanski, S Pietrzyk, B Eged, P Vago, F Wiedmann, 5GNOW: non-orthogonal, asynchronous waveforms for future mobile applications. IEEE Commun. Mag. **52**(2), 97–105 (2014)
23. HF Schepker, C Bockelmann, A Dekorsy, in *2013 IEEE 77th Vehicular Technology Conference (VTC Spring)*. Coping with cdma asynchronicity in compressive sensing multi-user detection (IEEE, Dresden, 2013), pp. 1–5
24. H Zhu, GB Giannakis, Exploiting sparse user activity in multiuser detection. IEEE Trans. Commun. **59**(2), 454–465 (2011)
25. AC Gilbert, JA Tropp, in *Proc. IEEE International Symposium on Information Theory (ISIT)*. Applications of sparse approximation in communications (IEEE, Adelaide, 2005)
26. JA Tropp, Just relax: convex programming methods for identifying sparse signals in noise. IEEE Trans. Inf. Theory. **52**(3), 1030–1051 (2006)
27. DL Donoho, A Maleki, A Montanari, Message-passing algorithms for compressed sensing. Proc. Natl. Acad. Sci. **106**(45), 18914–18919 (2009)
28. TM Cover, Broadcast channels. IEEE Trans. Inf. Theory. **18**(1), 2–14 (1972)
29. P Wang, J Xiao, L Ping, Comparison of orthogonal and non-orthogonal approaches to future wireless cellular systems. IEEE Veh. Technol. Mag. **1**(3), 4–11 (2006)
30. P Wang, J Xiao, L Ping, Correction to "Comparison of Orthogonal and Non-Orthogonal Approaches to Future Wireless Cellular Systems". EEE Veh. Technol. Mag. **1**(4), 42 (2006)
31. P Savazzi, P Gamba, L Favalli, An iterative algorithm for joint symbol timing recovery and equalization of short bursts. J. Commun. **3**(4), 34–40 (2008)
32. P Savazzi, P Gamba, Iterative symbol timing recovery for short burst transmission schemes. IEEE Trans. Commun. **56**(10), 1729–1736 (2008)

ATRAS: adaptive MAC protocol for efficient and fair coexistence between radio over fiber-based and CSMA/CA-based WLANs

Kaito Funabiki[1*], Takayuki Nishio[1], Masahiro Morikura[1], Koji Yamamoto[1], Daisuke Murayama[2] and Katsuya Nakahira[2]

Abstract

In this paper, we propose a medium access control (MAC) protocol to allow a radio over fiber-based wireless local area network (RoF-based WLAN) to coexist with legacy carrier sense multiple access with collision avoidance (CSMA/CA)-based WLANs. In RoF-based WLANs, there are long propagation delays between access points (APs) and stations (STAs). When an RoF-based WLAN uses the conventional CSMA/CA protocol and coexists with legacy WLANs, the propagation delay causes unexpected frame collisions and unfairness between the RoF-based WLAN and legacy WLANs. The proposed protocol reduces unexpected frame collisions and improves fairness between the RoF-based WLAN and legacy WLANs in environments where they coexist. It is only necessary to apply the proposed protocol to the AP in the RoF-based WLAN (RoF AP); there is no need to modify STAs, which is an advantage of the protocol. In the proposed scheme, the RoF AP transmits frames during the transmission and reception of frames in the legacy WLAN, such that the frame transmitted by the AP arrives at a destination STA one short interframe space (SIFS) period after the channel became idle, and thus the RoF AP's frames do not collide with other frames. As a result, the proposed method decreases the likelihood of frame collisions and increases the throughput of the RoF-based WLAN. In addition, the proposed method provides adaptive adjustment of the transmission probability, which enables the RoF-based WLAN and the legacy WLANs to fairly share wireless channels. The proposed protocol has been investigated for IEEE 802.11a/b/g WLANs. Numerical analysis and simulation evaluations show that the proposed scheme increases the TCP throughput of the RoF-based WLAN up to the same level as that in legacy WLANs.

Keywords: Radio over fiber, Long distance WLAN, Coexistence, MAC protocol, Propagation delay

1 Introduction

Wireless communication is becoming increasingly common, and many wireless standards are now available for wireless local area networks (WLANs). Accordingly, network operators must prepare dedicated wireless access points (APs) for each system, every time a new wireless standard emerges. Existing WLAN standards, such as IEEE 802.11b/g/a/n/ac, are utilized in the 2.4 and 5 GHz bands. Moreover, owing to increasing demand for more capacity, WLAN standards using the millimeter-wave (mmWave) band, such as IEEE 802.11ad and IEEE 802.11ay using 60 GHz, are actively being discussed for the provision of broadband wireless access [1]. In addition, the demand for Internet of Things (IoT) applications has drawn attention to sub-1 GHz wireless standards, such as IEEE 802.15.4 [2]. The upcoming IEEE 802.11ah standard has the potential to serve wide-area and low-cost wireless networks for IoT devices [3].

Access networks are required to support various communication standards with different physical (PHY) and medium access control (MAC) layer parameters by a single AP. Network function virtualization (NFV) is a promising method with which to economically realize such flexible wireless access networks [4].

One of the key enablers of NFV for wireless access networks is radio over fiber (RoF) technology. In wireless systems with RoF technology, only radio frequency (RF) units connected to a central office are located in remote

*Correspondence: nishio@i.kyoto-u.ac.jp
[1]Graduate School of Informatics, Kyoto University, Kyoto, Japan
Full list of author information is available at the end of the article

areas and the central office operates the PHY signal processing and MAC functionality [5–9]. The RF units, which are connected to the central office via an optical fiber, convert the optical signals to electrical signals before transmitting them over the radio channel, and vice versa. Using RoF for wireless systems enables the centralization of the wireless signal processing functions that are currently performed by each AP. The benefit of this approach is that WLANs or wireless personal area networks (WPANs) are able to support any wireless standard by only changing the processing functions at the central office.

An RoF-based access architecture for integrating multiple wireless systems was discussed in [8, 9]. However, the integration of the conventional carrier sense multiple access with collision avoidance (CSMA/CA)-based wireless standard with an RoF-based architecture requires research because the long propagation delays in the optical fiber links may cause the CSMA/CA protocol to behave unexpectedly, resulting in a decrease in the system throughput. When designing a MAC protocol for RoF-based WLANs, it is important to take into account backward compatibility with IEEE 802.11 standard to cover existing devices. Moreover, it is also important to take into account how best to support coexistence with legacy CSMA/CA-based systems. WLANs are used widely and many APs are densely deployed. The dense environment is typical for apartment buildings and offices. Because of the limited number of orthogonal frequency channels in 2.4 and 5 GHz band, situations where the neighboring WLANs utilizing the same channel can occur. This paper focuses on a MAC layer solution for IEEE 802.11 WLANs that employ an RoF architecture (RoF-based WLAN) and addresses the coexistence problems between the RoF-based and legacy WLANs.

In WLANs that have an RoF-based architecture, delays longer than one half of a slot time cause unexpected frame collision patterns and increase the probability of frame collisions when the WLAN operates using a conventional MAC protocol. In order to solve the severe frame collision problems caused by large propagation delays, several alternative MAC protocols have been proposed [10–12]. For example, Deronne et al. proposed a method for optimizing the slot time value so that unexpected collisions do not occur [10]. However, the use of large slot times increases the overhead, which decreases the system throughput. Moreover, if an RoF-based WLAN coexists with legacy WLANs, the RoF-based WLAN stations that use a large slot time are only able to transmit a few DATA frames because their slot time is much larger than that of the legacy WLAN stations. Other schemes described in [11, 12] improve the throughput of RoF-based WLANs by using the frame aggregation and BlockACK mechanisms. However, these schemes do not solve the frame collision problems caused by long propagation delay because frame aggregation does not reduce the number of frame collisions, not to mention that there are still open issues concerning the coexistence of RoF and legacy WLANs.

In this paper, we propose a new MAC protocol for an AP, called "advanced transmission for RoF-based WLAN system" scheme. The ATRAS scheme has been investigated for IEEE 802.11a/b/g WLANs. The ATRAS scheme solves the unfairness and TCP throughput degradation problems by increasing the number of transmission opportunities for an AP in RoF-based WLANs (RoF AP). Furthermore, the ATRAS scheme is able to control the throughput performance of an RoF-based WLAN by changing the transmission probability of the RoF AP. We also provide a mathematical throughput model of the coexistence environment, as well as an adaptive transmission probability control scheme for the RoF AP using the ATRAS scheme.

The contributions of this paper are as follows: (1) a design for an adaptive MAC protocol for RoF APs that enables fair and efficient coexistence without modification of the STAs and legacy WLANs; as far as we know, there is no MAC protocol suitable for the coexistence environment, (2) a mathematical throughput model of the proposed scheme and parameters determined based on the throughput model, and (3) numerical evaluations using the network simulator QualNet [13].

This paper is organized as follows. Section 2 introduces a system model of the coexistence environment and provides a problem statement. Section 3 describes related works. In Section 4, we describe how the proposed MAC protocol improves the throughput of an RoF-based WLAN, and then introduce a mathematical throughput model of the same. Section 5 shows the results of numerical evaluations performed by computer simulations, and our conclusions provided in Section 6.

2 RoF-based WLAN

2.1 System model

In this paper, we consider an environment where an RoF-based WLAN coexists with legacy WLANs, as shown in Fig. 1. The IEEE 802.11 standard is used for the WLANs and the WLANs are operated in the same frequency channel in the 2.4 or 5 GHz band.

The WLAN STAs associated with the legacy AP are denoted as legacy STAs, while the WLAN STAs associated with the RoF AP are denoted as RoF STAs. Furthermore, we define "local nodes" as the legacy APs, legacy STAs, and RoF STAs. All STAs operate according to the IEEE 802.11 standard MAC protocol [14].

The RoF AP in the RoF-based WLAN consists of an RF module in a premises of the customer and a processing unit located at a central office. The RF module is composed of an antenna, an amplifier, an analog-to-digital (AD)/digital-to-analog (DA) converter, and an interface

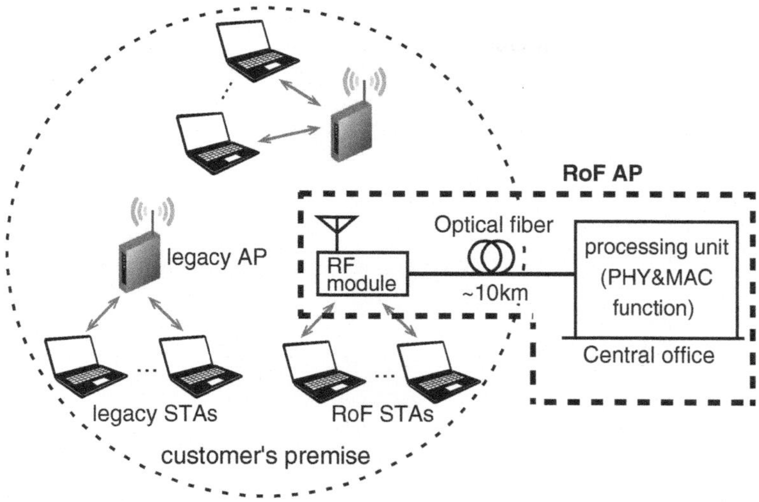

Fig. 1 Coexistence of RoF-based WLAN and legacy WLANs. Legacy APs and STAs, RoF STAs, and RF unit of RoF AP are deployed densely

circuit, and is connected to a centralized processing unit via an optical fiber. The processing unit handles the PHY and MAC functions for an AP. In this system, the frames transmitted by an STA are received by the RF module, and then amplified, down-converted, digitized, and converted into optical signals before being transmitted through the optical fiber to the processing unit located at the central office. We assume that distance between the RF module of the RoF AP and the other nodes is several meters to several hundred meters and the propagation delay of wireless signal is negligible.

It should be noted that there is a non-negligible propagation delay δ_{RoF} between the processing unit and STAs due to the presence of the optical links. In Japan, most telephone subscribers are located within 5 km of a telecommunication office. In order to accommodate most customers, our simulation assumed a maximum fiber length of 10 km, which corresponds to a one-way propagation delay of 50 μs. This delay can be derived from the refractive index of single-mode optical fiber, i.e., 1.5 [15].

2.2 Delay-induced issues on CSMA/CA

In this section, we discuss some of the issues that impair the throughput of RoF-based WLANs and degrade the fairness between an RoF-based WLAN and legacy WLANs in environments where they coexist.

2.2.1 Increase in collision between frames transmitted by RoF AP and local nodes

In the RoF-based WLAN, frames transmitted by the RoF AP or STAs are received after δ_{RoF}. The long propagation delay causes a large displacement in the carrier-sense timing at the receive side. This phenomenon causes the carrier-sense mechanism to underperform, causes unexpected collisions, and increases the probability of collisions.

In the RoF-based WLAN, frames transmitted in different slot times could collide while collisions between frames only occur if multiple nodes transmit in the same slot time in an ordinary WLAN with a small propagation delay. Figure 2 depicts a frame collision event between an RoF AP and local nodes. In the figure, a local node transmits an ACK frame, and after the ACK frame transmission, an RoF AP transmits a DATA frame based on a conventional distributed coordination function (DCF) in the kth slot time. The frames transmitted by the RoF AP in the kth slot time could collide with a frame transmitted by a local node in the kth and other slot times, the $(k+1)$th or $(k+2)$th slot time in the figure. In the $(k+3)$th slot time, local nodes can detect signal transmitted by the RoF AP, thus local nodes suspend their transmission. This phenomenon increases the frame collision probability.

When TCP downlink traffic is the dominant type of traffic in the RoF-based and legacy WLANs, TCP-DATA frames transmitted by the RoF AP could suffer from the frame collisions. Accordingly, this severely degrades the TCP throughput performance of the RoF-based WLAN.

2.2.2 ACK frame collision

In a conventional DCF, acknowledge (ACK) frames never collide with other frames because the ACK frames are transmitted after the short interframe space (SIFS) period. However, the ACK frames may collide with other frames in the RoF-based WLAN owing to the long propagation delays.

Figure 3 illustrates a collision between an ACK frame from an RoF AP and a frame transmitted by a local node, where the RoF AP replies ACK frame to an RoF STA and

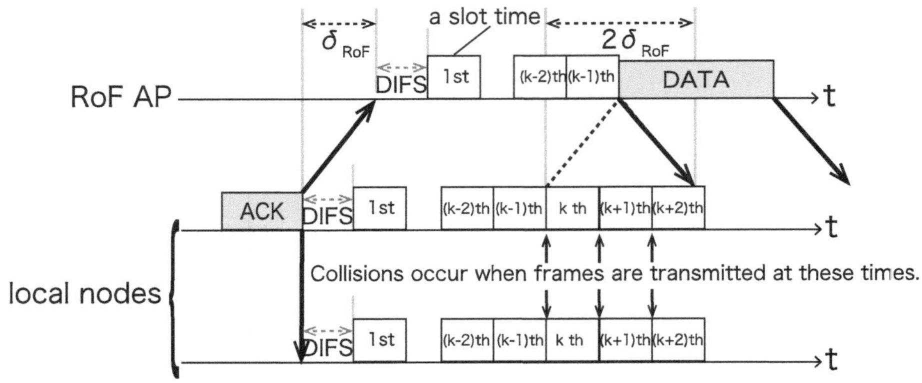

Fig. 2 Frame collision between RoF AP and local nodes

a local node transmits a DATA frame before the ACK is received by the RoF STA, which collides with the ACK frame. Although local nodes wait for a network allocation vector (NAV) period indicated in the DATA frame transmitted by the RoF STA before transmitting their frames, long propagation delays cause the NAV period to expire and the local noses could transmit frames before the RoF STA receives the ACK frame from the RoF AP. The ACK collision increases the number of retransmission in the RoF-based WLAN and degrades the throughput performance of the RoF-based WLAN.

3 Related works

3.1 Slot time adjustment scheme

One method to address this problem is to optimize the slot time value to account for the propagation delay. Deronne et al. proposed a method for optimizing the slot time value of RoF-based WLANs [10]. The original slot time value σ is defined in the IEEE 802.11 standard [14] as *aCCATime* + *aRxTxTurnaroundTime* + *aAirPropagationTime* + *aMACProcessingDelay*, where *aCCATime* is the time required to determine the state of the channel, *aRxTxTurnaroundTime* is the time required to change from receive to transmit, *aAirPropagationTime* is the roundtrip air propagation time, and *aMACProcessingDelay* is the time needed for computations in the MAC layer. In [10], the slot time value was adapted according to the following formula: Slottime $= \sigma + 2(\delta_{RoF} - 4.5\,\mu s)$. Adapting the slot time value decreases the number of unexpected frame collisions mentioned in Section 2.2.1 since the carrier-sense mechanism still functions normally.

However, extending the slot time increases the overhead duration in CSMA/CA, which leads to unfairness between an RoF-based WLAN and a legacy WLAN. For example, when the propagation delay is 50 μs, the slot time value must be greater than 100 μs, which is 11 times longer than the original slot time of 9 μs in the IEEE 802.11a standard [14]. Moreover, when an RoF-based WLAN uses an extended slot time value but a coexisting legacy WLAN employs the usual slot time, the nodes in the RoF-based WLAN are not able to transmit enough frames because their backoff period is much longer than that in the legacy WLAN, as shown in Fig. 4. This results in serious throughput degradation in the RoF-based WLAN.

3.2 Piggy back access scheme

In our previous work [16], we presented a transmission scheme to increase the number of successful RoF AP frame transmissions by applying a piggy back access (PBA) scheme [17] to an RoF AP in an environment where only

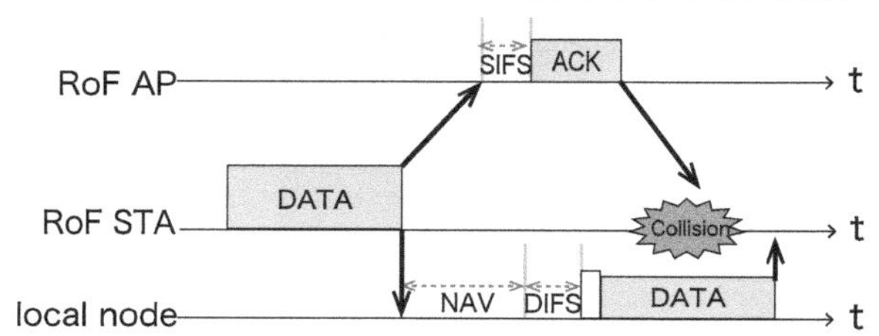

Fig. 3 ACK frame collision. An ACK frame transmitted by an RoF AP collides with a DATA frame transmitted by a local node

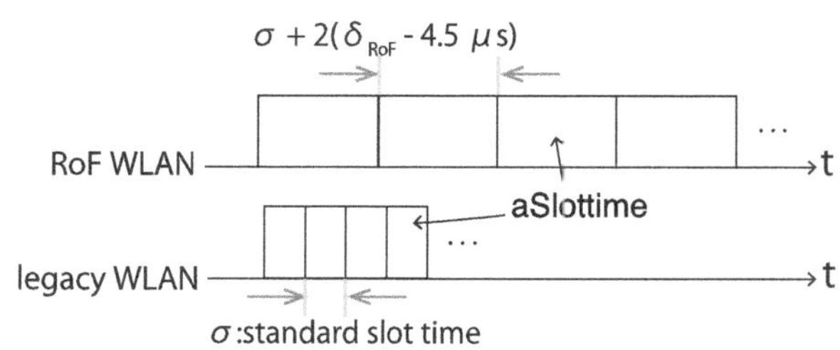

Fig. 4 Backoff time in slot time adjustment scheme. Backoff time of the RoF-based WLAN becomes much larger than that of the legacy WLAN

RoF-based WLANs exist. In that case, the RoF AP transmits a DATA frame after a SIFS period instead of after a DIFS period plus the backoff time when the RoF AP is transmitting an ACK frame to an STA. The RoF STA receives the DATA frame one SIFS period after the RoF STA received the ACK frame so that the RoF AP's DATA frame does not collide.

This PBA scheme works well in a bidirectional traffic scenario where RoF STAs frequently transmit frames; however, since TCP downlink traffic is dominant in actual wireless Internet access [18], where STAs do not have full buffer traffic, the throughput performance of the PBA scheme may decrease in a real-world wireless Internet access environment.

3.3 NAV extension method

To solve the ACK frame collision problem in Section 2.2.2, we also proposed extending the NAV period by using the request-to-send (RTS)/ clear-to-send (CTS) procedure for an RoF-based WLAN [16], as illustrated in Fig. 5. When an RoF AP receives an RTS frame, the RoF AP replies with a CTS frame that includes the extended NAV duration. Specifically, the NAV duration value of the CTS frame transmitted by the RoF AP T_{exNAV} is defined as $T_{\mathrm{exNAV}} = T_{\mathrm{NAV,st}} + 3\delta_{\mathrm{RoF}}$, where $T_{\mathrm{NAV,st}}$ is the conventional NAV duration of a CTS frame defined in the IEEE 802.11a standard [14]. This method prevents ACK frame collisions and is backwards compatible.

3.4 Collision avoidance using TDMA fashion

Collision free MAC protocols exploiting a TDMA fashion have been studied to solve collision problems in WLANs [19–25]. In these protocol, there are periods to transmit frames in a TDMA fashion, where APs and STAs can transmit frames without collision. Allocating TDMA timeslots to RoF APs taking into account the large delay could avoid delay induced collisions. However, these approaches require major modifications to a MAC protocol of the STAs, which means that conventional devices can not be used in the WLANs using the protocol exploiting a TDMA fashion. Our protocol focuses on RoF-based WLAN with backward compatibility with IEEE 802.11 standard, where existing WLAN devices are available. Therefore, these approaches exploiting TDMA are not suitable for our target application.

4 Proposed scheme (ATRAS)

In this section, we describe the proposed ATRAS method and its application only to the RoF AP. In this method, we assume that the RoF STAs use the conventional DCF procedure except with respect to the values of an ACK timeout interval and a CTS timeout interval. To prevent frame transmission failure due to timeout, it is necessary to extend the ACK and CTS timeout intervals of the RoF STAs. The timeout interval T_{timeout} should be set to the value obtained from $T_{\mathrm{timeout}} = T_{\mathrm{st}} + 2\delta_{\mathrm{RoF}}$, where T_{st} is a conventional timeout interval used in IEEE 802.11. The values of the ACK and CTS timeout intervals can be changed by setting or updating driver software [26, 27].

In this section, we describe the transmission procedures of the RoF AP in the ATRAS scheme and the theoretical model of the total throughput for UDP bidirectional full buffer traffic. We also explain the adaptive control scheme employing the transmission probability α ($0 < \alpha \leq 1$), the initial value of which is derived from the theoretical throughput model.

4.1 Basic procedure of the ATRAS scheme

In order to solve the RoF AP frame collision problem shown in Fig. 2 in the environment where an RoF-based WLAN and legacy WLANs coexist, we propose the ATRAS scheme. The procedures in the ATRAS scheme are performed when the RoF AP overhears a CTS frame transmitted by a node in a legacy WLAN, or when the RoF AP overhears a DATA frame transmitted by a node in a legacy WLAN not using RTS/CTS. Figure 6 depicts the ATRAS scheme when a legacy WLAN uses RTS/CTS. The following sections explain and evaluate the proposed scheme for a case where legacy WLANs use RTS/CTS. The RoF AP receiving a CTS frame transmitted by a legacy

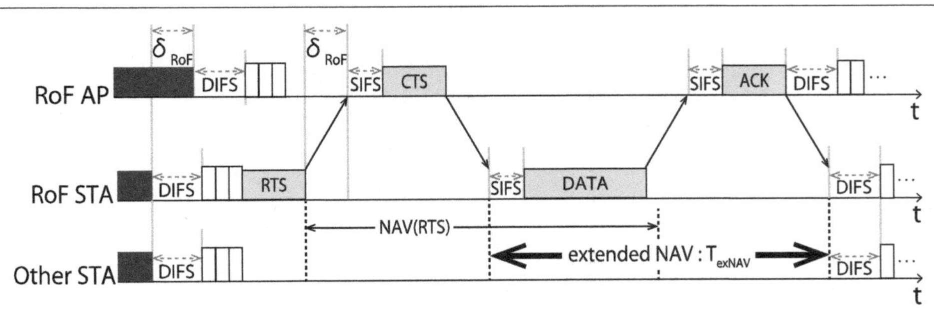

Fig. 5 NAV extension method in an RoF-based WLAN. A CTS frame transmitted by the RoF AP extends NAV at STAs

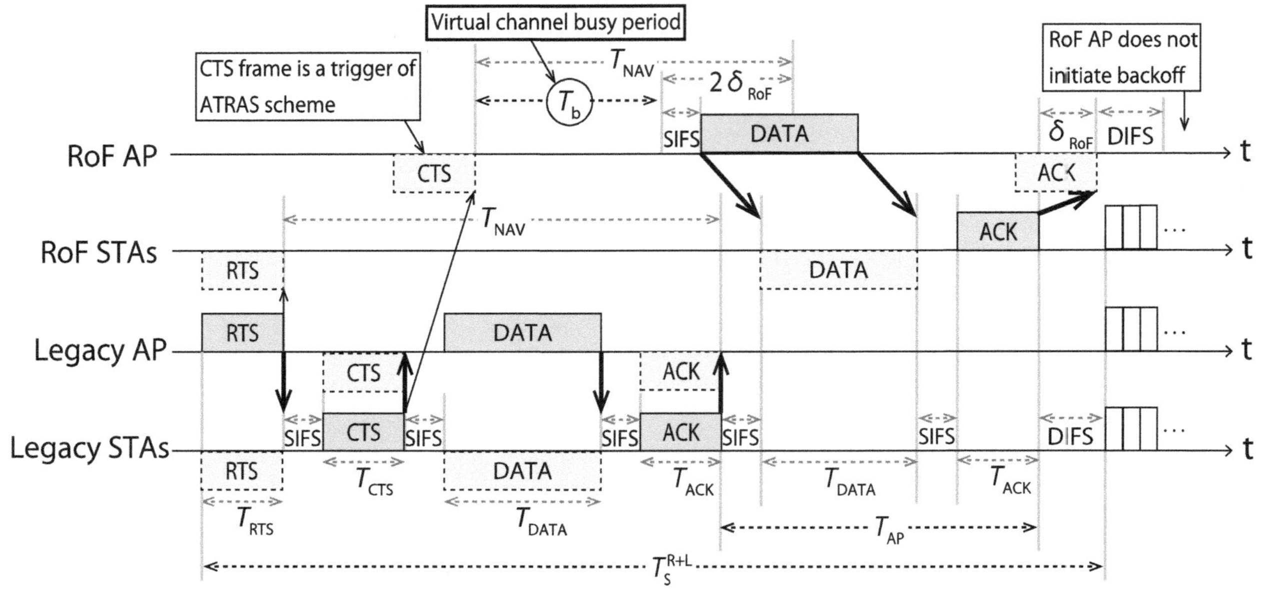

Fig. 6 Transmission procedure of ATRAS scheme when RTS/CTS is used in legacy WLANs. The frame transmission of the RoF AP is triggered by CTS frame transmitted by a legacy STA

STA sets a virtual channel busy period, the duration T_b of which is defined as $T_b = T_{NAV} - 2\delta_{RoF}$, where T_{NAV} is the duration of the NAV period. The virtual channel busy period is introduced to avoid collision between frames transmitted by a local node and the RoF AP. If the RoF AP transmits a frame in the virtual channel busy period, the frame collides with a frame transmitted by a local node. The RoF AP transmits its DATA frame SIFS period after the virtual channel busy period with the probability α, which is introduced to control throughput of the RoF AP. DATA frames transmitted by the RoF AP using the ATRAS scheme are received by local nodes SIFS period after the channel actually becomes idle. Therefore, the frame collision mentioned in Section 2.2.1 does not occur.

The parameter α represents the frame transmission probability, and the WLAN operator at the central office can control the downlink throughput of the RoF-based WLAN by adjusting this probability so that the channel is fairly shared with the RoF-based WLAN and the legacy WLAN. Introducing transmission probability to control throughput is commonly used approach in WLANs [28]. With the probability α, the RoF AP transmits its DATA frame SIFS period after the virtual channel busy period when the RoF AP receives a CTS frame from a legacy AP or STA. With the probability $(1 - \alpha)$, the RoF AP does not set a virtual channel busy period but waits for its NAV period T_{NAV}. We will describe the method to adjust the probability in Section 4.3.

The number of transmission opportunities for the RoF AP using the ATRAS scheme depends on α and the traffic volume of the legacy WLANs. When the traffic of the legacy WLANs is low, the RoF AP is not able to obtain many transmission opportunities by the ATRAS scheme. Using backoff or/and piggyback access scheme with the ATRAS scheme could increase transmission opportunity of the RoF AP. For example, it is conceivable that the RoF AP could measure the traffic volume of the legacy WLANs by overhearing frames transmitted by them and determine a certain threshold value, and when the traffic volume exceeded the threshold value, the RoF AP would use the ATRAS scheme. However, such a protocol design of adaptive usage of multiple medium access schemes is out of focus of this paper and we would like to include it to our future work.

4.2 Theoretical throughput model

We now develop a total throughput model for legacy WLANs and an RoF-based WLAN with the proposed scheme when they coexist. For simplicity, we provide a theoretical model for UDP bidirectional full buffer traffic, and we determine the initial value of α in the presence of TCP traffic by using an approximated model based on UDP bidirectional full buffer traffic, the details of which are described in the next section. Our model is an extension of a well-known analytical model for conventional WLANs [29].

The theoretical system throughput was derived for a conventional WLAN in [29]:

$$S = \frac{P_s P_{tr} E[P]}{(1 - P_{tr})\sigma + P_{tr} P_s T_s + P_{tr}(1 - P_s) T_c}, \quad (1)$$

where P_s and P_{tr} are the probability of successful transmission, and the probability of transmission, respectively, $E[P]$ is the average packet payload size of the DATA frames, σ is the slot time size, and T_s and T_c are the average periods during which the channel is busy due to the successful transmission and collision of frames, respectively.

$$T_s = T_{RTS} + \text{SIFS} + \delta + T_{CTS} + \text{SIFS} + \delta + T_{DATA} + \text{SIFS} + \delta + T_{ACK} + \delta + \text{DIFS}, \quad (2)$$

$$T_c = T_{RTS} + \text{EIFS}, \quad (3)$$

where T_{RTS}, T_{CTS}, T_{DATA}, and T_{ACK} are the required periods for transmitting an RTS frame, a CTS frame, a DATA frame, and an ACK frame, respectively, SIFS, DIFS, and EIFS are the SIFS period, the DIFS period, and the extended inter frame space (EIFS) period defined in [14], respectively, and δ is the one-way propagation delay in the WLAN.

Based on (1)–(3), we formulate the total system throughput S^{R+L} in an environment in which an RoF-based WLAN using the ATRAS scheme coexists with a legacy WLAN, as follows:

$$S^{R+L} = \frac{P_s P_{tr}(1+\alpha')E[P]}{(1-P_{tr})\sigma + P_{tr}P_s T_s^{R+L} + P_{tr}(1-P_s)T_c^{R+L}}, \quad (4)$$

$$T_s^{R+L} = T_{RTS} + \text{SIFS} + T_{delay} + T_{CTS} + \text{SIFS} + T_{delay} + T_{DATA} + \text{SIFS} + T_{delay} + T_{ACK} + T_{delay} + \text{DIFS} + T_{AP}, \quad (5)$$

$$T_{AP} = \alpha'(SIFS + T_{DATA} + \text{SIFS} + T_{ACK}), \quad (6)$$

$$T_c^{R+L} = T_c = T_{RTS} + \text{EIFS}. \quad (7)$$

In contrast to the conventional model, the term α' is added into (4), where α' is the ratio between the number of additional frame transmissions that use the ATRAS scheme and the number of frame transmissions that use the backoff procedure, as follows:

$$\alpha' = \frac{N_A^L + N_S^L}{N_A^L + N_S^L + N_S^R}\alpha, \quad (8)$$

where N_S^R, N_A^L, and N_S^L are the numbers of RoF STAs, legacy APs, and legacy STAs, respectively. The RoF AP transmits a DATA frame only when it overhears that a CTS frame has been transmitted by a legacy AP or STA. Thus, the larger ratio of the legacy AP and STAs to the RoF STAs increases the throughput of the RoF AP. When the number of legacy nodes is small, α should be large to obtain large throughput in the RoF AP.

Moreover, in (5), T_{AP} is the time required for the transmission from the RoF AP, as shown in Fig. 6, and T_{delay} is the average propagation delay for frames transmitted in the RoF and legacy WLANs. T_{delay} is calculated using the following equation:

$$T_{delay} = \frac{(N^R - 1)}{N}\delta_{RoF} + \frac{N^L}{N}\delta, \quad (9)$$

where N is the total number of local nodes, N^R is the total number of RoF APs and RoF STAs, and N^L is the total number of legacy APs and STAs, i.e., $N = N_A^L + N_S^L + N_S^R$, $N^R = N_A^R + N_S^R$, where N_A^R is the number of RoF APs, and $N^L = N_A^L + N_S^L$.

In the ATRAS scheme, the RoF AP does not transmit DATA or RTS frames using the backoff procedure but only transmits DATA frames using the ATRAS scheme. In addition, the frames transmitted by the RoF AP do not collide with other frames from local nodes. Therefore, the AP's transmissions do not affect the procedures in the local nodes in the DCF, and P_s and P_{tr} can be obtained by using the method that was described in [29]. The total system throughput can be calculated using (4)–(9).

4.3 Adaptive transmission probability control

The throughput of an RoF AP can be controlled by adjusting the transmission probability α. We control α such that the throughput of the RoF-based WLAN and legacy WLAN converge to the same value in full buffer traffic conditions. In this section, we describe the decision procedure to determine α. In this procedure, the initial value of α is determined based on the analytical throughput model; however, the initial value deviates from the optimal value because of traffic fluctuations. Thus, the proposed procedure adaptively adjusts the value of α after the initial setup.

4.3.1 Objective of the adaptive control

The objective of the adaptive transmission probability control is that the throughput of an RoF-based WLAN and a legacy WLAN converge to the same amount. We therefore consider the following minimization problem:

$$\underset{\alpha}{\text{minimize}} \quad |S^R - S^L| \qquad \text{subject to} \quad 0 \le \alpha \le 1. \quad (10)$$

S^R and S^L are the throughput of the RoF-based WLAN and legacy WLAN, respectively. From (4), the throughput of an RoF AP S_A^R can be derived from the following equation:

$$S_{\mathrm{A}}^{\mathrm{R}} = \frac{P_{\mathrm{s}} P_{\mathrm{tr}} \alpha' \mathrm{E}[P]}{(1-P_{\mathrm{tr}})\sigma + P_{\mathrm{tr}} P_{\mathrm{s}} T_{\mathrm{s}}^{\mathrm{R+L}} + P_{\mathrm{tr}}(1-P_{\mathrm{s}}) T_{\mathrm{c}}^{\mathrm{R+L}}}. \tag{11}$$

The total throughput of the local nodes can be calculated by subtracting $S_{\mathrm{A}}^{\mathrm{R}}$ from the total system throughput $S^{\mathrm{R+L}}$. When the local nodes have the same number of transmission opportunities, the average throughput s_i of local node i is shown in (12).

$$s_i = \left(S^{\mathrm{R+L}} - S_{\mathrm{A}}^{\mathrm{R}}\right)/N. \tag{12}$$

Therefore, the values of S^{R} and S^{L} can be derived from:

$$S^{\mathrm{R}} = S_{\mathrm{A}}^{\mathrm{R}} + \sum_{i \in N_{\mathrm{S}}^{\mathrm{R}}} s_i, \tag{13}$$

$$S^{\mathrm{L}} = \sum_{i \in N^{\mathrm{L}}} s_i, \tag{14}$$

where S^{R} is a summation of the throughput of RoF AP derived from (11) and the throughput of RoF STAs, each of which is derived from (12), and S^{L} is a summation of throughput of legacy AP and STAs, each of which is also derived from (12).

4.3.2 *Initial value of α*

Here, we attempt to find an initial value of α :α_0 based on (10) for the case of UDP and TCP traffic.

α_0 under UDP traffic: To accomplish our objective of adaptive control based on (10), we need to determine α_0 by solving the following equation:

$$|S^{\mathrm{R}} - S^{\mathrm{L}}| = 0. \tag{15}$$

If the above equation is satisfied, then the throughput of the RoF and legacy WLANs are identical. By solving (15) using (8) and (11)–(14), we obtain

$$\alpha_0 = \frac{N_{\mathrm{A}}^{\mathrm{L}} + N_{\mathrm{S}}^{\mathrm{L}} - N_{\mathrm{S}}^{\mathrm{R}}}{N_{\mathrm{A}}^{\mathrm{L}} + N_{\mathrm{S}}^{\mathrm{L}}}. \tag{16}$$

Accordingly, α_0 is determined by the number of local nodes sending UDP traffic. Since $1 \geq \alpha \geq 0$, when $N_{\mathrm{A}}^{\mathrm{L}} + N_{\mathrm{S}}^{\mathrm{L}} \leq N_{\mathrm{S}}^{\mathrm{R}}$, (15) cannot be satisfied. In the case, the number of RoF STAs should be limited or other fairness criteria should be selected, which is beyond the scope of this paper.

α_0 under TCP traffic: Bruno et al. reported that TCP traffic results in very low contention in a WLAN because of the TCP flow control algorithm [30]. In the presence of persistent TCP upload and download connections, the average number of STAs that compete for transmission has an upper bound of less than two on average, regardless of the total number of STAs in a WLAN. Based on this result, we can assume that the number of STAs competing for transmission in the RoF-based WLAN is equal to that in the legacy WLAN, $N_{\mathrm{S}}^{\mathrm{R}} = N_{\mathrm{S}}^{\mathrm{L}}$. Thus, the estimated value of α satisfying (15), which should be used for initial value of α, is obtained from (16) as

$$\alpha_0 = \frac{N_{\mathrm{A}}^{\mathrm{L}}}{N_{\mathrm{A}}^{\mathrm{L}} + N_{\mathrm{S}}^{\mathrm{L}}}, \tag{17}$$

which is a ratio of the number of legacy APs to the total number of nodes in the legacy WLANs. Since CSMA/CA allows each contending node to obtain the same amount of transmission opportunity and the ATRAS scheme is triggered by frames transmitted by legacy nodes, the RoF AP using the ATRAS scheme can obtain $\alpha C(N_{\mathrm{A}}^{\mathrm{L}} + N_{\mathrm{S}}^{\mathrm{L}})$ transmission opportunity, where C is a transmission opportunity obtained by a node. Thus, when using α_0 obtained from (17), the RoF AP obtains $CN_{\mathrm{A}}^{\mathrm{L}}$ transmission opportunity, which is equal to the total number of transmission opportunity obtained by legacy APs.

4.3.3 *Adaptive control procedure by α*

We now describe the adaptive control procedure and α. The term α_{t+1}, which represents the value of α at time $t+1 (t \geq 1)$, can be derived as follows:

$$\alpha_{t+1} = \alpha_t - \Delta\alpha_t, \tag{18}$$

$$\Delta\alpha_t = \frac{A}{1 + \mathrm{e}^{-g(S^{\mathrm{R}}(t) - S^{\mathrm{L}}(t))}} - \frac{A}{2}, \tag{19}$$

where $\Delta\alpha_t$ refers to a variation of α. (19) uses the sigmoid function [31]; $S^{\mathrm{R}}(t)$ and $S^{\mathrm{L}}(t)$ represent the estimated throughput of an RoF and legacy WLAN from $t-1$ to t, the units of $S^{\mathrm{R}}(t)$ and $S^{\mathrm{L}}(t)$ are bit/s, g indicates the gain of the sigmoid function, and A is a constant that determines the range of $\Delta\alpha_t$, i.e., $-A/2 < \Delta\alpha_t < A/2$. The throughput of both WLANs can be estimated by the RoF AP, which overhears the frame headers and counts the frames transmitted in each WLAN. The RoF AP regularly updates the value of α using (18) and (19) based on the estimated throughput. The details of this procedure are given in Algorithm 1.

Algorithm 1 Adaptive transmission probability control

1. **Input**: A, g
2. **Initialize**: Estimate the current number of APs and STAs by overhearing transmitted frames, and set $\alpha \leftarrow \alpha_0$ calculated from (17).
3. **repeat** (for each certain update period):
4. Estimate the current throughput of RoF and legacy WLANs $S^{\mathrm{R}}(t)$ and $S^{\mathrm{L}}(t)$ by overhearing frames.
5. $\Delta\alpha \leftarrow \frac{A}{1+\mathrm{e}^{-g(S^{\mathrm{R}}(t)-S^{\mathrm{L}}(t))}} - \frac{A}{2}$
6. $\alpha \leftarrow \alpha - \Delta\alpha$
7. **until** (the states of WLANs change significantly)

We now explain how to determine the value of g. When $|S^{R}(t) - S^{L}(t)|$ is an infinitesimal value, then $|\Delta\alpha_t|$ should be 0. The wireless throughput may fluctuate for various reasons, and thus we do not want to change α in response to small throughput fluctuations. If we presume that the throughput fluctuations are less than 1 kbit/s, we set g to a value, such as 10^{-5} so that $e^{-g \cdot 10^3} \simeq 1$. In contrast, when $|S^{R}(t) - S^{L}(t)|$ is large, $|\Delta\alpha_t|$ should be set to a large value in order to quickly approach the target value.

In this context, A determines the range of $\Delta\alpha_t$ ($-A/2 < \Delta\alpha_t < A/2$). The value of A should be selected so as to satisfy $|A| < 2$ because $0 < \alpha < 1$. The optimal value of A depends on g, the number of nodes in the RoF and legacy WLAN, and the traffic offered to each node. In this paper, the optimal value of A was determined empirically based on the simulation results.

Since the RoF AP changes the transmission probability based on the measured throughput values, even when the number of users changes, it is possible to control the system such that the throughputs of the RoF WLAN and legacy WLAN are equal. In an asymmetric system configuration, such as an unequal number of users in the RoF WLAN and legacy WLAN, fairness is improved by changing the control target to the ratio of the number of users.

5 Performance evaluation

5.1 Simulation setup

We evaluated the performance of the ATRAS scheme using the network simulator QualNet [13], which is a reliable simulator used in many literature to evaluate wireless network performance. The simulation model consisted of an RoF-based WLAN and a legacy WLAN, and each had an AP with nine associated STAs, i.e., $N_A^R = N_A^L = 1$ and $N_S^R = N_S^L = 9$. These WLANs were operated based on the IEEE 802.11a standard and the PHY/MAC parameters shown in Table 1. We also assumed that there were no hidden nodes. We employed the TCP SACK algorithm using a delayed ACK and the Nagle Algorithm, and FTP was chosen as the application protocol. The maximum segment size was 146 bytes, and the sending/receiving buffer size was 64 Kbytes. The ACK/CTS timeout value $T_{timeout}$ was defined as $T_{timeout} = T_{st} + 2\delta_{RoF}$. The maximum optical fiber length was simulated at 10 km, which corresponded to a δ_{RoF} of 50 μs. The propagation delay between the local nodes δ was negligibly small, so δ was set to zero.

In the ATRAS scheme, the RoF STAs and legacy WLAN nodes transmitted frames using the RTS/CTS exchange protocol, and the RoF-based WLAN employed the NAV extension method mentioned in Section 3.3 to solve the ACK frame collision problem in 2.2.2. In the PBA scheme, the RoF AP transmitted the DATA frames using the PBA scheme instead of the ATRAS scheme. In the conventional scheme, the local nodes and RoF AP transmitted frames using the standard CSMA/CA protocol with the RTS/CTS exchange option, and the RoF-based WLAN employed the NAV extension method. In the slot time adjustment scheme, the RoF-based WLAN used an extended slot time and the legacy WLAN used the standard slot time defined

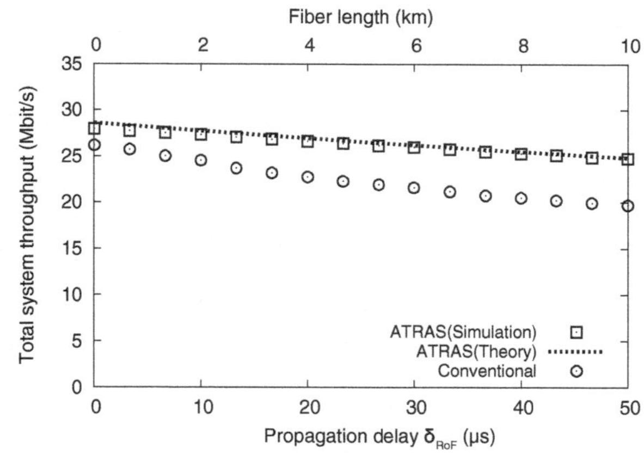

Fig. 7 Total system throughput vs. δ_{RoF} with UDP bidirectional full buffer traffic when $\alpha = 1$

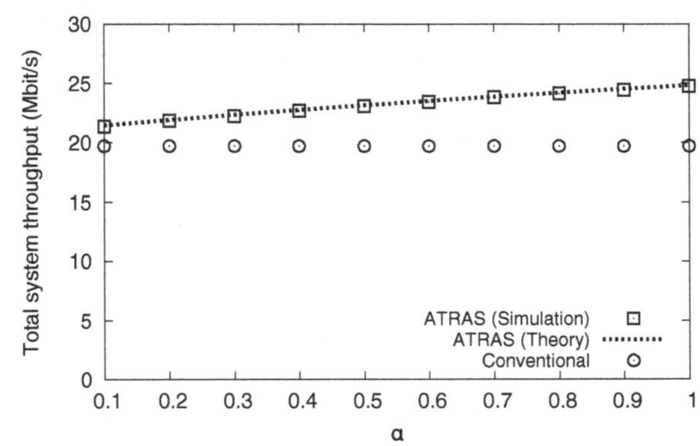

Fig. 8 Total system throughput vs. α with UDP bidirectional full buffer traffic when $\delta_{RoF} = 50\ \mu s$

Table 1 PHY/MAC parameters

Parameters	Values
SIFS	16 μs
DIFS	34 μs
Slot time	9 μs
Data rate	54 Mbit/s
MAC payload	1500 bytes
CWmin	15
CWmax	1023
Retry limit	7

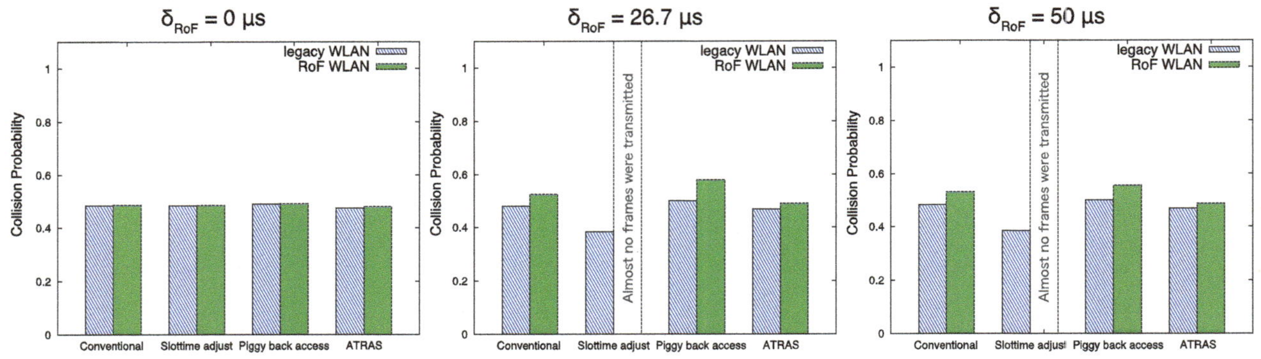

Fig. 9 Collision probability of the RoF WLAN and legacy WLAN when $\alpha = 1$

in the IEEE 802.11a standard. The extended slot time was $2\delta_{\mathrm{RoF}} + 4\mu$s.

5.2 Simulation and numerical results

We evaluated the performance of the MAC and transport layers. In the MAC layer performance evaluation, we considered UDP bidirectional full buffer traffic and confirmed that the theoretical throughput performance was in good agreement with the simulation results. We also compared the frame collision probability in the MAC layer performance evaluation. When evaluating the performance of the transport layer, we compared the performance of the ATRAS scheme with that of the conventional CSMA/CA scheme, slot time adjustment scheme, and the PBA scheme with TCP downlink traffic.

5.2.1 MAC-layer performance

Figure 7 shows the total system throughput of the RoF-based WLAN and legacy WLAN as a function of the propagation delay δ_{RoF} with $\alpha = 1$. When $\alpha = 1$, the RoF AP transmits a frame every time the RoF AP overhears a CTS frame transmitted by a legacy AP or a legacy STA. The ATRAS scheme achieved a higher throughput than the conventional scheme. In the both schemes, the total throughput decreases as the propagation delay δ_{RoF} increases since the required time to exchange frames in RoF-based WLAN increases. In addition to the throughput degradation, increase in frame collision degrades throughput in the conventional scheme, which is the reason why the throughput gain of the proposed scheme increases as δ_{RoF} increases.

Figure 8 shows the total system throughput as a function of transmission probability α, where $\delta_{\mathrm{RoF}} = 50$ μs. The ATRAS scheme can control the throughput of the RoF-based WLAN by adjusting α. Moreover, in both of the figures, the simulation results of the ATRAS scheme are in agreement with the theoretical values.

Figure 9 shows the collision probability of frames as a function of the propagation delay δ_{RoF} when $\alpha = 1$ in the ATRAS scheme. Since the RoF WLAN with slot time adjustment scheme cannot transmit frames when δ_{RoF} is longer than 26.7 μs due to long DIFS and backoff, the collision probability of the RoF WLAN with slot time adjustment is not shown in the figure. When the ATRAS scheme is used, the collision probability of the RoF WLAN does not increase as δ_{RoF} increases, while the probabilities in the other schemes increase. The average collision probability of both WLANs also increases in the scheme

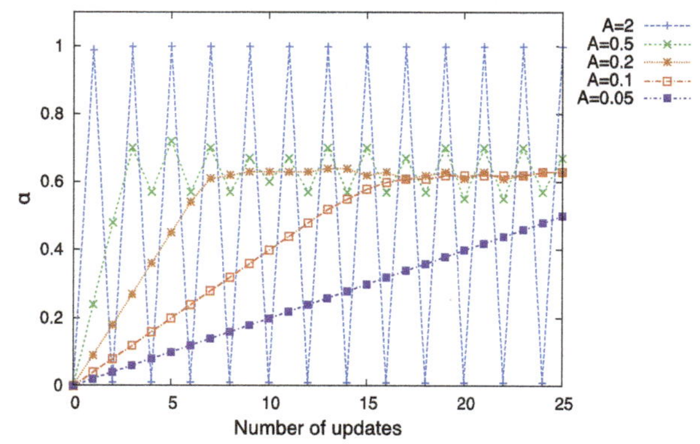

Fig. 10 Time transition of α parameterized by A when $\delta_{\mathrm{RoF}} = 50$ μs

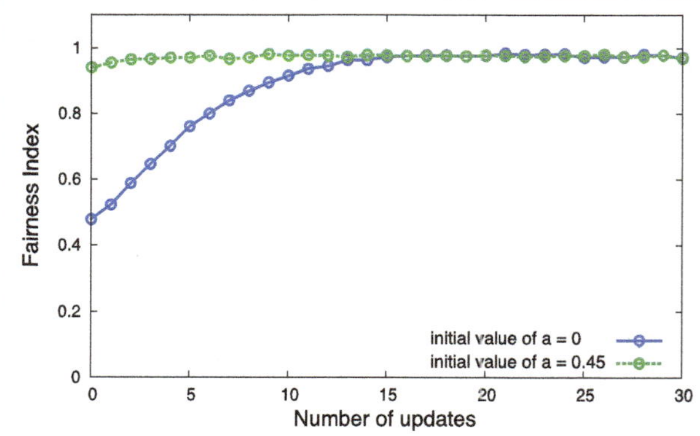

Fig. 11 Fairness index of TCP downlink flow

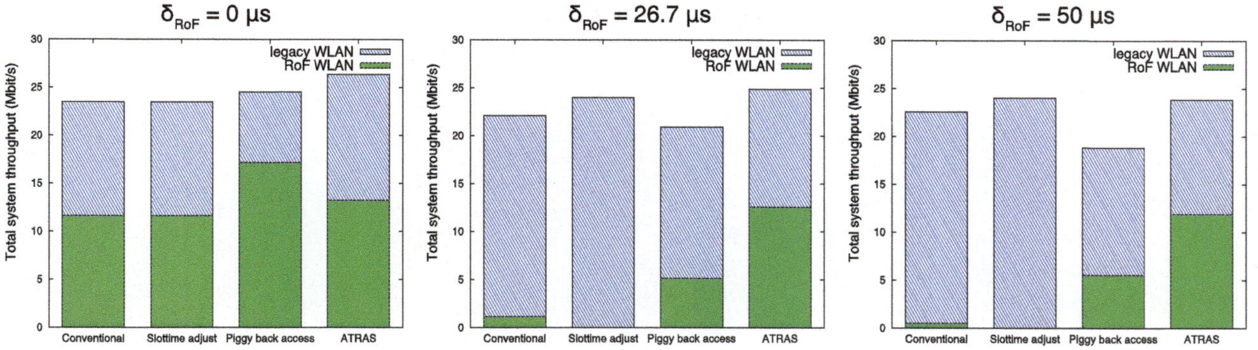

Fig. 12 Total throughput of the RoF-based WLAN and legacy WLAN

other than the ATRAS scheme as δ_{RoF} increases. This is because the frame collision due to delay mentioned in Section 2.2.1 occurs frequently when δ_{RoF} is long. In the ATRAS scheme, the RoF AP can transmit frames without collision, which enables WLANs to maintain lower collision probability when δ_{RoF} is long.

5.2.2 Transport layer performance

In this scenario, we considered the TCP downlink traffic in both the RoF and legacy WLANs. First, we evaluated the convergence of α with the adaptive control scheme. We determined the gain of the sigmoid function g and the constant value A in (19). The update of α was performed every time the RoF AP received 100,000 TCP-DATA packets transmitted by the legacy AP and legacy STAs. As we discussed in 4.3.3, g was set to $g = 10^{-6}$ in the simulation.

Figure 10 shows the time dependence of α parameterized by A. The horizontal axis indicates a number of updates to α. In the cases where $A = 2$ and $A = 0.5$, the value of α fluctuated owing to the large value of A. On the other hand, if we set the value of A to be small, such as $A = 0.05$, then α converged slowly. Based on these simulation results, we set $A = 0.1$.

Next, we evaluated the convergence time of α with the adaptive control scheme in order to clarify the effectiveness of our initial value determination. As was mentioned above, we set $g = 10^{-6}$ and $A = 0.1$ in (19). From (17), α_0 was set to 0.45 in this scenario. We compared the proposed scheme to a scheme where $\alpha_0 = 0$.

Figure 11 shows the Jain's fairness index [32], which is defined as

$$F = \frac{\left(\sum_{i=1}^{n} \theta_i\right)^2}{n \sum_{i=1}^{n} \theta_i^2}, \tag{20}$$

where n is the total number of downlink flows in the RoF and legacy WLANs, and θ_i is the TCP throughput of the ith connection. As shown in Fig. 11, the fairness index is close to one as a result of using the proposed adaptive control procedure. Moreover, it takes less time to achieve fairness by using the initial value $\alpha_0 = 0.45$ derived from (17) as compared to the time required when $\alpha_0 = 0$.

Next, we compared the throughput of the ATRAS scheme with that of the conventional CSMA/CA scheme, the slot time adjustment scheme, and the PBA scheme. Figures 12 and 13 show the system throughput of each WLAN and the fairness index based on the value of δ_{RoF}. The ATRAS scheme was able to increase the throughput of the RoF-based WLAN and achieve fairness regardless of the δ_{RoF} value. This is because the ATRAS scheme gives collision-free transmission opportunity to the RoF AP, and the adaptive transmission probability control adjusts α to increase the throughput of the RoF AP up to the same level as the legacy WLAN. Moreover, the total system throughput in the proposed scheme was higher than in the other schemes because the number of frame collisions caused by the propagation delay was reduced. Using the CSMA/CA scheme, the throughput of the RoF-based WLAN decreased and the fairness index also decreased as the δ_{RoF} increased. When the RoF-based WLAN used the slot time adjustment scheme, the throughput of the RoF-based WLAN approached zero if $\delta_{\mathrm{RoF}} \geq 10\ \mu$s. This is because the nodes in the RoF-based WLAN have much

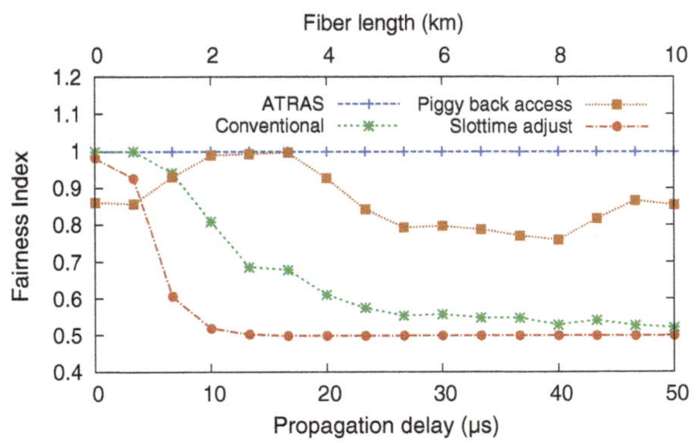

Fig. 13 Fairness Index when the RoF AP used each scheme

longer backoff periods than those in the legacy WLAN, which means that they cannot transmit enough frames for the reasons described in Section 3.1. The use of the PBA scheme was able to improve the throughput of the RoF-based WLAN compared to the throughput in the conventional CSMA/CA scheme, because the RoF AP was able to transmit its frames using both the CSMA/CA and PBA schemes. The throughput of the RoF-based WLAN, however, decreased based on δ_{RoF} as a consequence of the increase in frame collisions.

6 Conclusions

In this paper, we proposed the ATRAS scheme to reduce unexpected frame collisions and improve fairness between an RoF-based WLAN and legacy WLANs in a coexisting environment. The ATRAS scheme enables the RoF AP to transmit its frames in advance so that the frames do not collide with other frames, and the adaptive transmission probability control is able to achieve fairness between the RoF-based WLAN and legacy WLANs. Our simulation and numerical results demonstrated the performance of the proposed scheme, which are highlighted as follows: The proposed scheme increases MAC layer throughput by 12.7–24.6% and decreases frame collision probability compared to the conventional scheme. The proposed adaptive transmission probability control converges quickly within a few updates. The proposed scheme increases the TCP throughput of the RoF-based WLAN up to the same level as that in the legacy WLAN and achieves fairness regardless of propagation delay.

Our future work includes applying the proposed protocol to environments where IEEE 802.11n/ac is used for WLANs or where multiple RoF-based WLANs and legacy WLANs coexist and extending the proposed protocol to use multiple medium access schemes adaptively as mentioned in Section 4.1.

Competing interests

The authors declare that they have no competing interests.

Author details

[1]Graduate School of Informatics, Kyoto University, Kyoto, Japan. [2]NTT Network Innovation Laboratories, NTT Corporation, Yokohama, Japan.

References

1. SK Yong, C-C Chong, An overview of multigigabit wireless through millimeter wave technology: potentials and technical challenges. EURASIP J. Wirel. Commun. Netw. **2007**(1), 1–10 (2006)
2. JA Gutierrez, M Naeve, E Callaway, M Bourgeois, V Mitter, B Heile, IEEE 802.15. 4: a developing standard for low-power low-cost wireless personal area networks. IEEE Netw. **15**(5), 12–19 (2001)
3. E Khorov, A Lyakhov, A Krotov, A Guschin, A survey on IEEE 802.11 ah: an enabling networking technology for smart cities. Comput. Commun. **58**, 53–69 (2015)
4. R Riggio, A Bradai, D Harutyunyan, T Rasheed, T Ahmed, Scheduling wireless virtual networks functions. IEEE Trans. Netw. Serv. Manag. **13**(2), 240–252 (2016)
5. H Al-Raweshidy, S Komaki, *Radio over fiber technologies for mobile communications networks*. (Artech House, London, 2002)
6. L Maksymiuk, M Kowalczyk, J Siuzdak, Multimode fiber passive optical network for IEEE 802.11 signal distribution. J. Opt. Commun. Netw. **6**(8), 743–753 (2014)
7. D Novak, RB Waterhouse, A Nirmalathas, C Lim, PA Gamage, TR Clark, ML Dennis, JA Nanzer, Radio-over-fiber technologies for emerging wireless systems. IEEE J. Quantum Electron. **52**(1), 1–11 (2016)
8. M Zhu, L Zhang, J Wang, L Cheng, C Liu, G-K Chang, Radio-over-fiber access architecture for integrated broadband wireless services. J. Light. Technol. **31**(23), 3614–3620 (2013)
9. J Beas, G Castanón, I Aldaya, A Aragón-Zavala, G Campuzano, Millimeter-wave frequency radio over fiber systems: a survey. IEEE Commun. Surv. Tutor. **15**(4), 1593–1619 (2013)
10. S Deronne, V Moeyaert, S Bette, in *Proc. IEEE SCVT*. Analysis of the MAC performances in 802.11g radio-over-fiber systems, (Ghent, Belgium, 2011), pp. 1–5
11. S Deronne, V Moeyaert, S Bette, in *Proc. ONDM*. WiFi transmission in radio-over-fiber systems: performance of the IEEE 802.11n aggregation mechanism, (Brest, France, 2013), pp. 167–172
12. S Zhang, D Franklin, in *Proc. IEEE ICCC*. Feasibility study on the implementation of IEEE 802.11 on cloud-based radio over fibre architecture, (Sydney, Australia, 2014), pp. 2891–2896
13. QualNet . http://web.scalable-networks.com/qualnet-network-simulator-software. Accessed 7 July 2017
14. IEEE Std. IEEE Standard for Information technology–Telecommunications and information exchange between systems Local and metropolitan area networks–Specific requirements - Part 11: Wireless LAN Medium Access Control (MAC) and Physical Layer (PHY) Specifications, in IEEE Std 802.11-2016, pp.1–3534, doi:10.1109/IEEESTD.2016.7786995. http://ieeexplore.ieee.org/document/7786995/
15. GP Agrawal, *Nonlinear fiber optics*. (Academic press, Massachusetts, 2007)
16. T Nishio, K Funabiki, M Morikura, K Yamamoto, D Murayama, K Nakahira, MAC protocol for improving throughput and balancing uplink/downlink throughput for wireless local area networks with long propagation delays. IEICE Trans. Commun. **E100.B**(5), 874–883 (2017)
17. Z Zeng, Y Gao, K Tan, PR Kumar, in *Proc. IEEE INFOCOM*. CHAIN: introducing minimum controlled coordination into random access MAC, (Shanghai, China, 2011), pp. 2669–2677
18. F Wamser, R Pries, D Staehle, K Heck, P Tran-Gia, Traffic characterization of a residential wireless internet access. Telecommun. Syst. **48**(1-2), 5–17 (2011)
19. Q Wei, I Aad, L Scalia, J Widmer, P Hofmann, L Loyola, E-MAC: an elastic MAC layer for IEEE 802.11 networks. Wirel. Commun. Mob. Comput. **13**(4), 393–409 (2013)
20. Y Khan, M Derakhshani, S Parsaeefard, T Le-Ngoc, in *Proc. IEEE Globecom Workshops*. Self-organizing TDMA MAC protocol for effective capacity improvement in IEEE 802.11 WLANs, (California, 2015), pp. 1–6
21. G Jakllari, M Neufeld, R Ramanathan, in *Proc. IEEE MASS*. A framework for frameless tdma using slot chains, (Nevada, 2012), pp. 56–64
22. P Djukic, P Mohapatra, Soft-TDMAC: a software-based 802.11 overlay TDMA MAC with microsecond synchronization. IEEE Trans. Mob. Comput. **11**(3), 478–491 (2012)
23. M Tuysuz, HA Mantar, A beacon-based collision-free channel access scheme for IEEE 802.11 WLANs. Wirel. Pers. Commun. **75**(1), 155–177 (2014)
24. MF Tuysuz, HA Mantar, Exploiting the channel using uninterrupted collision-free MAC adaptation over IEEE 802.11 WLANs. Wirel. Commun. Mob. Comput. **15**(5), 889–909 (2015)
25. J Barcelo, B Bellalta, C Cano, M Oliver, in *Proc. Eunice Open European Summer School*. Learning-BEB: avoiding collisions in wlans, (Brest, France, 2008), pp. 1–8
26. NetComm Wireless HS1200N. http://www.netcommwireless.com/sites/default/files/HS1200N-User-Manual.pdf. Accessed 7 July 2017

27. ath9k driver. https://wireless.wiki.kernel.org/en/users/drivers/ath9k. Accessed 7 July 2017
28. M Derakhshani, T Le-Ngoc, in *Proc. IEEE Globecom*. Adaptive access control of CSMA/CA in wireless lans for throughput improvement, (Georgia, USA, 2013), pp. 2951–2955
29. G Bianchi, Performance analysis of the IEEE 802.11 distributed coordination function. IEEE J. Sel. Areas Commun. **18**(3), 535–547 (2000)
30. R Bruno, M Conti, E Gregori, in *Proc. ACM MSWiM*. Performance modelling and measurements of TCP transfer throughput in 802.11-based WLAN (Malaga, Spain, 2006), pp. 4–11
31. WA Little, *The existence of persistent states in the brain*. (Science & Business Media, US, 1974), pp. 145–164
32. R Jain, D-M Chiu, WR Hawe, A quantitative measure of fairness and discrimination for resource allocation in shared computer system. DEC Research Report TR-301 (1984). https://jwcn-eurasipjournals.springeropen.com/articles/10.1186/s13638-015-0340-3

7

Subcarrier allocation based simultaneous wireless information and power transfer for multiuser OFDM systems

Zhenyu Na[1], Xiaotong Li[1], Xin Liu[2*] and Zhian Deng[1]

Abstract

Most existing algorithms of simultaneous wireless information and power transfer (SWIPT) for orthogonal frequency division multiplexing (OFDM) systems are studied based on power splitting or time splitting, which lead to time delay and the decrease of subcarrier utilization. In this paper, a multiuser OFDM system with multichannel is established and the subcarriers are divided into two parts. One part is used for information decoding and the other part is used for energy harvesting. We maximize the sum rate of the users under the constraint of energy harvesting by optimizing the channel allocation and power allocation. By means of iterative calculation, an efficient subcarrier allocation algorithm is proposed. Simulation results demonstrate that the proposed algorithm converges fast and can achieve higher sum rate than the conventional algorithm.

Keywords: Simultaneous wireless information and power transfer (SWIPT), OFDM, Subcarrier allocation, Power allocation

1 Introduction

Orthogonal frequency division multiplexing (OFDM) is the viable air interface for providing the high spectral efficiency and ubiquitous communication services because of its ability to combat frequency selective fading and flexibility in resource allocation. However, the power-hungry circuitries and the limited energy supplies in portable devices lead to the bottlenecks in prolonging the lifetime of networks and guaranteeing quality of service (QoS). As a result, the energy-efficient mobile communication has drawn much attention from both the industry and the academia [1–4].

Traditionally, the energy has been harvested from natural renewable energy sources, such as solar, wind and geothermal heat, thereby substantially reducing the reliance on the energy supply from conventional energy sources. In this context, simultaneous wireless information and power transfer (SWIPT) has emerged and attracted the widespread concern. Varshney put forward the concept of transmitting information and energy simultaneously and defined the capacity energy function for the first time [5]. Pulkit Grover proposed the model based on the electromagnetic induction principle and analyzed the noise coupling circuit with SWIPT [6–10] maximized the energy efficiency in spectrum sensing. Two transmission protocols based on power splitting relaying and mode adaptation were proposed in OFDM relaying SWIPT systems [11]. Liang Liu proposed a classical transmission solution [12]. Specifically, the received signal is divided into two circuits: one is used for information decoding, and the other one is used for energy harvesting. As two classical models, time switching (TS) model and power switching (PS) model were put forward in [13–15], respectively. In TS model, the receiver switches to energy harvesting mode or information mode within one transmission period. In PS model, the receiver splits the power into two parts with some ratios of which one part is used for information decoding and another part is used for energy harvesting. Javer Rubio combined SWIPT with multiple-input-single-output (MISO) in [16] where a transmitter with multi-antennas transmits the same information to several banks of single antenna simultaneously. Various types of SWIPT systems including two-user MIMO broadcast channels, two-way communication links, and point-to-point links assisted by passive relays have been formulated and optimized in [17–20] proposed an optimal algorithm of power splitting based on downlink

* Correspondence: liuxinstar1984@dlut.edu.cn
[2]School of Information and Communication Engineering, Dalian University of Technology, Dalian, China
Full list of author information is available at the end of the article

OFDMA by means of iterative algorithm. A tradeoff between TS and PS was proposed in [21].

As the promising technology for improving spectrum efficiency, cognitive radio (CR) has been investigated a lot recently [22–24]. SWIPT in CR networks with multiusers was studied in [25]. Two SWIPT schemes were proposed [26] for CR networks with a single SU. The secure beamforming schemes for SWIPT in a MISO broadcast channel was investigated in [27].

Different from PS and TS models, we study a subcarrier allocation algorithm based on SWIPT for the OFDM system with multiusers which is without a splitter at the receiver. The subcarriers of each user are separated into the information decoding part and the energy harvesting part. On the basis, we address the problem of maximizing the sum rate of users under the condition of enough energy harvesting. The problem is non-convex, and an iterative algorithm is used to solve it.

2 System model

We consider a wireless OFDM downlink system consisting of one cognitive base station (CBS) and k users. As shown in Fig. 1, each user is only equipped with one antenna. Let K denote the set of k users. The OFDM system bandwidth is assumed to be equally divided into n ($n \geq k$) channels. The set of subcarriers is denoted as N ($N = \{1,2,\cdots n\}$). Each subcarrier must be allocated to only one user. Parts of subcarriers are used for energy harvesting, while the others are utilized for information decoding simultaneously. We suppose that the channel power gain on each subcarrier is always constant in one transmission period provided by the base station. Let $h_{k,n}$ represent the gain of the k-th user on the n-th subcarrier. Then, the noise power of each subcarrier is modeled as an additive white Gaussian noise (AWGN) random variable with zero mean and variance σ^2. The total transmission power is limited to the power budget P. Therefore, the power allocated on the n-th subcarrier is denoted as P_n. Let S^P represent the subcarriers used for energy harvesting to power transfer. Accordingly, the other subcarriers used for information decoding are denoted by S^I. Hence, S^I_K represents the subcarriers of K-th user for information transfer. Since one subcarrier cannot be used for energy harvesting and transfer information simultaneously, we have $S^I \cap S^P = \varnothing$and $S^I \cup S^P = N$.

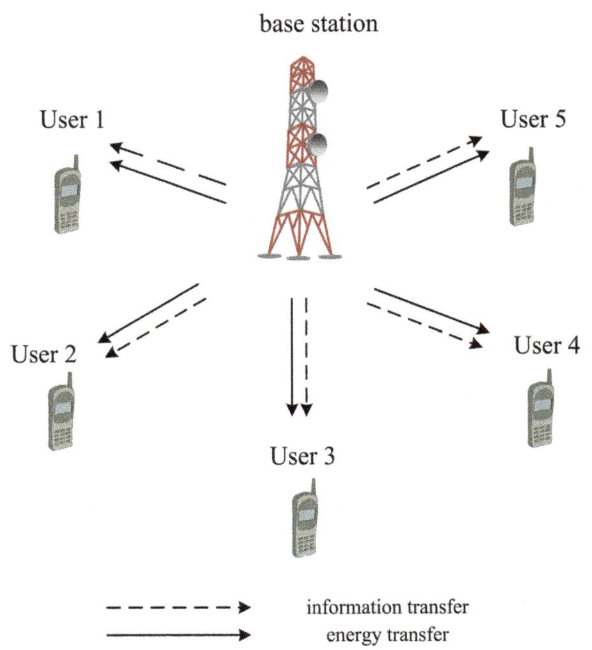

Fig. 1 System model

3 Problem formulation

Our aim is to maximize the sum rate of the OFDM downlink under the constraint of the minimum harvested energy for each user. Let B_k represent the minimum harvested energy of the k-th user. Since one subcarrier can only be allocated to one user, we use $\alpha_{n,k}$ to stand for a binary channel allocation index. In other words, $\alpha_{n,k} = 1$ means that the subcarrier n is only allocated to the user k, while $\alpha_{n,k} = 0$ is determined on other terms. Thus, it is written as:

$$\sum\nolimits_{k=1}^{K} \alpha_{k,n} = 1, \forall n \in N \tag{1}$$

The sum rate can be formulated as:

$$\sum\nolimits_{k=1}^{K} \sum\nolimits_{n \in S^I} \alpha_{k,n} \log\left(1 + \frac{h_{k,n} P_n}{\sigma^2}\right) \tag{2}$$

Here, $n \in S^I$. With energy harvesting efficiency ε, the harvested energy during one transmission block for user k is determined by:

$$\sum\nolimits_{n \in S^P} \left(\varepsilon h_{k,n} P_n + \sigma^2\right) \tag{3}$$

For $\forall k \in K$. Therefore, the optimization model of maximum sum rate can be expressed as:

$$\begin{aligned} &\max_{\alpha_{n,k}, S^I, P_n} \sum\nolimits_{k=1}^{K} \sum\nolimits_{n \in S^I} \alpha_{k,n} \log\left(1 + \frac{h_{k,n} P_n}{\sigma^2}\right) \\ &\text{s.t. } \sum\nolimits_{n \in N} P_n \leq P \ \ (P_n \geq 0) \\ &S^P \cup S^I = N \\ &S^P \cap S^I = \varnothing \\ &\sum\nolimits_{k=1}^{K} \alpha_{k,n} = 1, \quad \forall n \in N \\ &\alpha_{k,n} \in \{0,1\}, \quad \forall k \in K, n \in N \end{aligned} \tag{4}$$

4 Optimal solution

Since the problem is non-convex, it is impossible to obtain the optimal solution directly. In this section, a sub-optimal algorithm is proposed for solving the non-convex problem.

The optimization task can be decomposed into three steps: to optimize $\alpha_{k,n}$ with given P_n and $S^I(S^P)$, to optimize $P_n$with given $\alpha_{k,n}$ and $S^I(S^P)$, as well as to optimize $S^I(S^P)$ with given $\alpha_{k,n}$ and P_n.

Firstly, since P_n and $S^I(S^P)$ are determined, $\alpha_{k,n}$is optimized as follows:

$$\max_{\alpha_{n,k}} \sum_{k=1}^{K} \alpha_{k,n} \log\left(1+\frac{h_{k,n}P_n}{\sigma^2}\right), \quad n \in S^I$$
$$\text{s.t.} \sum_{k=1}^{K} \alpha_{k,n} = 1, \quad \forall n \in N$$
$$\alpha_{k,n} \in \{0,1\}, \quad \forall k \in K, n \in N \tag{5}$$

The problem above is regarded as allocating the subcarrier n to the assigned user for obtaining the maximum sum rate. In other words, the subcarrier n ($n \in S^I$) is allocated to the user k which can get the maximum $h_{k,n}P_n$, i.e., $\alpha_{k^*,n} = 1$, $k^* = \arg\max_{k\in K} h_{k,n}P_n$ and $\alpha_{k,n} = 0, \forall k \neq k^*, k \in K$.

Secondly, P_n is optimized by $\alpha_{k,n}$ and $S^I(S^P)$. In this proposition, the problem can be rewritten as:

$$\max_{P_n} \sum_{n\in S_k^I} \log\left(1+\frac{h_{k^*,n}P_n}{\sigma^2}\right) \tag{6}$$
$$\text{s.t.} \quad \sum_{n\in S^P}\left(\varepsilon h_{k^*,n}P_n+\sigma^2\right) \geq B_k$$
$$\sum_{n\in N} P_n \leq P \quad (P_n \geq 0)$$

Note that $\alpha_{k^*,n} = 1, \alpha_{k,n} = 0, \forall k \neq k^*, k \in K$. The converted problem is satisfied with convex model. Therefore, the Lagrange dual decomposition is adopted to solve this problem. The Lagrange dual function is as follows:

$$g(\beta_1,,\beta_2) = \max_{\{P_n\}} L(P_n) \tag{7}$$

Where, β_1 and β_2 are the Lagrange multipliers and they are determined by the sub-gradient method. Meanwhile, $L(P_n)$ is expressed as:

$$L(P_n) = \sum_{n\in S_k^I} \log\left(1+\frac{h_{k^*,n}P_n}{\sigma^2}\right) + \beta_1\left\{\sum_{n\in S^P}\left(\varepsilon h_{k^*,n}P_n+\sigma^2\right)-B_k\right\} + \beta_2\left(P-\sum_{n\in N}P_n\right) \tag{8}$$

Then, the dual problem can be simplified as follows:

$$\min_{\beta_1,\beta_2} g(\beta_1,,\beta_2) \tag{9}$$
$$\text{s.t} \quad \beta_1,\beta_2 \geq 0$$

Because the dual problem is differentiable, it can be solved by the classic sub-gradient method, which solves the optimal problem based on the gradient and the suitable step size [28]. The result is presented as below:

$$\Delta\beta_1 = \sum_{n\in S^P} \varepsilon h_{k^*,n}P_n + \sigma^2 - B_k \tag{10}$$

$$\Delta\beta_2 = P - \sum_{n\in N} P_n \tag{11}$$

For the given β_1 and β_2, the optimal power $P_n (n \in S^I)$ is obtained according to KKT conditions using mathematical manipulation:

$$P_n = \left(\frac{1}{\beta_2} - \frac{\sigma^2}{h_{k^*,n}}\right)^+ \tag{12}$$

Where, $(.)^+$denotes max(.,0).

Similarly, the allocated power P_n used for energy harvesting is determined as:

$$P_n = \begin{cases} P_{\max} \beta_1 h_{k^*,n}\varepsilon > \beta_2 \\ P_{\min} \beta_1 h_{k^*,n}\varepsilon \leq \beta_2 \end{cases} \tag{13}$$

Where, $P_{\max}$ and $P_{\min}$ represent the maximum and minimum power constraints on information decoding, respectively.

Algorithm 1: Proposed algorithm for the problem (6)

1. **Initialize**: $\beta_1(0)$, $\beta_2(0)$, $t = 0$.
2. **Repeat**
3. Calculate$P_n(n \in S^I)$ according to (11)
4. Calculate$P_n(n \in S^P)$ according to (12)
5. $\beta_1(t+1) = \beta_1(t) -$ stepsize1$(\sum_{n\in S^P} \varepsilon h_{k^*,n}P_n + \sigma^2 - B_k)$
6. $\beta_2(t+1) = \beta_2(t) -$ stepsize2$(P - \sum_{n\in N} P_n)$
7. $t = t+1$
8. **Until** $|\beta_1(t) - \beta_1(t-1)| \leq \varepsilon$ and $|\beta_2(t) - \beta_2(t-1)| \leq \varepsilon$

Where, $\varepsilon(>0)$ denotes the error tolerance and t denotes the times of iteration.

According to P_n and $\alpha_{n,k}$, $S^I(S^P)$ can be obtained by substituting (11) and (12) into (8). Consequently, the Lagrange dual function can be rewritten as below:

$$L(S^P) = \sum_{k=1}^{K}\sum_{n\in N}\alpha_{n,k}\log\left(1+\frac{h_{k,n}P_n}{\sigma^2}\right) - \sum_{k=1}^{K}\sum_{n\in S^P}\alpha_{n,k}\log\left(1+\frac{h_{k,n}P_n}{\sigma^2}\right) + \beta_1\sum_{n\in S^P}\left(\varepsilon h_{k,n}P_n+\sigma^2\right) - \beta_1 B_k + \beta_2 P - \beta_2\sum_{n\in N}P_n$$
$$= \sum_{n\in S^P}\left\{\beta_1\left(\varepsilon h_{k,n}P_n+\sigma^2\right) - \sum_{k=1}^{K}\alpha_{n,k}\log\left(1+\frac{h_{k,n}P_n}{\sigma^2}\right)\right\} + \sum_{n\in N}\left\{\sum_{k=1}^{K}\alpha_{n,k}\log\left(1+\frac{h_{k,n}P_n}{\sigma^2}\right) - \beta_2 P_n\right\} - \beta_1 B_k + \beta_2 P$$
$$= \sum_{n\in S^P}F_n + \sum_{n\in N}\left\{\sum_{k=1}^{K}\alpha_{n,k}\log\left(1+\frac{h_{k,n}P_n}{\sigma^2}\right) - \beta_2 P_n\right\} - \beta_1 B_k + \beta_2 P \tag{14}$$

Where,

$$F_n = \beta_1\left(\varepsilon h_{k,n}P_n + \sigma^2\right) - \sum\nolimits_{k=1}^{K} \alpha_{n,k} \log\left(1 + \frac{h_{k,n}P_n}{\sigma^2}\right) \tag{15}$$

When we observe the formula (13), only the first item on the right side is about to S^P. Thus, the optimal S^P can be achieved by maximizing the item F_n, i.e.,

$$S^{P*} = \arg\ \max_{S^P} \sum\nolimits_{n \in S^P} F_n^* \tag{16}$$

S^{P*} can be easily obtained by substituting all the n into F_n to find the ones which make F_n positive. As a result, the rest of the set N belongs to S^{I*}.

The proposed algorithm to solve the optimal problem is presented as below:

Algorithm 2: Proposed algorithm for the joint optimizationproblem

1. **Initialize:** $P_n(0) = 0$, $\alpha_{n,k^*}(0)$, $t = 0$
2. Given$P_n(0)$ and $\alpha_{n,k^*}(0)$, obtain $S^I(0)$ according to formula (16)
3. $R_k(0) = \sum_{k=1}^{K} \alpha_{n,k}(0)\log\left(1 + \frac{h_{k,n}P_n(0)}{\sigma^2}\right)$, $n \in S^I$
4. **Repeat**
5. Given $P_n(0)$ and $S^I(t)$, set $\alpha_{n,k^*}(t) = 1$, $k^* = \arg\max_{k \in K}(h_{k,n}P_n)$ and $\alpha_{n,k^*}(t) = 0$, $\forall k \neq k^*$, $k \in K$
6. Given $\alpha_{n,k^*}(t+1)$ and $S^I(t)$, obtain $P_n(t+1)$ for all $n \in N$, according to formulas (11) and (12)
7. Given $\alpha_{n,k^*}(t+1)$ and $P_n(t+1)$, obtain $S^I(t+1)$
8. $R_k(t+1) = \sum_{k=1}^{K} \alpha_{n,k}(t+1)\log\left(1 + \frac{h_{k,n}P_n(t+1)}{\sigma^2}\right)$, $n \in S^I$
9. $t = t + 1$
10. **Until** $|R_k(t+1) - R_k(t)| \leq \varepsilon$

Where, $\varepsilon(>0)$ denotes the error tolerance and t denotes the times of iteration.

5 Simulation results

In this section, the performance of the proposed subcarrier allocation algorithm based on SWIPT for the OFDM system with multiusers is demonstrated by simulation results.

We denote all the channels involved follow the Rayleigh distribution. For simplicity, we suppose that the minimum harvested energy limits for all the users are the same, i.e., $B_k = B$. In addition, we set N = 16, K = 5, σ^2=1, $P_{\max}$ = 3 W, $P_{\min}$ = 0, ε = 1, and the bandwidth is equal to 1 MHz.

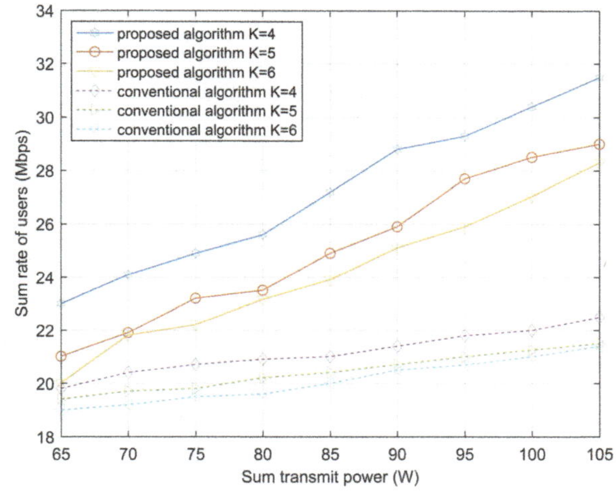

Fig. 3 Sum rate of users vs. sum transmission power

Figure 2 shows the convergence behavior of the proposed algorithm. It can be seen that the proposed algorithm converges fast. It indicates that the proposed algorithm can be implemented practically.

Figure 3 presents the comparison between the proposed optimization algorithm and the conventional algorithm of subcarrier allocation based on [29]. It can be observed that the proposed algorithm performs better compared with the conventional algorithm.

When the conventional algorithm allocates N subcarriers to K users, all the subcarriers are used for information decoding and the consumed energy comes from the system. Since the system can not produce energy by itself, the energy comes from the finite battery. Therefore, the constraint of minimum energy harvesting can not be

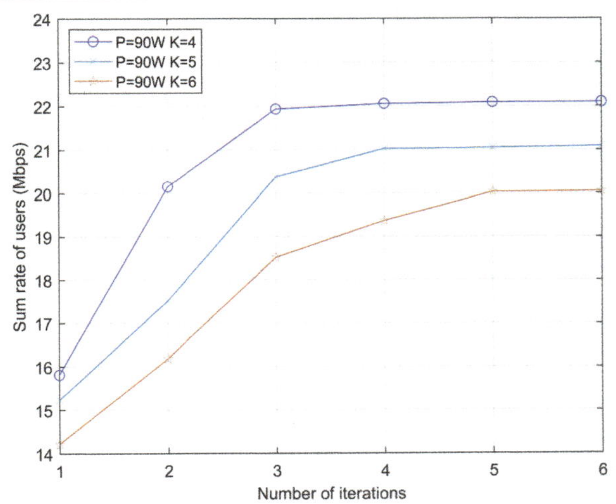

Fig. 2 Convergence behavior of the proposed algorithm

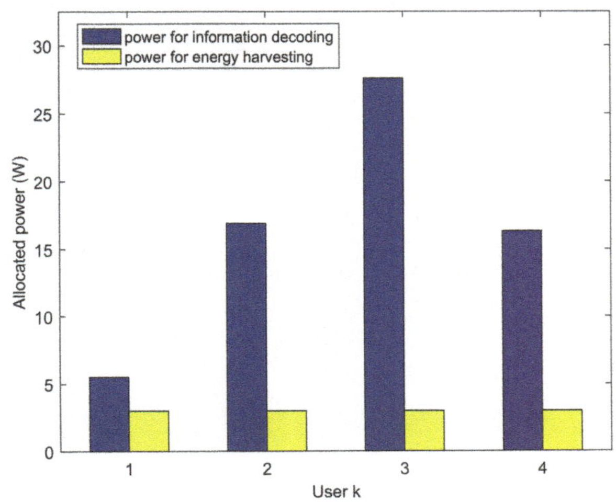

Fig. 4 The allocated power for user k

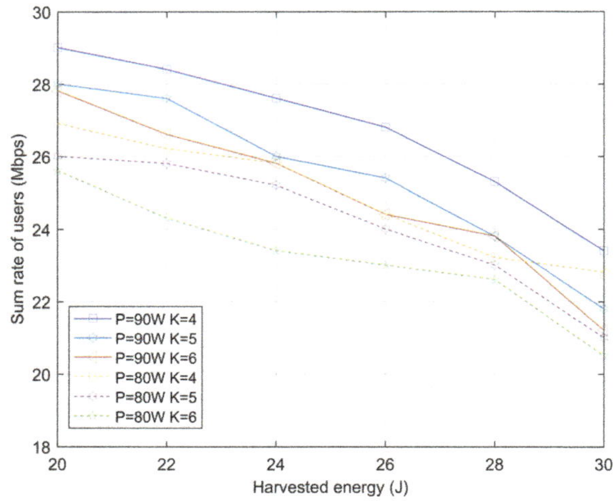

Fig. 5 Sum rate of users vs. harvested energy

satisfied more often than not. Since the water-filling approach is used for the power allocation of the conventional algorithm, it would cause some power waste. What is more, all the subcarriers allocated to information decoding lead to the energy consumption and less power is used for information decoding. Figure 2 also demonstrates that the sum rate of users increases with the increase of sum transmit power P. It can be interpreted as that the increase of sum transmit power brings about the more power allocated to information decoding with the same target harvested energy.

Figure 4 shows that the total transmit power used for information decoding of user k. It can be seen that the user 3 is allocated the most power, while the user 1 is the least. This is because, in our simulation, the user 3 has the best channel condition so that it can achieve higher sum rate.

Figure 5 shows that the sum rate of users versus the minimum energy harvesting limit B_k under the fixed power budget. It is obvious that the sum rate varies with the constraint of the harvested energy. When B_k increases, the sum rate decreases. This is because when the constraint of energy harvesting increases, more subcarriers are allocated to satisfy the constraint of the harvested energy. It leads to less resources that are allocated to the information decoding. It can also be observed that, with regard to the same constraint of energy harvesting, the sum rate decreases as the number of users increases. That is expected due to the fact that under the same power budget, the increasing number of users causes the decrease of the average allocated power for each subcarrier. As a result, more subcarriers are allocated to energy harvesting in order to satisfy the same constraint of the required energy, as well as the sum rate of users increases.

Figure 6 shows the allocated power to each subcarrier under the condition of the power budget equal to 80 W. It also illustrates the average channel gain for each subcarrier. As shown in Fig. 6, when the subcarrier has the good channel condition, it is always used for energy harvesting. On the other hand, when the subcarrier has the bad channel condition, it is always used for information decoding and less power is allocated. The phenomenon appears because to allocate the good channel for energy harvesting leads to less subcarriers used for energy harvesting and more subcarriers used for information decoding to achieve maximum sum rate.

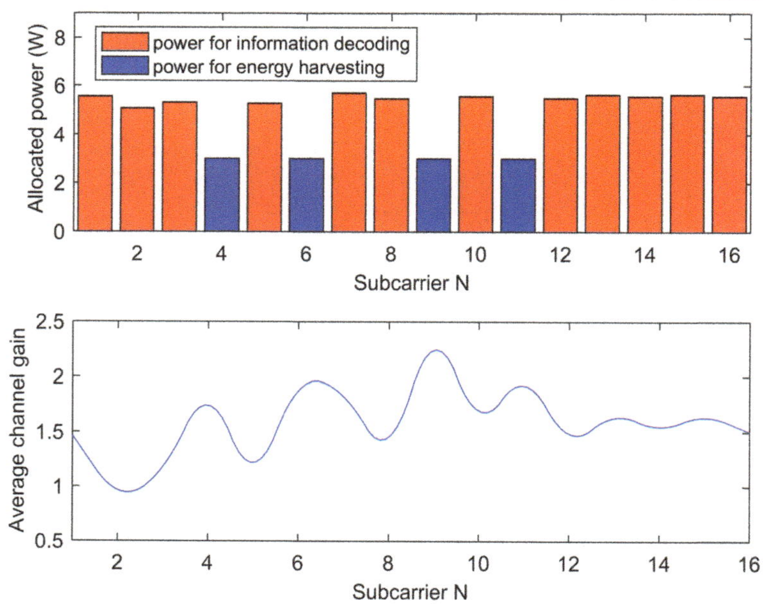

Fig. 6 The allocated power and channel condition of subcarrier n

6 Conclusions

A subcarrier allocation algorithm based on SWIPT for the OFDM system with multiusers was proposed in this paper. Traditionally, SWIPT is used for single user in OFDM systems more often than not. However, we proposed a joint optimization algorithm used for multiusers. Specifically, the OFDM subcarriers of each user are divided into two parts whose one part is used for information decoding and the other part is used for energy harvesting. Therefore, in contrast to the conventional time or power splitter at the receiver, the enough information rate can be obtained on the premise of the harvested energy we require. Simulation results show that the proposed algorithm converges fast and performs better than the conventional algorithm in the rate of information decoding.

Funding

This work was supported by the National Natural Science Foundations of China under Grant No. 61301131 and 61601221, the Natural Science Foundations of Jiangsu Province under Grant No. BK20140828, the China Postdoctoral Science Foundations under Grant No. 2015M580425 and the Fundamental Research Funds for the Central Universities under Grant No. 3132016347 and DUT16RC(3)045.

Authors' contributions

ZN designed the main body of the algorithm and completed the first simulation. XL gave the guidance on the ideas of the paper, optimized the parameters, and discussed the related problems with other authors. XL completed the remaining simulations, analyzed data, and processed simulation results. All authors read and approved the final manuscript.

Authors' information

[1] Zhenyu Na and Xiaotong Li are from the School of Information Science and Technology of Dalian Maritime University (DMU), Dalian, China. Zhenyu Na is the associate professor of DMU. His research interests include OFDM communications, non-orthogonal multicarrier techniques, satellite communications, and networking. Xiaotong Li is the postgraduate of DMU whose research field is the OFDM communications and systems, information and energy transfer optimization in wireless multicarrier systems.

[2] Xin Liu is the associate professor of the School of Information and Communication Engineering of Dalian University of Technology (DUT), Dalian, China. His research interests include cognitive radio, SWIPT, joint resource management in mobile communication systems, satellite communications, and space laser communications.

Competing interests

The authors declare that they have no competing interests.

Author details

[1]School of Information Science and Technology, Dalian Maritime University, Dalian, China. [2]School of Information and Communication Engineering, Dalian University of Technology, Dalian, China.

References

1. T. Chen, Y. Yang, H. Zhang, H. Kim, K. Horneman, Network energy saving technologies for green wireless access networks. IEEE Wirel. Commun. **18**(5), 30–38 (2011)
2. F. Zhao, X. Sun, H. Chen, R. Bie, Outage performance of relay-assisted primary and secondary transmissions in cognitive relay networks. EURASIP J. Wirel. Commun. Netw. **2014, 60**(1) (2014)
3. M. Jia, L. Wang, Q. Guo, A low complexity detection algorithm for fixed up-link SCMA system in mission critical scenario. IEEE Int Things J **1**(1), 99 (2017)
4. M. Jia, Z. Gao, Q. Guo, Fast compressive tracking with robust example selection based on multiple instance learning in smart and autonomous systems. Pattern Recognition. **69**, 325-335(2017)
5. L.R. Varshney, *Transporting Information and Energy Simultaneously. IEEE International Symposium on Information Theory. IEEE* (2016), pp. 1612–1616
6. P. Grover, A. Sahai, *Shannon meets Tesla: wireless information and power transfer* (2010), pp. 2363–2367
7. J. Guo, N. Zhao, F.R. Yu, X. Liu, Exploiting adversarial jamming signals for energy harvesting in interference networks. IEEE Trans. Wirel. Commun. **16**(2), 1267–1280 (2017)
8. W. Zhong, K. Chen, X. Liu, Joint optimal energy-efficient cooperative spectrum sensing and transmission in cognitive radio. IEEE Wirel. Commun. **14**(1), 98–110 (2014)
9. M. Jia, X. Liu, X. Gu, Joint cooperative spectrum sensing and channel selection optimization for satellite communication systems based on cognitive radio. Int. J. Satell. Commun. Netw. 23(3), 139-150(2015)
10. M. Jia, X. Gu, Q. Guo, Broadband hybrid satellite-terrestrial communication systems based on cognitive radio toward 5G. IEEE Wirel. Commun. **23**(6), 96–106 (2013)
11. Y. Liu, X. Wang, Information and energy cooperation in OFDM relaying: Protocols and optimization. IEEE Trans. Veh. Technol. **65**(7), 5088–5098 (2016)
12. L. Liu, R. Zhang, K.C. Chua, Secrecy wireless information and power transfer with MISO beamforming. IEEE Trans. Signal Process. **62**(7), 1850–1863 (2014)
13. L. Liu, R. Zhang, K.C. Chua, Wireless information transfer with opportunistic energy harvesting. IEEE Trans. Wirel. Commun. **12**(1), 288–300 (2012)
14. F. Zhao, L. Wei, H. Chen, Optimal time allocation for wireless information and power transfer in wireless powered communication systems. IEEE Trans. Veh. Technol. **65**(3), 1830–1835 (2016)
15. L. Liu, R. Zhang, C.K. Ho, Wireless information and power transfer: a dynamic power splitting approach. IEEE Trans. Wirel. Commun. **61**(9), 3990–4001 (2013)
16. D.W.K. Ng, R. Schober, Spectral efficient optimization in OFDM systems with wireless information and power transfer. IEEE Wirel. Commun. **6**(2), 1–5 (2013)
17. R. Zhang, C. Ho, MIMO broadcasting for simultaneous wireless information and power transfer. IEEE Trans. Wirel. Commun. **12**(5), 1989–2001 (2013)
18. F. Zhao, B. Li, H. Chen, X. Lv, Joint beamforming and power allocation for cognitive MIMO systems under imperfect CSI based on game theory. Wirel. Pers. Commun. **73**(3), 679–694 (2013)
19. F. Zhao, W. Wang, H. Chen, Q. Zhang, Interference alignment and game-theoretic power allocation in MIMO heterogeneous sensor networks communications. Signal Process. **126**(C), 173–179 (2016)
20. X. Zhou, R. Zhang, C.K. Ho, Wireless information and power transfer in multiuser OFDM systems. IEEE Trans. Wirel. Commun. **13**(4), 2282–2294 (2014)
21. D.W.K. Ng, E.S. Lo, R. Schober, Wireless information and power transfer: energy efficiency optimization in OFDMA systems. IEEE Trans. Wirel. Commun. **12**(12), 6352–6370 (2013)
22. X. Liu, M. Jia, X.Z. Tan, Threshold optimization of cooperative spectrum sensing in cognitive radio network. Radio Sci. **48**(1), 23–32 (2013)
23. X. Liu, G.A. Bi, M. Jia, Y.L. Guan, Joint optimization of sensing threshold and transmission power in wideband cognitive radio. Radio Sci. **48**(4), 359–370 (2013)
24. F. Zhao, H. Nie, H. Chen, Group buying spectrum auction algorithm for fractional frequency reuses cognitive cellular systems. Ad Hoc Netw. **58**, 239–246 (2017)
25. L. Sibomana, H.J. Zepernick, H. Tran, *Signal Processing and Communications Systems (ICSPCS), 2014 8th international conference on*, Wireless information and power transfer in an underlay cognitive radio network (2014), pp. 1–7
26. Z. Gao, B. Wang, *Wireless Communication & Signal Processing (WCSP), 2015 International Conference on. User cooperation in OFDM-based cognitive radio networks with simultaneous wireless information and power transfer* (2015), pp. 1–6
27. D.W.K. Ng, R. Schober, *2003 IEEE GlobeconWorkshops (GC Wkshps), Resource allocation for secure communication in systems with wireless information and power transfer* (2013), pp. 1251–1257
28. A. Nedic, A. Ozdaglar, Distributed subgradient methods for multi-agent optimization. IEEE Trans. Autom. Control **54**(1), 48–61 (2007)
29. W. Lu, Y. Gong, J. Wu, Simultaneous wireless information and power transfer based on joint subcarrier and power allocation in OFDM systems. IEEE Access **1**(1), 99 (2017)

8

Hardware and software design of BMW system for multi-floor localization

Mu Zhou[1,2], Bin Wang[1*], Zengshan Tian[1] and Liangbo Xie[1]

Abstract

Although the Micro Electro Mechanical System (MEMS) sensors are capable of providing short-term high positioning accuracy, every positioning result significantly depends on the historical ones, which inevitably leads to the long-term error accumulation. The Bluetooth Low Energy (BLE) is independent of the accumulative error, but the positioning accuracy is suffered by the irregular jump error resulted from the Received Signal Strength Indicator (RSSI) jitter. Considering the requirement of accurate, seamless, and consecutive positioning by the existing commercial systems, we propose a new integrated BLE and MEMS Wireless (BMW) system for multi-floor positioning. In concrete terms, first of all, the way of fingerprint database construction with the reduced workload is introduced. Second, the fingerprint database is denoised by the process of affinity propagation clustering, outlier detection, and RSSI filtering. Third, the robust M estimation-based extended Kalman filter is applied to estimate the two-dimensional coordinates of the target on each floor. Finally, the barometer data are used to calculate the height of the target. The extensive experimental results show that the proposed system can not only restrain the accumulative error caused by the MEMS sensors but also eliminate the irregular jump error from the BLE RSSI jitter. In an actual multi-floor environment, the proposed system is verified to be able to achieve the Root Mean Square (RMS) positioning error within 1 m.

Keywords: Multi-floor positioning, Data fusion, Robust M estimation, BLE, MEMS

1 Introduction

At present, the indoor positioning has broad application such as searching the cars and elevators in underground parking lot and pushing advertisements or discounts to the customers in large shopping malls. The Global Navigation Satellite System (GNSS) [1–3] well meets the precision requirement of outdoor positioning, but in indoor environment, its performance may drastically deteriorate due to the serious signal blocking and multipath effect. In response to this compelling problem, a batch of researchers put forward a variety of indoor positioning systems based on the Bluetooth Low Energy (BLE) [4], Ultra Wideband (UWB) [5], Radio Frequency Identification (RFID) [6], Micro Electro Mechanical System (MEMS) sensors [7], and Wireless Local Area Network (WLAN) [8, 9].

The hardware cost, range limitation, and long-term error accumulation limit the development of the conventional positioning systems. Among them, the MEMS sensors can be used to perform the Pedestrian Dead Reckoning (PDR) [7, 10] by using the inertia and heading information though it contains the accumulative error. The BLE Received Signal Strength Indicator (RSSI)-based positioning system can meet the requirement of low power, low cost, and no accumulative error though the RSSI jitter caused by multipath effect may seriously decrease the positioning accuracy.

In recent decade, various integrated systems have been significantly concerned to improve positioning accuracy [11, 12]. The systems [13–15] using Kalman and particle filter to achieve the WLAN/MEMS fusion positioning reduce the accumulative error of MEMS sensors, but they fail to constrain the WLAN RSSI jitter. To solve the RSSI jitter problem, the authors in [16] build the observation equation of the RSSI according to the signal propagation model, which is difficult to be constructed in the complex indoor environment. In [17], the authors design an indoor positioning system by fusing the WLAN and Magnetic

*Correspondence: wangbin-cq@foxmail.com
[1]Chongqing Key Lab of Mobile Communications Technology, Chongqing University of Posts and Telecommunications, Chongqing 400065, China
Full list of author information is available at the end of the article

Angular Rate and Gravity (MARG) data to solve the problems of RSSI jitter and accumulative error. However, this system cannot achieve the three-dimensional positioning, which is significantly required by the current commercial applications.

This paper presents a new integrated BMW system for multi-floor positioning. First of all, the fingerprint database is optimized by using the affine propagation clustering, outlier detection, and RSSI filtering algorithms. Specifically, the fingerprint denoising is performed to reduce the probability of large errors of BLE positioning, and then, the gait detection approach is used to estimate the walking speed and heading angle of the target based on the extended Kalman filter (EKF). Second, the robust EKF is designed to restrain the accumulative error caused by the MEMS sensors, as well as eliminate the irregular jump error from the BLE RSSI jitter. Finally, according to the barometer data and geographical location information, the height of the target is calculated to achieve the multi-floor positioning.

2 Related work

In this section, we will briefly introduce some related work in three aspects of fast fingerprint database construction, fusion positioning, and multi-floor positioning and also address the corresponding limitations.

Fast fingerprint database construction Although the fingerprint positioning has been studied for a long time, it still cannot be applied widely since its offline phase generally spends a huge amount of time on fingerprint database construction. By using the conventional approaches, the target area is calibrated with a batch of equally spaced grids, and then, the fingerprints are collected at the grids point-by-point [18, 19]. The involved time cost rises significantly with the increase of environmental size, which hinders the development of fingerprint positioning. In [20], the authors propose to construct an incomplete fingerprint database with realistic coverage gaps, and meanwhile study the performance of several interpolation and extrapolation approaches used for recovering the missing fingerprints. In [21], according to the distribution of Reference Points (RPs) with respect to each Wi-Fi Access Point (AP), the signal propagation model is constructed as a function of spatial structure, which can be used to construct the fingerprint database quickly.

Fusion positioning At present, the common single positioning systems are difficult to adapt to the complex indoor environment. In response to this compelling problem, a variety of fusion positioning systems are designed to compensate for the shortcomings of each single one. In [22], the data fusion from both the proprioceptive and exteroceptive sensors, like the odometer, Global Positioning System (GPS), Light Detection and Ranging (LIDAR), and vision, as well as the knowledge of road map is considered to perform the fusion positioning. In [23], the authors conduct data fusion by integrating the RSSI and Time difference of Arrival (TDOA) measurements to estimate the superior locations of the target. Specifically, by employing the nonparametric estimation approach, which is robust to the variations of measurement noise and quantization, it is addressed that the fusion positioning is more robust and higher accurate and has lower implementation cost.

Multi-floor positioning The mainstream of indoor positioning systems mainly focuses on the horizontal coordinate estimation, whereas little research has been done on the vertical coordinate estimation. In [24], the authors propose a WiFi-based indoor positioning system that takes both the characteristics of trilateration and scene analysis into account. The authors in [25] rely on the path loss model to construct a light fingerprint radio map to find the target floor. However, these systems are applied to only determine the floor on which the target is most probably located, but they cannot estimate the accurate locations of the target.

3 System description

3.1 System framework

As shown in Fig. 1, our system contains four modules, BLE fingerprint positioning module, speed and heading calculation module, EKF module, and height calculation module. According to the output of accelerometer, gyroscope, and magnetometer, the gait detection and quaternion calculation are carried out to estimate the walking speed and heading angle of the target, which will be sent to the PDR to perform localization. Then, the localization results by the PDR and Weighted K Nearest Neighbor (WKNN) are selected as the input of the EKF to obtain the two-dimensional coordinates of the target. After that, based on the output of barometer, two-dimensional positioning result, and geographical location information, the height of the target is inferred.

3.2 BLE fingerprint positioning

BLE fingerprint positioning includes two phases, offline phase and online phase. The main tasks of offline phase are the coordinate calibration and fingerprint database construction with respect to the target environment. In addition, the affinity propagation clustering, outlier detection, and RSSI filtering are carried out to eliminate the large errors as well as optimize each sub-database. The main task of online phase is to match the newly collected RSSI data against the fingerprint database to obtain the positioning result. This process is shown in Fig. 2.

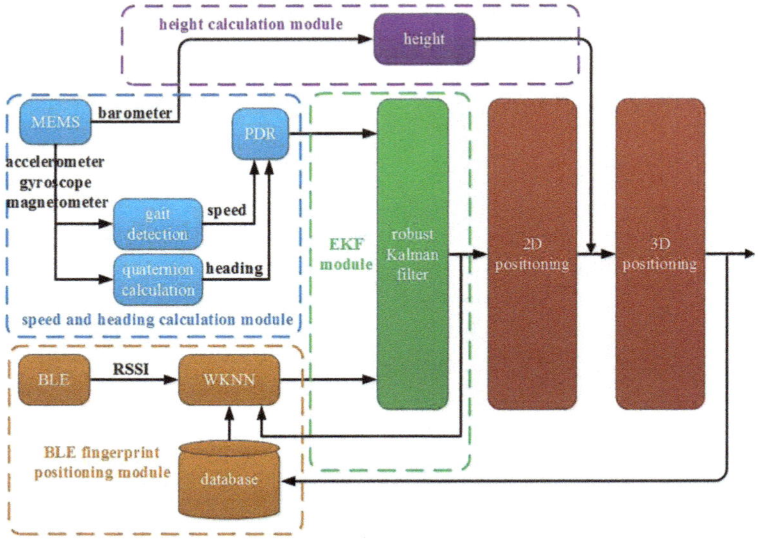

Fig. 1 Framework of BLE/MEMS fusion positioning system

3.2.1 Fast fingerprint database construction

The data in the conventional fingerprint database are collected point-by-point, which limits the application of fingerprint positioning due to the requirement of huge time cost. In response to this compelling problem, we propose a new fast fingerprint database construction approach. First of all, the floor plan of the target environment is imported into the terminal as a map. Second, the map is represented by a batch of straight lines (labeled with dash blue lines) with the starting and ending points (labeled with red triangles), as shown in Fig. 3.

Supposing that the pixel size of the map is m_{mapx} by m_{mapy} and the actual size of the target environment is M_x by M_y, we can obtain the relations of the pixels and actual locations in (1).

$$\begin{cases} X_i = x_{\text{map}_i} * \dfrac{M_x}{m_{\text{map}}} + X_0 \\ Y_i = y_{\text{map}_i} * \dfrac{M_y}{m_{\text{map}}} + Y_0 \end{cases} \quad (1)$$

where (X_0, Y_0) is the actual origin location, which is represented by $(0, 0)$ in the map. $(x_{\text{map}_i}, y_{\text{map}_i})$ and (X_i, Y_i) are the coordinates of the ith point in the map (with the length m_{mapx} and width m_{mapy}) and actual environment (with the length M_x and width M_y) respectively.

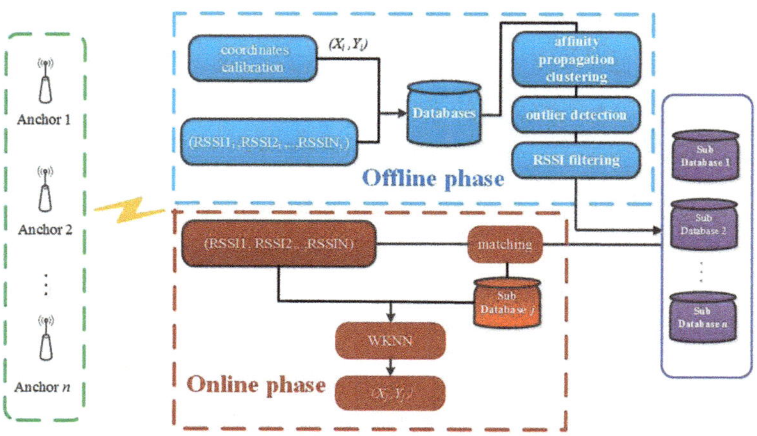

Fig. 2 Process of BLE fingerprint positioning

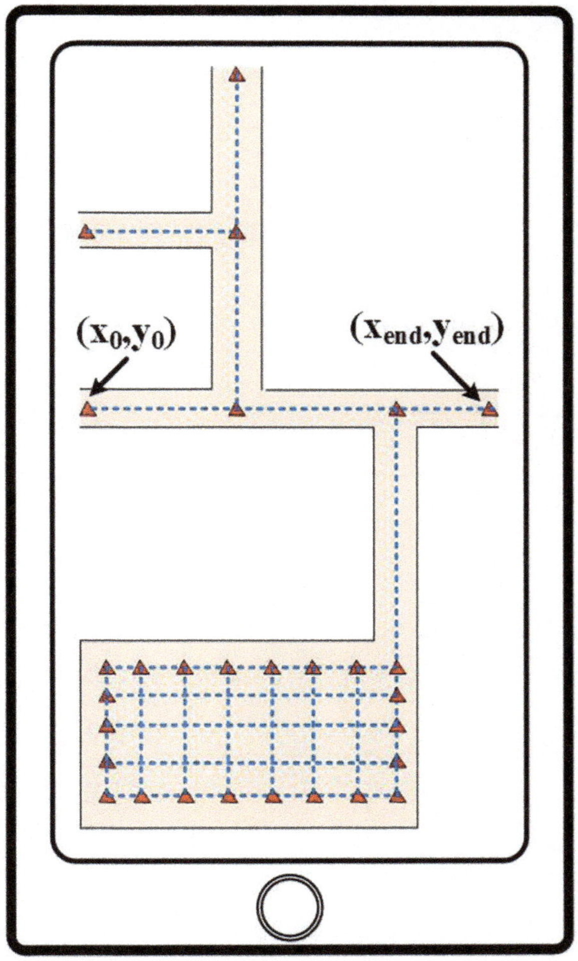

Fig. 3 Floor plan imported into the terminal

We construct the standard fingerprint database by

$$\begin{cases} X_i = X_0 + L_{\text{step\_}x} * i i \in 1, \cdots, \frac{|X_{\text{end}} - X_0|}{L_{\text{step\_}x}} \\ Y_j = Y_0 + L_{\text{step\_}y} * j j \in 1, \cdots, \frac{|Y_{\text{end}} - Y_0|}{L_{\text{step\_}y}} \end{cases} \quad (2)$$

where $L_{\text{step\_}x}$ and $L_{\text{step\_}y}$ are the physical intervals in the X and Y directions, respectively, in the actual environment, which are normally set as $L_{\text{step\_}x} = L_{\text{step\_}y}$.

By taking the walking path from (x_0, y_0) to $(x_{\text{end}}, y_{\text{end}})$ (in Fig. 3) as an example, a volunteer starts walking after labeling (x_0, y_0) on the screen of the terminal and then labels $(x_{\text{end}}, y_{\text{end}})$ on the screen of the terminal when he/she stops walking. During the walk, (x_i, y_i) is notated as the ith location, which is calculated by

$$\begin{cases} x_i = x_0 + \sum_{n=1}^{i} v_n * \sin(\theta_n) + \varepsilon_{xi} \\ y_i = y_0 + \sum_{n=1}^{i} v_n * \cos(\theta_n) + \varepsilon_{yi} \end{cases} \quad (3)$$

where v_n and θ_n are the walking speed and heading. $\varepsilon_{xi} = \frac{v_i \sin(\theta_i)}{\sum v_i \sin(\theta_i)}(L \sin(\Psi_1) - L_{pdr} \sin(\Psi_2))$ and $\varepsilon_{yi} = \frac{v_i \cos(\theta_i)}{\sum v_i \cos(\theta_i)}(L \cos(\Psi_1) - L_{pdr} \cos(\Psi_2))$ are the error compensation in the X and Y directions respectively. L and $L_{pdr} = \sum_{i=1}^{N} v_i$ are the actual distance and the estimated one by the PDR from the starting to ending locations. Ψ_1 is the heading of $(x_{\text{end}}, y_{\text{end}})$ relative to (x_0, y_0). Ψ_2 is the heading of (x_N, y_N) relative to (x_0, y_0).

Since the RSSI in standard fingerprint database may be lost when the walking speed is too fast, we propose to rely on the interpolation approach to estimate the lost RSSI by propagation modeling, as shown in Fig. 4.

We use the propagation model below to describe the relations of the RSSI and distance from each anchor to the terminal.

$$P_d = -10N_t \log d + P_{d_0} \quad (4)$$

where P_{d_0} is the RSSI with 1-m distance from the anchor. d is the distance from the anchor to terminal. N_t is the path loss exponent.

Figure 5 shows two groups of test points indicating the variation of RSSI with respect to the distance from an anchor to the terminal. As can be seen from Fig. 6, the constructed propagation model matches the variation of RSSI well when the distance is within 10 m.

We randomly select N_m terminal locations to train the path loss exponent for each anchor by

$$\begin{cases} P_{d_1} = -10N_1 \log \|D_1 - D_0\|_2 + P_{d_0} \\ P_{d_2} = -10N_2 \log \|D_2 - D_0\|_2 + P_{d_0} \\ \vdots \\ P_{d_i} = -10N_i \log \|D_i - D_0\|_2 + P_{d_0} \\ \vdots \\ P_{d_m} = -10N_m \log \|D_m - D_0\|_2 + P_{d_0} \end{cases} \quad (5)$$

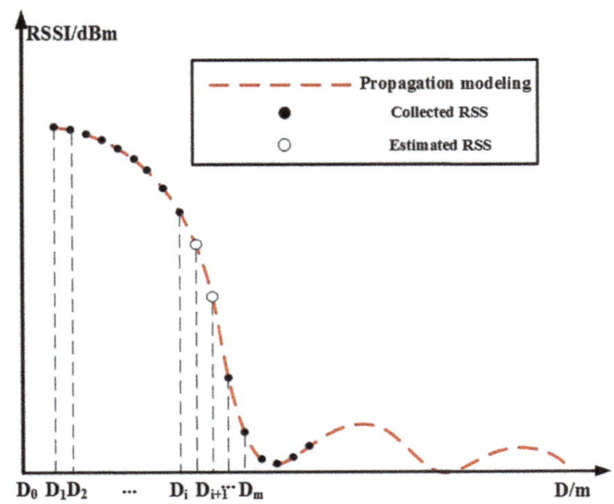

Fig. 4 RSSI interpolation

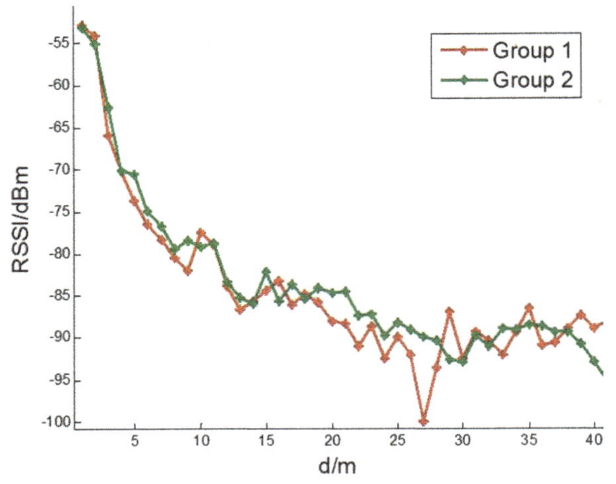

Fig. 5 Variation of RSSI with respect to the distance

where the notation "$\|\bullet\|_2$" represents the 2-norm operation. D_0 is the anchor location. D_i is the ith location of the terminal. P_{d_i} is the RSSI at D_i. We calculate $\overline{N} = \sum N_i/m$ as the path loss exponent. Thus, the lost RSSI at D_{i+1} is estimated by

$$P_{d_{i+1}} = -10\overline{N}\log\|D_{i+1} - D_0\|_2 + P_{d_0} \tag{6}$$

3.2.2 *Outlier detection*

We rely on the affinity propagation clustering [26] to classify the off-line RSSI data and then construct each cluster as a sub-database, as shown in Table 1. In online phase, we match the newly collected RSSI data to the cluster center of each sub-database and then select the sub-database with the best matching to perform positioning. To achieve this goal, we define density(x, k) and relative_density(x, k) as the density and relative density [27] of the newly collected RSSI x with respect to its k nearest neighbors respectively.

$$\text{density}(x,k) = \left(\frac{\sum\limits_{y \in N(x,k)} \text{distance}(x,y)}{|N(x,k)|} \right)^{-1} \tag{7}$$

$$\text{relative\_density}(x,k) = \frac{\text{density}(x,k)}{\sum\limits_{y \in N(x,k)} \frac{\text{density}(y,k)}{|N(x,k)|}} \tag{8}$$

where distance(x, y) is the distance between x and y. $N(x, k)$ is the set of k nearest neighbors with respect to x. $|N(x, y)|$ is the number of RSSI data in $N(x, k)$.The pseudo-code of sub-database construction is shown in Table 1.

cluster$_i$ is the ith cluster. point$_j$ is the jth point. *Threshold* is the threshold of relative density.

Table 1 Pseudo-code of sub-database construction

Algorithm: Sub-database construction

Data: Raw fingerprint database

Result: sub-databases

```
1 cluster ← AP_Cluster (input);
2 for each cluster_i ∈ clusters do
3   for each point_j ∈ cluster_i do
4     if relative_density (point_j) >Threshold
5       Outliers are reassigned to the nearest neighbor;
6       back to Step 1;
7     end
8   end
9   Obtain the cluster center;
10 end
11 Construct each cluster as a sub-database;
```

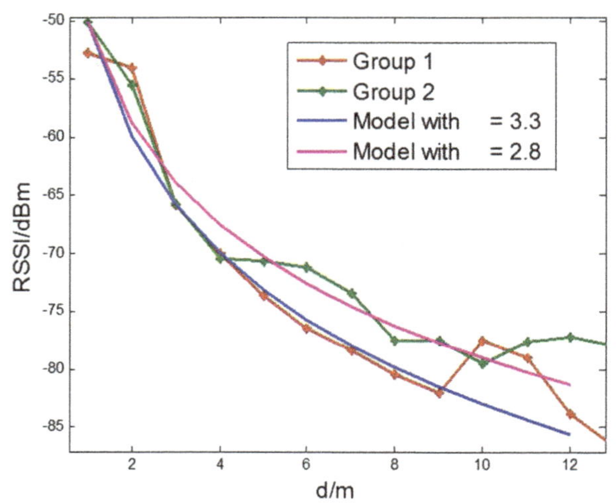

Fig. 6 Result of propagation modeling

3.3 Speed and heading estimation

We integrate the data from the 3-axis accelerometer, 3-axis gyroscope, and 3-axis magnetometer to estimate the walking speed and heading of the target (Fig. 7). First of all, the median and mean filters are fused to eliminate the device noise. Second, the low-pass filtering is performed on the output of 3-axis accelerometer to detect the gait and consequently estimate the walking speed. Finally, the data from the 3-axis accelerometer, 3-axis gyroscope, and 3-axis magnetometer are combined with the posture matrix [28] to estimate the heading.

3.3.1 *Speed estimation*

We calculate the step size of the kth step by $P_k = C_{Sk}\sqrt[4]{A_{\text{norm}_{\max}} - A_{\text{norm}_{\min}}}$, where the maximum and minimum acceleration respectively are $A_{\text{norm}_{\max}}$ and $A_{\text{norm}_{\min}}$.

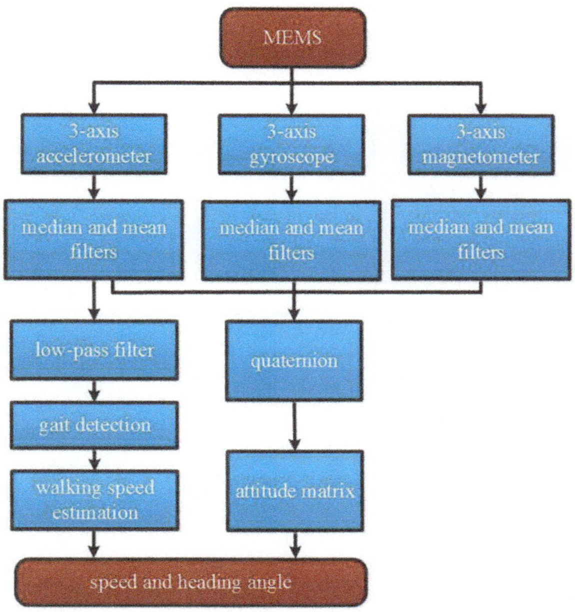

Fig. 7 Process of speed and heading estimation

The parameter C_{Sk} is trained by using the Back Propagation (BP) neural network [29, 30]. Specifically, the step frequency F_{Sk} and height H_{S_k} are selected as the input and C_{Sk} is selected as the output to train the coefficients V_{ij} and W_j. The number of neurons on hidden layer is 11, as shown in Fig. 8.

With the pedestrian walking, the output of accelerometer $A_{\text{norm}} = \sqrt{a_{x_b}^2 + a_{y_b}^2 + a_{z_b}^2}$ generally changes in sinusoid mode [28], where a_{x_b}, a_{y_b}, and a_{z_b} are the output of accelerometer in X, Y, and Z directions respectively. A step is detected when A_{norm} reaches a local maximum higher than a given threshold A_0. By setting the sampling frequency as f_s and number of samples between two adjacent maximum as ΔN, the time duration of the kth step is $t_k = \frac{\Delta N}{f_s}$ and the corresponding estimated speed is

$$v_k = \frac{P_k}{t_k} = \frac{P_k f_s}{\Delta N} \tag{9}$$

3.3.2 Heading estimation

We define the carrier and geographic reference coordinate systems as x-y-z and ENU, where E, N, and U axes point to the geographic east, north, and sky directions, respectively, as shown in Fig. 9.

The coordinate transformation from the carrier to geographic reference coordinate systems is described as

$$\begin{bmatrix} x_b \\ y_b \\ z_b \end{bmatrix} = T_n^b(q) \begin{bmatrix} x_n \\ y_n \\ z_n \end{bmatrix} \tag{10}$$

where $T_n^b(q)$ is the posture matrix. $q = q_0 + q_1 i + q_2 j + q_3 k$ is the quaternion. $\begin{bmatrix} x_b & y_b & z_b \end{bmatrix}^T$ and $\begin{bmatrix} x_n & y_n & z_n \end{bmatrix}^T$ are the coordinates of the target in the carrier and geographic reference coordinate systems respectively. According to the relations of the posture matrix and quaternion [31], we have

$$T_n^b(q) = \begin{bmatrix} q_0^2 + q_1^2 - q_2^2 - q_3^2 & 2(q_1 q_2 + q_0 q_3) & 2(q_1 q_3 - q_0 q_2) \\ 2(q_1 q_2 - q_0 q_3) & q_0^2 - q_1^2 + q_2^2 - q_3^2 & 2(q_2 q_3 + q_0 q_1) \\ 2(q_1 q_3 + q_0 q_2) & 2(q_2 q_3 - q_0 q_1) & q_0^2 - q_1^2 - q_2^2 + q_3^2 \end{bmatrix} \tag{11}$$

Using the rigid body angular differential equation, we obtain

$$\frac{dq}{dt} = \frac{1}{2} q \otimes w \tag{12}$$

where $w = 0 + w_x i + w_y j + w_z k$ is the angular velocity rotation quaternion. w_x, w_y, and w_z are the angular velocities

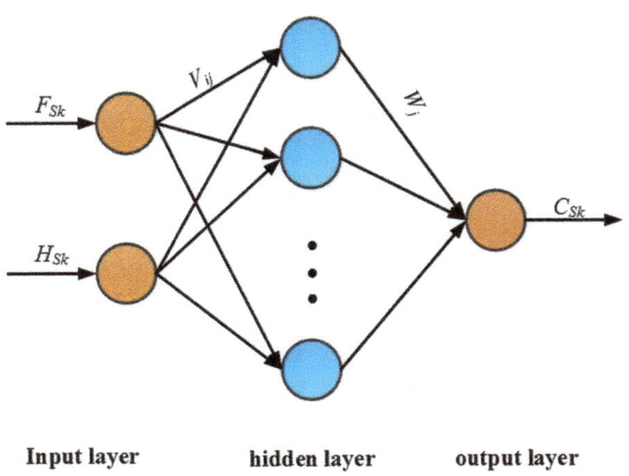

Fig. 8 Structure of BP neural network

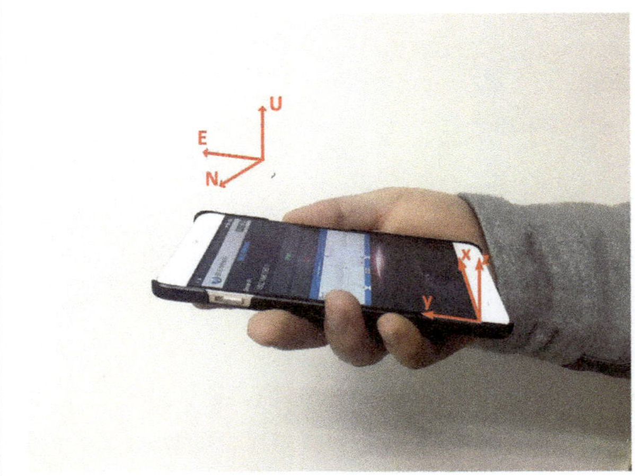

Fig. 9 Geographic reference coordinate system

in X, Y, and Z directions in the carrier coordinate system. The notation "$\otimes$" represents the quaternion multiplication, which is also used in [31]. We rewrite the formula above in matrix form as

$$\frac{dq}{dt} = \frac{1}{2}\begin{bmatrix} 0 & -w_x & -w_y & -w_z \\ w_x & 0 & w_z & -w_y \\ w_y & -w_z & 0 & w_x \\ w_z & w_y & -w_x & 0 \end{bmatrix} \times \begin{bmatrix} q_0 \\ q_1 \\ q_2 \\ q_3 \end{bmatrix} = \frac{1}{2}\Omega(w)\vec{q} \tag{13}$$

We discretize the formula above as

$$\begin{cases} \vec{q}_{m+1} = \left(I + \frac{1}{2}\Omega(wT_s)\right)\vec{q}_m, \ m = 0, 1, \cdots \\ \vec{q}_0 = \vec{q}(0) \end{cases} \tag{14}$$

where T_s is the sampling interval. I is the unit matrix. $\vec{q}(0)$ is the initial quaternion. m is the index of samples. We normalize the quaternion as

$$\vec{q}_m = \frac{\vec{q}_m}{\|\vec{q}_m\|} = \frac{\vec{q}_m}{\sqrt{q_0^2 + q_1^2 + q_2^2 + q_3^2}} \tag{15}$$

The posture matrix can be updated based on the quaternion obtained in (12). However, the quaternion may not be accurate due to the error of the gyroscope. To solve this problem, we construct the quaternion equations of status and observation as follows.

$$\begin{cases} \vec{q}_{m+1} = A_m\vec{q}_m + W_m \\ Y_{m+1} = h(\vec{q}_{m+1}, V_{m+1}) \end{cases} \tag{16}$$

where $A_m = \left(I + \frac{1}{2}\Omega(wT_s)\right)$ is the transition matrix of status. W_m is the process noise. V_{m+1} is the observation noise. The observation variable Y_{m+1} is calculated by

$$Y_{m+1} = \begin{bmatrix} a_{m+1} \\ c_{m+1} \end{bmatrix} = \begin{bmatrix} T_n^b(\vec{q}_{m+1}) & 0 \\ 0 & T_n^b(\vec{q}_{m+1}) \end{bmatrix}\begin{bmatrix} g \\ L \end{bmatrix} + V_{m+1} \tag{17}$$

where $a_{m+1} = \left[a_{x_b}\ a_{y_b}\ a_{z_b}\right]^T$ and $c_{m+1} = \left[c_{x_b}\ c_{y_b}\ c_{z_b}\right]^T$ are the output of accelerometer and magnetometer, respectively, in the carrier coordinate system. $g = \left[0\ 0\ 1\right]^T$ is the normalized output of accelerometer when the target is static in the geographic reference coordinate system. $L = \left[0\ b_y\ b_z\right]^T$ is the output of magnetometer in the geographic reference coordinate system, in which $b_y = \sqrt{c_{x_b}^2 + c_{y_b}^2}$ and $b_z = c_{z_b}$. Based on (14), we can calculate that

$$\Phi_{m+1} = \frac{\partial h(\vec{q}_{m+1}, V_{m+1})}{\partial \vec{q}_{m+1}} = \begin{bmatrix} -2q_2 & 2q_3 & -2q_0 & 2q_1 \\ 2q_1 & 2q_0 & 2q_3 & 2q_2 \\ 2q_0 & -2q_1 & -2q_2 & 2q_3 \\ 2(q_3b_y - q_2b_z) & 2(q_2b_y + q_3b_z) & 2(q_1b_y - q_0b_z) & 2(q_0b_y + q_1b_z) \\ 2(q_0b_y + q_1b_z) & -2(q_1b_y - q_0b_z) & 2(q_2b_y + q_3b_z) & -2(q_3b_y - q_2b_z) \\ -2(q_1b_y - q_0b_z) & -2(q_0b_y + q_1b_z) & 2(q_3b_y - q_2b_z) & 2(q_2b_y + q_3b_z) \end{bmatrix} \tag{18}$$

Finally, the heading of the target is estimated by

$$\varphi = \arctan\left(-\frac{2(q_1q_2 + q_0q_3)}{q_0^2 + q_1^2 - q_2^2 - q_3^2}\right) \tag{19}$$

3.4 Robust EKF

3.4.1 EKF

We choose the geographic east and north directions, walking speed, and heading to construct the equation of status below.

$$X_t = f(X_{t-1}, W_{t-1}) = \begin{bmatrix} 1 & 0 & \sin(\varphi_{t-1}) & 0 \\ 0 & 1 & \cos(\varphi_{t-1}) & 0 \\ 0 & 0 & 1 & 0 \\ 0 & 0 & 0 & 1 \end{bmatrix} \times \begin{bmatrix} E_{t-1} \\ N_{t-1} \\ v_{t-1} \\ \varphi_{t-1} \end{bmatrix} + W_{t-1} \tag{20}$$

where $X_t = [E_t\ N_t\ v_t\ \varphi_t]^{\mathrm{T}}$. E_{t-1} and N_{t-1} are the geographic east and north directions, respectively, at moment $t-1$. v_{t-1} and φ_{t-1} are the walking speed and heading, respectively, in the geographic reference coordinate system at moment $t-1$. W_{t-1} is the Gaussian white noise with zero mean at moment $t-1$. $E\left[W_iW_j^{\mathrm{T}}\right] = Q(i,j)\delta_{ij},\ i,j = 1,\cdots,o$, in which o is the number of estimation variables, δ_{ij} is the Kronecker function, and Q is the covariance matrix of process noise.

The output of BLE fingerprint positioning and the estimated speed and heading are selected as the observation variable to construct the equation of observation below.

$$Z_t = h(X_t, V_t) = \begin{bmatrix} 1 & 0 & 0 & 0 \\ 0 & 1 & 0 & 0 \\ 0 & 0 & 1 & 0 \\ 0 & 0 & 0 & 1 \end{bmatrix} \times \begin{bmatrix} E_t \\ N_t \\ v_t \\ \varphi_t \end{bmatrix} + V_t \tag{21}$$

where $Z_t = \left[E_t^{\text{ble}}\ N_t^{\text{ble}}\ v_t^{\text{mems}}\ \varphi_t^{\text{mems}}\right]^{\text{T}}$. E_t^{ble} and N_t^{ble} are the geographic east and north directions by the BLE fingerprint positioning at moment t. v_t^{mems} and φ_t^{mems} are the estimated speed and heading, respectively, by using the MEMS sensors at moment t. V_t is the Gaussian white noise with zero mean at moment t. $E\left[V(i)V^T(j)\right] = R(i,j)\delta_{ij}$, $i,j = 1, \cdots, n$, in which n is the number of observation variables and $R(R > 0)$ is the covariance matrix of observation noise.

3.4.2 *Robustness enhancement*

Considering that the change of the weighting factors of the IGG3 [32] is slight, we first define the IGG3 weight function below to enhance the robustness of the EKF. We first define the IGG3 weight function below.

$$\bar{p}_i = \begin{cases} p_i & |u_i| \le k_0 \\ p_i \frac{k_0}{|u_i|}\left(\frac{k_1 - |u_i|}{k_1 - k_0}\right)^2 & k_0 < |u_i| \le k_1 \\ 0 & |u_i| > k_1 \end{cases} \tag{22}$$

where p_i and $\bar{p}_i$ are the ith diagonal elements in the weight matrix $P = R^{-1}$ and equivalence weight matrix $\bar{P}$ respectively. $k_0 \in [1, 1.5]$ and $k_1 \in [2.5, 3]$ are the weights. The normalized residual vector u_i is calculated by

$$u_i = \frac{v_i}{\sigma_{v_i}} \tag{23}$$

where v_i is the ith element in the $n \times 1$-dimensional residual vector of observation V. $\sigma_{v_i} = \frac{\sigma_0}{\sqrt{q_{v_i}}}$, in which q_{v_i} is the reciprocal of v_i and σ_0 is the variance factor [32].

Then, we apply the M estimation approach to enhance the robustness of the EKF. The main difference between the conventional EKF and robust one is the updating of filter gain, as shown below.

$$\bar{K}(t) = P(t, t-1)H^{\text{T}}(t) \times \left[H(t)P(t,t-1)H^{\text{T}}(t) + \bar{R}(t)\right]^{-1} \tag{24}$$

where $\bar{K}(t)$ is the filter gain at moment t. $P(t, t-1)$ is the one-step prediction of the matrix of error covariance at moment t. $H(t)$ is the matrix of observation at moment t. $\bar{R}(t) = \bar{B}_t^{-1}$ is the covariance matrix of observation noise, in which $\bar{B}_t$ is the equivalent weight matrix.

3.5 Height estimation

3.6 Pressure measure

The principle of barometric pressure measure is that the atmospheric pressure in the gravitational field decreases as the height increases [33]. Thus, the altitude of the target can be estimated by using the barometric pressure measured by the barometer. When the height of gravity, H, is less than 11 km, the standard pressure formula can be described as

$$H = 44330.76\left[1 - \left(\frac{P_s}{101.325}\right)^{0.190255}\right] \tag{25}$$

where P_s is the pressure value.

Finally, the height of the target is estimated by

$$h = \frac{R \times H}{R - H} \tag{26}$$

where $R = 6356766$ m is the radius of earth. Since $R \gg H$, we can obtain $\frac{R}{R-H} \approx 1$ and $h \approx H$.

3.7 Algorithm design

The walking pattern in indoor environment can be simply divided into walking flat, upstairs, and downstairs. In our system, we rely on the output of barometer to determine the walking pattern of the target and then count the number of steps to estimate the height of the target, as shown in Table 2. In this table, h_{t-1} and h_{stair} stand for the height of the target at moment t and step height respectively.

3.8 System implementation

Figure 10 shows the schematic diagram of the proposed system. The BLE anchor broadcasts the packets in a real-time manner. A smartphone equipped with the BLE and MEMS modules is selected as the target. Both the BLE

Table 2 Pseudo-code of height estimation

Algorithm: Height estimation
Data: *Positioning_result* (x, y);
Result: Height of the target;
if (x, y) is located in the staircase
if Pressure value rises
$h_t = h_{t-1} + h_{stair}$;
else if Pressure value drops
$h_t = h_{t-1} - h_{stair}$;
else
$h_t = h_{t-1}$;
end
else
$h_t = h_{t-1}$;
end

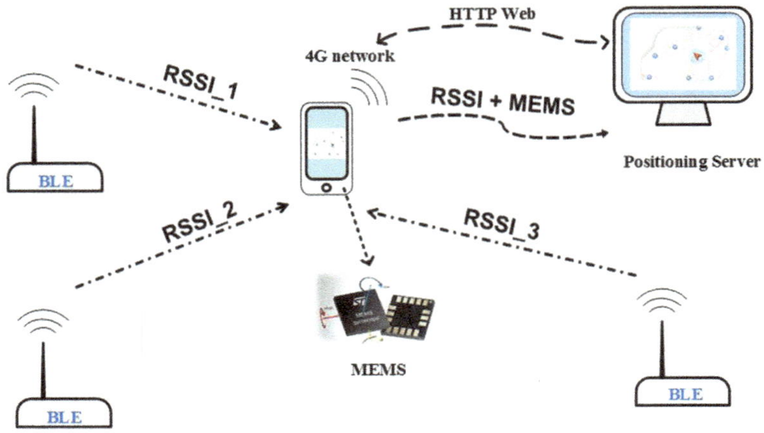

Fig. 10 Schematic diagram of the proposed system

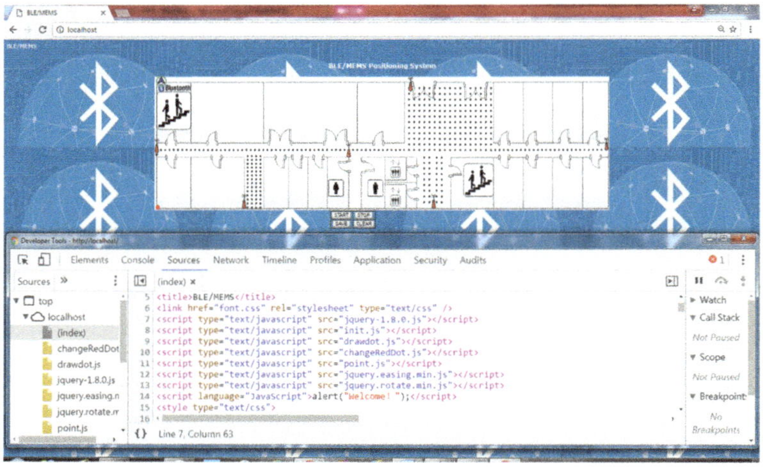

Fig. 11 Interface of positioning server

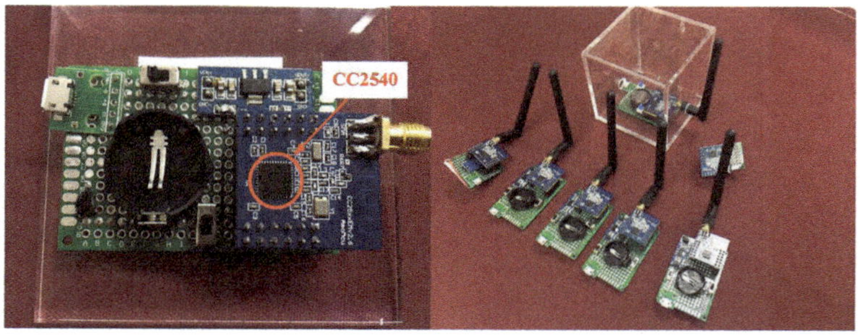

Fig. 12 BLE anchors

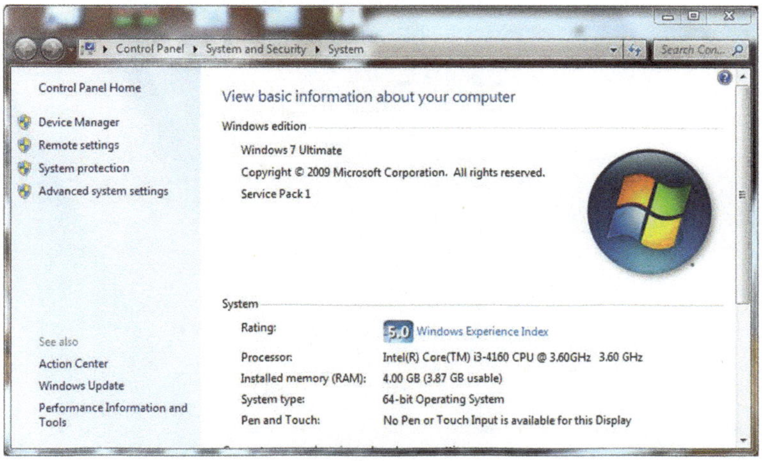

Fig. 13 Hardware configuration

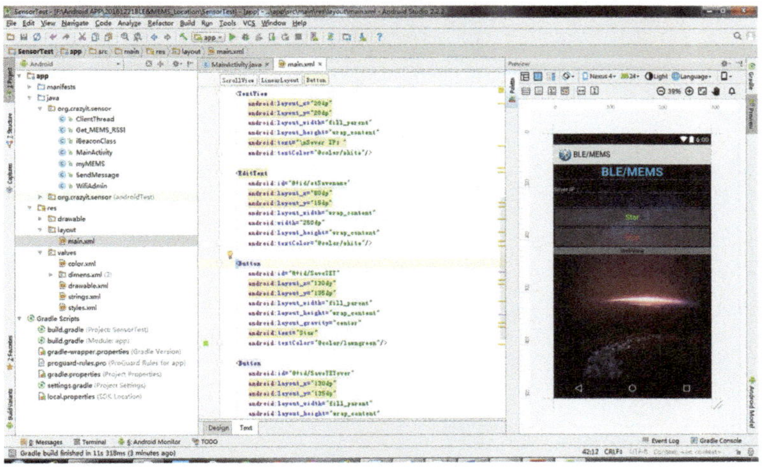

Fig. 14 Interface of Android Studio software

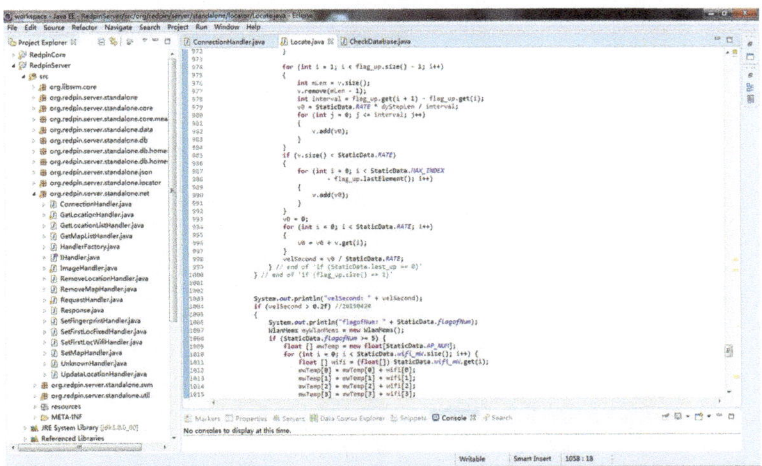

Fig. 15 Interface of Eclipse software platform

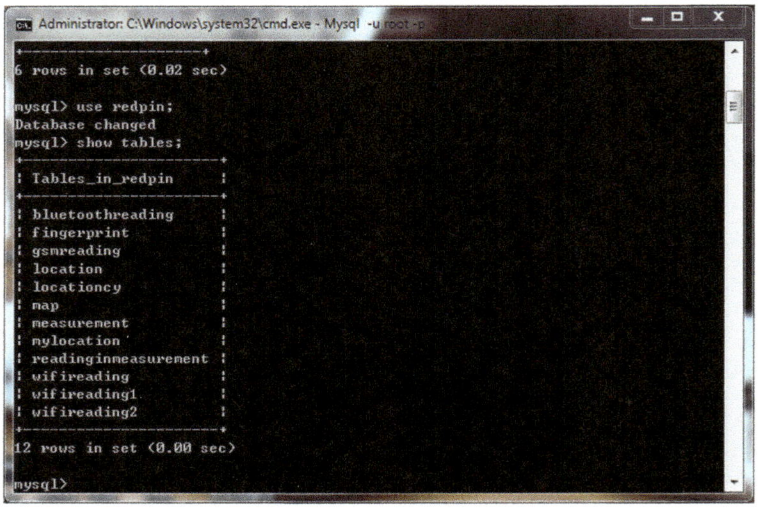

Fig. 16 Interface of database management

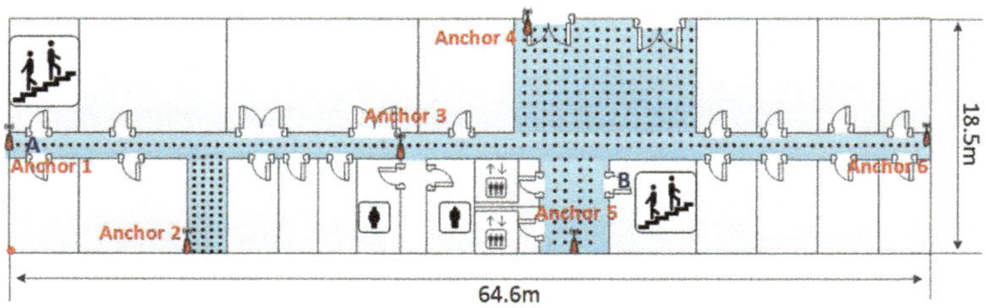

Fig. 17 Floor plan of the lower floor

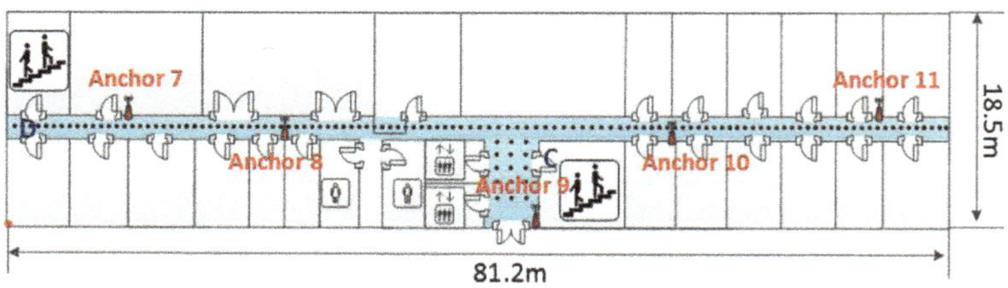

Fig. 18 Floor plan of the upper floor

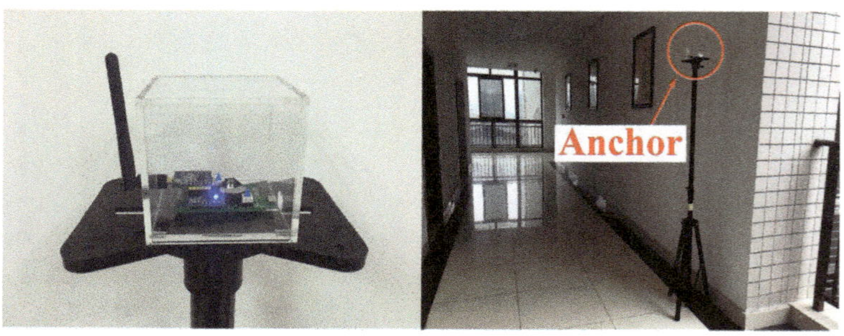

Fig. 19 Photos of the actual environment

Fig. 20 Display on terminal side

and MEMS data are packaged and transmitted to the positioning server through the 4G network, and then, the estimated locations of the target are displayed by the server, as shown in Fig. 11.

3.8.1 Hardware platform

The CC2540 chip of Texas Instruments (TI) is selected as the BLE anchor with the broadcast frequency 10 Hz and power 0 dBm, as shown in Fig. 12.

The target is installed with the Android 6.0 operating system and designed to support the BLE, accelerometer, magnetometer, and barometer modules. The positioning and web servers are based on the Windows 7 operating system with the processor of Intel (R) Core (TM) i3-416 CPU @ 3.60 GHz, as shown in Fig. 13.

3.8.2 Software platform

The APP used to collect the data from the BLE and MEMS sensors as well as access the positioning server is developed by Android Studio software, and the corresponding algorithms are written on the Eclipse software platform in JAVA (Figs. 14 and 15). The MySQL database and Apache Web server are selected to save the positioning result and show them through the PHP and JavaScript (Fig. 16).

4 Experimental results

Two adjacent floors in a building are selected as the experimental environment, as shown in Figs. 17 and 18. There are 11 BLE anchors, namely Anchor 1, · · · , 11, fixed in target environment and the 514 RPs (marked with bullet) are uniformly calibrated with the interval of 0.6 m. Figure 19 shows some photos of the actual environment. The positioning result can be displayed on both the terminal and server sides, as shown in Figs. 20 and 21.

Figures 22 and 23 show the result of fingerprints clustering without and with denoising. The process of denoising increases the probability of merging the physically adjacent RPs into the same cluster.

Although the irregular jump error of BLE positioning may exceed 10 m, the ultimate error by fusion positioning will converge into a small value as the time going on. By taking the result in Fig. 24 as an example, the error of initial positioning is over 13 m, but the one almost decreases to less than 2 m when the timestamp is over 100 s.

We continue to compare the performance of the proposed and conventional positioning algorithms by using the MEMS or BLE solely. Figure 25 shows the result of locations tracking by different algorithms on a floor. The real trajectory starts from point A, along Anchor 3 and Anchor 6, along Anchor 4 and Anchor 3, and back to point A (see Fig. 25).

The result of locations tracking on two different floors is also shown in Fig. 26. In this test, the real trajectory is

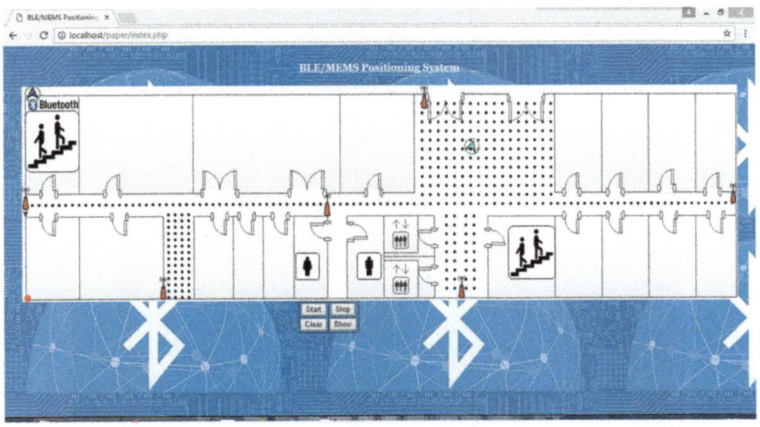

Fig. 21 Display on server side

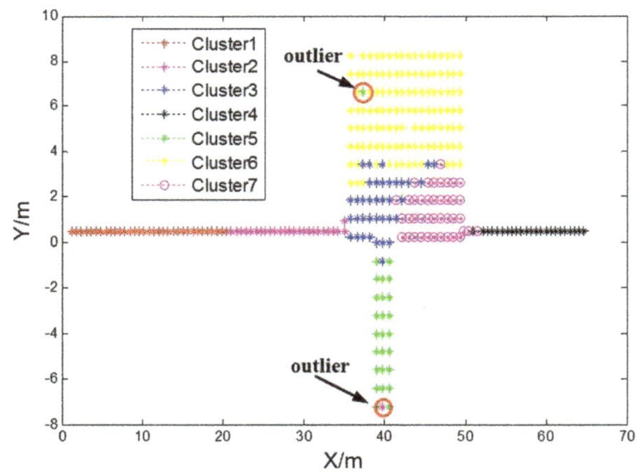

Fig. 22 Fingerprints clustering without denoising

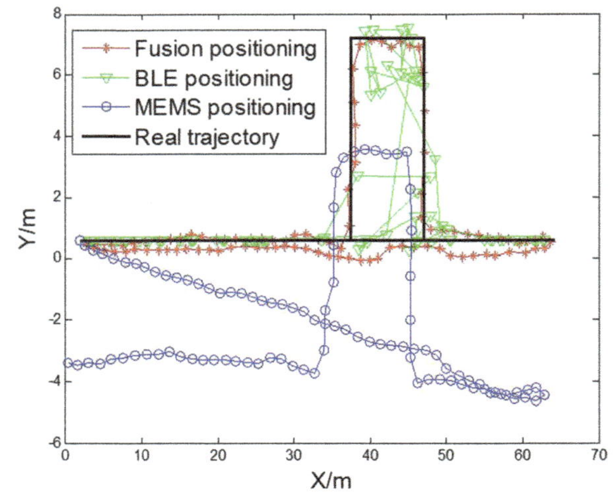

Fig. 25 Locations tracking on a floor

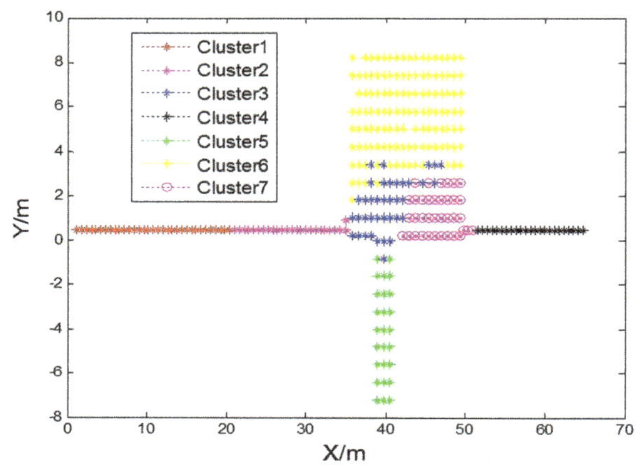

Fig. 23 Fingerprints clustering with denoising

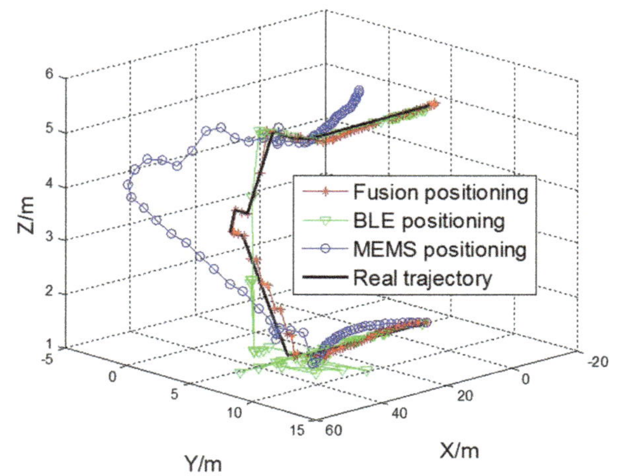

Fig. 26 Results in 3D view

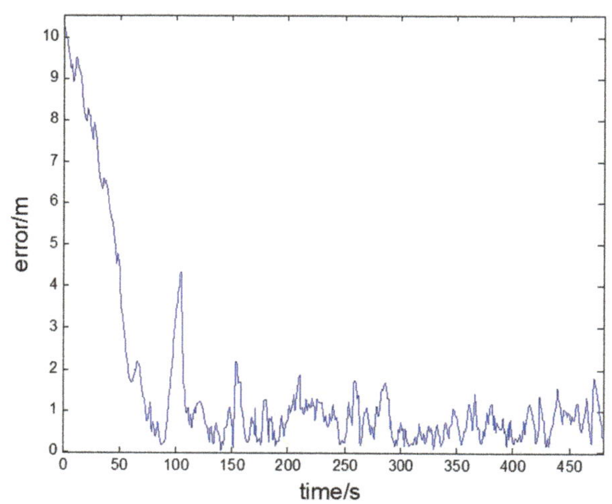

Fig. 24 Error variation of fusion positioning

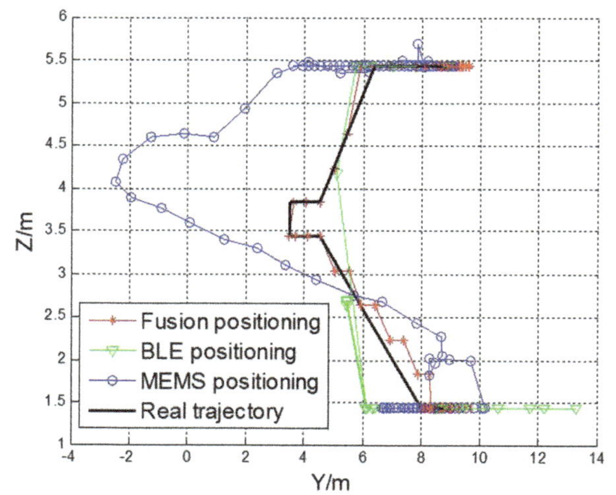

Fig. 27 Projection result in *Y-Z* plane view

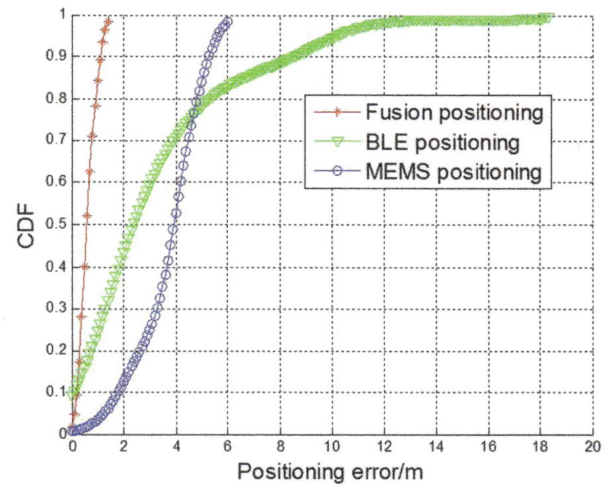

Fig. 28 CDF of positioning errors

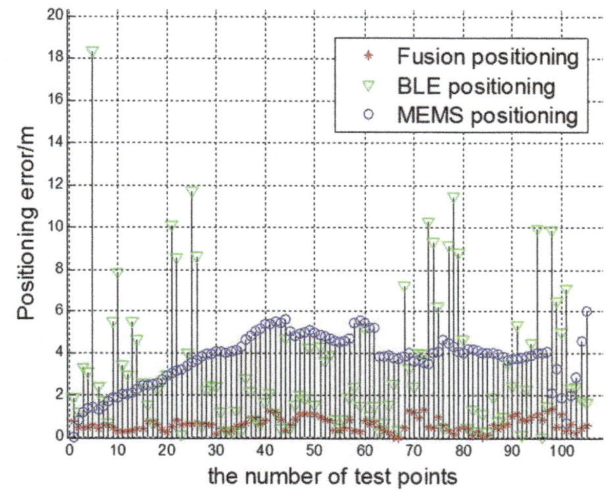

Fig. 29 Positioning error at each test point

selected from point A, along Anchor 3 and point B, going upstairs to point C, along Anchor 8 and Anchor 7, and ending at point D (see Figs. 26 and 27).

As can be seen from Fig. 26, the large accumulative error of MEMS positioning exists, while the BLE positioning is suffered by the irregular jump error and cannot be accurate enough when the target is located in the staircase. In contrast, the proposed fusion positioning algorithm can restrain the accumulative error of MEMS positioning as well as eliminate the irregular jump error of BLE positioning. In addition, it is also verified that the proposed algorithm is featured with good height resolution, which makes the system more robust to the actual indoor multi-floor positioning.

Figures 27 and 28 show the Cumulative Density Function (CDF) of errors and the error at each test point respectively. From these figures, we can find that the positioning errors by using the BLE or MEMS solely are much larger than the one of the proposed fusion positioning. In addition, different percentile errors of fusion, BLE, and MEMS positioning are also illustrated in Table 3.

Figure 29 shows the height error of each test point when the target is located in the staircase. From this figure, we can find that our system is capable of achieving better height resolution compared to the conventional ones using the MEMS or BLE solely.

Table 3 Different percentile values with respect to the positioning error

Percentile values (%)	Fusion positioning (m)	BLE positioning (m)	MEMS positioning (m)
50	<0.60	<3.80	<2.29
70	<0.77	<4.48	<3.37
90	<1.03	<5.26	<8.41

Finally, we investigate the stability of our system under the long-term testing. To achieve this goal, three shapes (square, linear, and irregular shapes) of trajectories are considered in Figs. 30, 31, 32, 33, 34, 35, and 36 respectively. Obviously, our system exhibits the best performance in terms of accumulative error constraint, irregular jump error elimination, and height resolution under different shapes of trajectories.

5 Conclusions

In this paper, both the hardware and software of BMW system are designed and implemented for indoor multi-floor positioning. Based on the extensive experimental results, it is demonstrated that the proposed system is capable of solving the problems of accumulative error constraint and irregular jump error elimination in MEMS

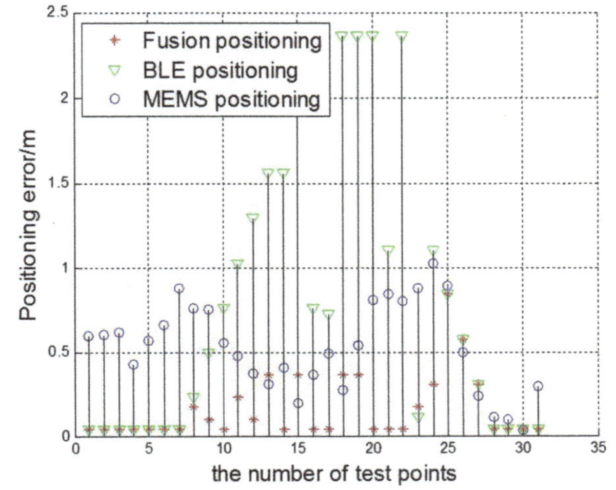

Fig. 30 Height error at each test point

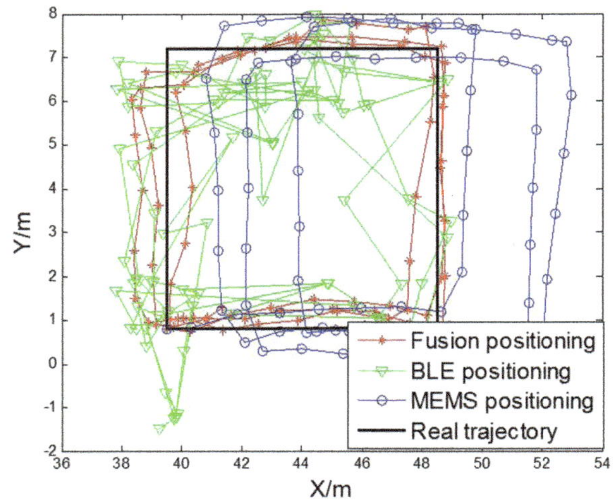

Fig. 31 Locations tracking under *square shape* of trajectory

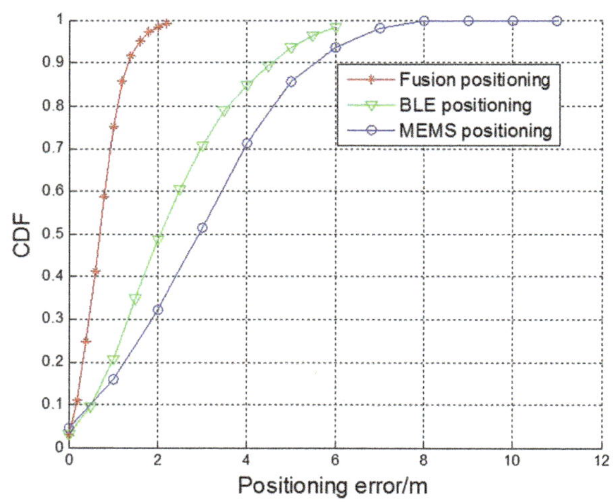

Fig. 32 CDF of positioning errors under *square shape* of trajectory

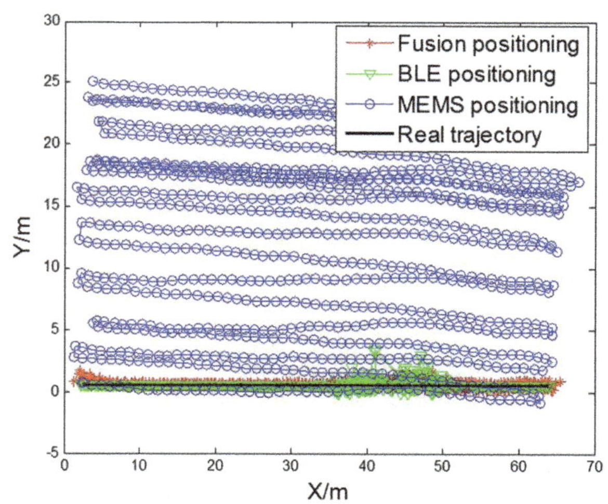

Fig. 33 Locations tracking under *linear shape* of trajectory

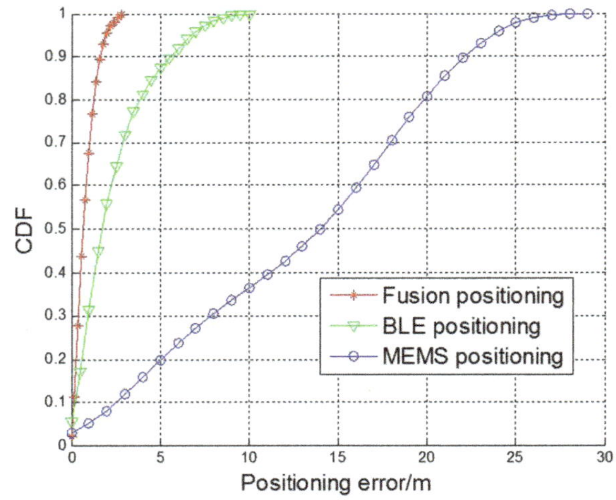

Fig. 34 CDF of positioning errors under *linear shape* of trajectory

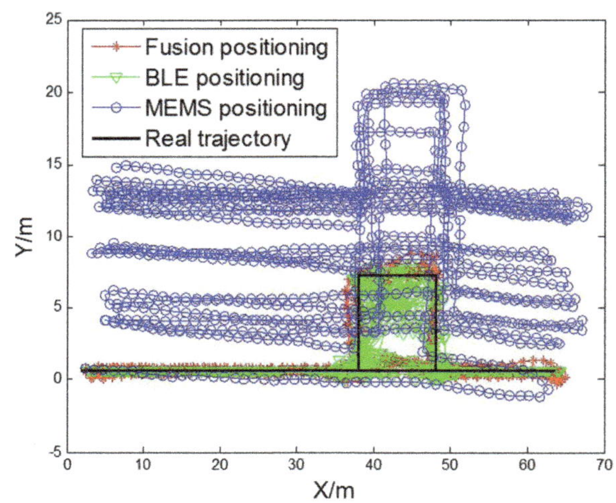

Fig. 35 Locations tracking under *irregular shape* of trajectory

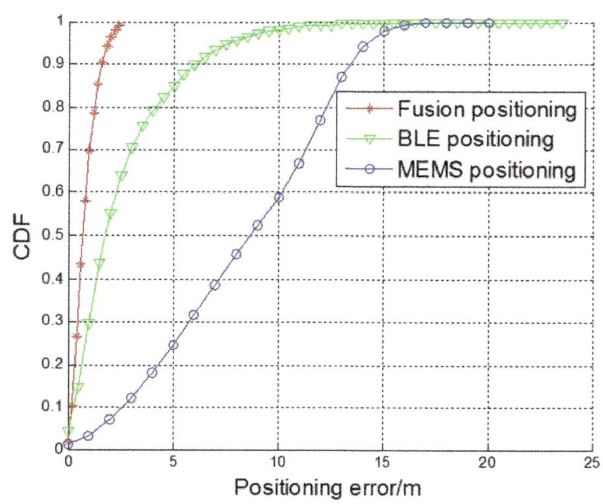

Fig. 36 CDF of positioning errors under *irregular shape* of trajectory

and BLE positioning respectively. In general, our system is featured with high positioning accuracy, good height resolution, and strong long-term stability.

Acknowledgements

This work was supported in part by the Program for Changjiang Scholars and Innovative Research Team in University (IRT1299), Special Fund of Chongqing Key Laboratory (CSTC), Fundamental and Frontier Research Project of Chongqing (cstc2017jcyjAX0380, cstc2015jcyjBX0065, cstc2015jcyjBX0085), and University Outstanding Achievement Transformation Project of Chongqing (KJZH17117).

Authors' contributions

The authors have contributed jointly to all parts on the preparation of this manuscript, and all authors read and approved the final manuscript.

Competing interests

The authors declare that they have no competing interests.

Author details

[1]Chongqing Key Lab of Mobile Communications Technology, Chongqing University of Posts and Telecommunications, Chongqing 400065, China. [2]Tianjin Key Laboratory of Wireless Mobile Communications and Power Transmission, Tianjin Normal University, Tianjin 300387, China.

References

1. M Jia, L Wang, Q Guo, X Gu, W Xiang, A low complexity detection algorithm for fixed up-link scma system in mission critical scenario. IEEE Internet Things J. **PP**(99), 1–1 (2017)
2. M Jia, X Gu, Q Guo, W Xiang, N Zhang, Broadband hybrid satellite-terrestrial communication systems based on cognitive radio towards 5g. IEEE Wirel. Commun. **23**(6), 96–106 (2016)
3. M Jia, L Wang, Z Yin, Q Guo, X Gu, A novel spread slotted aloha based on cognitive radio for satellite communications system. EURASIP J. Wirel. Commun. Netw. **2016**(1), 232 (2016)
4. ES Lohan, J Talvitie, PFE Silva, in *International Conference on Localization and GNSS*. Received signal strength models for WLAN and BLE-based indoor positioning in multi-floor buildings (IEEE, Gothenburg, 2015), pp. 1–6
5. K Yu, JP Montillet, A Rabbachin, UWB location and tracking for wireless embedded networks. Signal Process. **86**(9), 2153–2171 (2006)
6. YL Lai, J Cheng, A cloud-storage RFID location tracking system. IEEE Trans. Magn. **50**(7), 1–4 (2014)
7. L Zhang, J Liu, H Jiang, Senstrack: energy-efficient location tracking with smartphone sensors. IEEE Sensors J. **13**(10), 3775–3784 (2013)
8. N Alsindi, Z Chaloupka, N Alkhanbashi, An empirical evaluation of a probabilistic RF signature for WLAN location fingerprinting. IEEE Trans. Wirel. Commun. **13**(6), 3257–3268 (2014)
9. M Zhou, F Qiu, K Xu, Z Tian, Semi-supervised learning for indoor hybrid fingerprint database calibration with low effort. Comput. Commun. **86**(C), 57–74 (2016)
10. B Koo, S Lee, M Lee, in *International Conference on Indoor Positioning and Indoor Navigation*. PDR/fingerprinting fusion indoor location tracking using RSS recovery and clustering (IEEE, Busan, 2015), pp. 699–704
11. M Tanenhaus, D Carhoun, T Geis, in *IEEE/ION Position Location and Navigation Symposium*. Miniature IMU/INS with optimally fused low drift MEMS gyro and accelerometers for applications in GPS-denied environments (IEEE, South Carolina, 2012), pp. 259–264
12. K Kloch, P Lukowicz, C Fischer, in *International Symposium on Wearable Computers*. Collaborative PDR localisation with mobile phones (IEEE, San Francisco, 2011), pp. 37–40
13. W Xiao, W Ni, KT Yue, in *International Conference on Indoor Positioning and Indoor Navigation*. Integrated Wi-Fi fingerprinting and inertial sensing for indoor positioning (IEEE, Guimaraes, 2011), pp. 1–6
14. H Wang, H Lenz, A Szabo, in *Positioning, Navigation and Communication, 2007. Wpnc '07. Workshop On*. WLAN-based pedestrian tracking using particle filters and low-cost MEMS sensors (IEEE, Hannover, 2007), pp. 1–7
15. M Atia, M Korenberg, A Noureldin, in *IEEE International Symposium on Mechatronics and Its Applications*. A WiFi-aided reduced inertial sensors-based navigation system with fast embedded implementation of particle filtering (IEEE, Sharjah, 2012), pp. 1–5
16. V Malyavej, W Kumkeaw, M Aorpimai, in *International Conference on Electrical Engineering/Electrinics, Computer, Telecommunications and Information Technology*. Indoor robot localization by RSSI/IMU sensor fusion (IEEE, Krabi, 2013), pp. 1–6
17. X Liu, Z Wu, X Lin, WLAN/MARG/GPS integrated positioning system based on a self-adaptive weighted algorithm. Qinghua Daxue Xuebao/J. Tsinghua University. **53**(7), 955–960 (2013)
18. M Zhou, Q Zhang, Y Wang, Hotspot ranking based indoor mapping and mobility analysis using crowdsourced Wi-Fi signal. IEEE Access. **5**, 3594–3602 (2017)
19. M Zhou, Y Tang, Z Tian, Semi-supervised learning for indoor hybrid fingerprint database calibration with low effort. IEEE Access. **5**, 4388–4400 (2017)
20. J Talvitie, M Renfors, ES Lohan, Distance-based interpolation and extrapolation methods for RSS-based localization with indoor wireless signals. IEEE Trans. Veh. Technol. **64**(4), 1340–1353 (2015)
21. SS Jan, SJ Yeh, YW Liu, Received signal strength database interpolation by kriging for a Wi-Fi indoor positioning system. Sensors. **15**(9), 21377–21393 (2015)
22. J Laneurit, C Blanc, R Chapuis, IEEE, Columbus, in *Intelligent Vehicles Symposium*. Multisensorial data fusion for global vehicle and obstacles absolute positioning, (2003), pp. 138–143
23. M Mcguire, KN Plataniotis, AN Venetsanopoulos, Data fusion of power and time measurements for mobile terminal location. IEEE Trans. Mob. Comput. **4**(2), 142–153 (2005)
24. HH Liu, YN Yang, in *TENCON 2011 - 2011 IEEE Region 10 Conference*. WiFi-based indoor positioning for multi-floor environment (IEEE, Cebu, 2012), pp. 597–601
25. IH Alshami, NA Ahmad, S Sahibuddin, in *Malaysian Software Engineering Conference*. A light WLAN radio map for floor detection in multi-floor environment localization (IEEE, Seri Kembangan, 2015), pp. 135–139
26. X Wang, L Yang, L Wang, in *IEEE International Conference on Mobile Ad-Hoc and Sensor Networks*. Improved AP clustering algorithm based on target segmentation (IEEE, Dalian, 2013), pp. 539–541
27. IR King, Density data and emission measure for a model of the coma cluster. Astrophys. J. **174**(174), 123 (1972)
28. K Zhao, B Li, A Dempster, in *IEEE International Conference on Wireless Communication and Sensor Network*. A new approach of real time step length estimation for waist mounted PDR system (IEEE, Changsha, 2015), pp. 400–406
29. Y Zhang, L Wu, Stock market prediction of S&P 500 via combination of improved BCO approach and BP neural network. Expert Syst. Appl. **36**(5), 8849–8854 (2009)
30. S Jia, Q Qiu, J Li, in *IEEE International Conference on Information and Automation*. BP neural network based localization for a front-wheel drive and differential steering mobile robot (IEEE, Lijiang, 2015), pp. 2270–2274
31. Y Yang, W Gao, X Zhang, Robust Kalman filtering with constraints: a case study for integrated navigation. J. Geodesy. **84**(6), 373–381 (2010)
32. S Zhao, S Chan, in *IEEE International Symposium on Circuits and Systems*. A novel algorithm for mobile station location estimation with none line of sight error using robust least m-estimation (IEEE, Seattle, 2008), pp. 1176–1179
33. Y Kim, Y Hwang, S Choi, in *IEEE/ACM International Conference on Advanced Intelligent Mechatronics*. Height estimation scheme of low-cost pedestrian dead-reckoning system using Kalman filter and walk condition estimation algorithm (IEEE, Wollongong, 2013), pp. 1492–1497

9

Adaptive deactivation and zero-forcing scheme for low-complexity LDPC decoders

Taehyun Kim[1], Jonghyun Baik[2], Myeongwoo Lee[3] and Jun Heo[1*]

Abstract

A modified message propagation algorithm is proposed for a low-complexity decoder of low-density parity-check (LDPC) codes, which controls the information propagated from variable and check nodes.The proposed threshold-based node deactivation for variable nodes and zero-forcing scheme for check nodes remarkably reduce decoding complexity required for similar error performance. In the proposed scheme, different thresholds, which are determined from the base matrix of the LDPC codes, are applied for each type of variable node. In addition, thresholds for deactivating variable nodes are increased while the decoding process is operated for a reduction in decoding complexity without early error floor, which is a drawback of the conventional threshold-based deactivation scheme. Simulation results show that the proposed scheme enables normalized min-sum decoders to decode successfully with less complexity than the conventional threshold-based deactivating scheme.

Keywords: Low-density parity-check codes, Protograph-based extrinsic transfer chart, Normalized min-sum algorithm, Complexity reduction

1 Introduction

Low-density parity-check (LDPC) codes were first introduced by Gallager [1] in 1962 and rediscovered by Mackay [2] in 1999. LDPC codes have been in competition with turbo codes, which were proposed in 1993 [3], for error control in many applications. In the viewpoint of error performance, LDPC codes exhibit better behavior for high code rates compared to turbo codes, which the last are better for lower code rates. Complexity as well as error performance is one of the factors that determine the error correction code to use. Therefore, reducing the complexity of the decoding algorithm makes LDPC codes more applicable, especially to the energy limited applications, such as Internet of Things (IoT) and deep space communication. Since this paper is the research for LDPC codes, only schemes for LDPC codes are briefly introduced in this section, but there were various researches for practical usage of turbo codes [4–7]. The complexity of the LDPC decoding process is determined by three major factors: the computational complexity, the number of iterations, and the number of activated nodes.

Although the original iterative decoding algorithm for LDPC codes, known as the sum-product algorithm [2], shows good error correction performance, its computational complexity is quite high. The sum-product algorithm can be simplified using a mathematical approximation, which is called the min-sum algorithm [8]. The degraded performance from the approximation can be improved by simply scaling messages [9, 10]. On the other hand, Savin proposed a self-correction (SC) method for improving performance of min-sum algorithm [11, 12]. He checked the sign changes of the messages from variable nodes between two consecutive iterations to identify unreliable messages. Erasing unreliable messages yields better performance compared to other modified min-sum algorithms [9], especially in the error floor region [13]. As an additional effect, the SC method affects power efficiency due to a reduction in switching activity [14, 15]. The cost of the performance improvement is the additional memory required to store the signs of passing messages. This overhead can be relieved by changing the rules of the conventional SC method [16]. This scheme has memory requirements similar to those of the conventional min-sum algorithm.

The decoding complexity is proportional to the number of iterations. In order to reduce the number of

*Correspondence: junheo@korea.ac.kr
[1]The School of Electrical Engineering, Korea University, 145, Anam-ro, Seongbuk-gu, 02841 Seoul, Republic of Korea
Full list of author information is available at the end of the article

iterations, several serial scheduling methods have been proposed. Unlike classical decoding, iteratively generated information is updated following a certain sequence in a serial scheduling. Check node-wise and variable node-wise schedules have been known as layered decoding [17] and shuffled decoding [18], respectively. The required number of iterations can be reduced to half of that of parallel decoding while the performance is maintained. A stopping criterion can also reduce the average number of iterations. Recently, a stopping criterion for reducing not only the average number of iteration but also the number of calculations per iteration was proposed [19].

In addition, decoding complexity has close relationship with the number of activated nodes. An activated node is a node that calculates messages for increasing reliability. In contrast, a node that does not calculate any information, is called an deactivated node. The forced convergence (FC) method [20] reduces the number of activated nodes per iteration. During the decoding procedure, the reliability of some variable node messages become sufficiently high within a few iterations. The update of those reliable variable nodes can be skipped in the rest of the iterations [20]. Similarly, the check nodes can be deactivated when they satisfy some inequality conditions [21]. An important issue for achieving good error correction performance is proper selection of the threshold value. Sarajlc et al. tried to obtain optimized threshold values through an optimization problem [22, 23]. In addition, there exist simpler versions of the FC method [24–26] for operational efficiency.

In this paper, we introduce a low-complexity decoding scheme that reduces the number of activated nodes. Compared to FC scheme, the proposed scheme has lower complexity with less performance loss. The main contribution of this work can be presented as follows:

- Thresholds for the proposed scheme are determined by using information theoretic analysis based on a base matrix of LDPC codes.
- For the variable nodes, a threshold-based deactivation process is proposed with dynamically increasing thresholds.
- For the check nodes, the total amount of check-node operations is reduced by generating zero-forced log-likelihood ratios.

Different thresholds are applied to each variable node according to its connectivity. In addition, prematurely deactivated variable nodes are reactivated by dynamically increasing threshold values. The threshold increasing condition is checked with a simple estimator based on logical circuits. For the check nodes, the SC method is modified to make it useful for reducing check node calculations.

The rest of this paper is organized as follows: In Section 2, we explain precisely the classical decoding algorithm and conventional techniques that motivated our proposed scheme. Section 3 describes the proposed scheme. The simulation results of the proposed scheme are discussed in Section 4 and a conclusion is provided in Section 5.

2 Background

2.1 Decoding algorithms for LDPC codes

There are three major decoding algorithms for LDPC codes: maximum likelihood (ML) decoder, sum-product algorithm (SPA), and min-sum algorithm (MSA). The ML decoder is an alternative maximum a posteriori (MAP) decoder. It essentially selects the codeword that maximizes the likelihood of the received signal. Even though the ML decoder is optimum, it is impractical due to its high computational complexity. The SPA is a decoding method exchanging variable-to-check (V2C) and check-to-variable (C2V) information iteratively. The C2V message, $U_{mn}^{(itr)}$, which propagates from check node m to variable node n at the itr-th iteration, is generated as follows:

$$U_{mn}^{(itr)} = 2\tanh^{-1}\left(\prod_{n'\in N(m)\setminus n} \tanh\left(\frac{1}{2}V_{mn'}^{(itr-1)}\right)\right), \quad (1)$$

where $V_{mn}^{(itr-1)}$ denotes the V2C log-likelihood ratio (LLR) from variable node n to check node m at the $(itr-1)$-th iteration. The term $n' \in N(m) \setminus n$ denotes the neighbors of check node m except the variable node n. The V2C message, $V_{mn}^{(itr)}$, is generated using

$$V_{mn}^{(itr)} = U_{n,ch} + \sum_{m'\in M(n)\setminus m} U_{m'n}^{(itr)}. \quad (2)$$

The term $U_{n,ch}$ denotes the channel LLR of variable node n, and $M(n) \setminus m$ denotes the neighbors of variable node n except the check node m. Along the decoding procedure, the a posteriori probability LLR (APP-LLR) of variable node n at the itr-th iteration, $\beta_n^{(itr)}$, is updated as follows:

$$\beta_n^{(itr)} = U_{n,ch} + \sum_{m\in M(n)} U_{mn}^{(itr)} \quad (3)$$

Although SPA is more efficient to implement than ML, the complexity is still high due to the numerous multiplications of soft values in each iteration. MSA is a low-complexity algorithm that uses min operations instead of using multiplications. The simplified version of the check node updating rule can be represented as follows:

$$U_{mn}^{(itr)} = \prod_{n'\in N(m)\setminus n} \operatorname{sgn}\left(V_{mn'}^{(itr-1)}\right) \min_{n'\in N(m)\setminus n} \left|V_{mn'}^{(itr-1)}\right| \quad (4)$$

The drawback of MSA is degraded performance compared with SPA. For performance improvement, the normalized MSA was used in this study. The performance of MSA can be improved by simply scaling the absolute value of the C2V message. Although optimum scaling factor is an real value [10], the multiplication of real value is sublated in implementation due to the high complexity. Therefore, for the sake of simplicity and practical usefulness, we used a scaling factor of 0.75, which is close to the optimized value and can be implemented by several shift registers.

2.2 Forced convergence

The FC scheme forces variable nodes, which have highly reliable information, to stop updating their information. Specifically, variable nodes whose absolute value of APP-LLR is greater than a threshold t are deactivated during the decoding process as follows:

$$\left|\beta_n^{(itr-1)}\right| \geq t \tag{5}$$

In order to reduce the complexity stem from the variable nodes, the deactivation process should be operated before the variable node operation is started. Therefore, the APP-LLR, which was calculated in the previous iteration, is used as a criterion for deactivation. Deactivated variable nodes do not update their V2C messages and propagate the last updated messages. In order not to experience performance degradation due to hastily deactivated variable nodes, the decoder reactivates the deactivated variable nodes when they are found in the unsatisfied parity-check equation. The behavior of the FC scheme is depicted in Fig. 1. Hereafter, "FC scheme" or "conventional FC scheme" both mean the scheme proposed in [20].

2.3 Self-correction for the min-sum algorithm

The SC method is proposed to recover the performance loss of MSA and achieves a performance gain by erasing unreliable information. It identifies unreliable information by comparing the predicted V2C message with the message at the previous iteration. When the sign is changed, the decoder regards that message as unreliable information. Therefore, the process of the SC decoder can be defined as

$$V_{mn}^{(itr)} = \begin{cases} 0 & V_{mn}^{(itr)} V_{mn}^{(itr-1)} < 0 \\ V_{mn}^{(itr)} & V_{mn}^{(itr)} V_{mn}^{(itr-1)} \geq 0 \end{cases} \tag{6}$$

The erased message can be represented by assigning a zero value, which means the bit states are equiprobable. Whenever the message at the previous iteration is erased, the predicted message is sent without considering its sign. The behavior of the SC scheme is depicted in Fig. 2.

3 Adaptive deactivation and zero-forcing scheme

In this section, calculation reduction techniques for variable and check nodes are proposed. We were motivated by the FC and SC schemes that are explained in Section 2. The proposed node deactivation process for variable nodes is conceptually similar to the conventional FC scheme in the viewpoint of reducing the operational complexity of variable nodes with high reliability. The conventional FC scheme uses an equal threshold value for all variable nodes. In contrast, we decided a different threshold value for each variable node. Because the increment of reliability per iteration during the decoding process is different according to the connectivity, the use of various threshold is more efficient for complexity reduction. In addition, not fixed but variable thresholds are used for preventing early error floor, which is a drawback of the FC scheme.

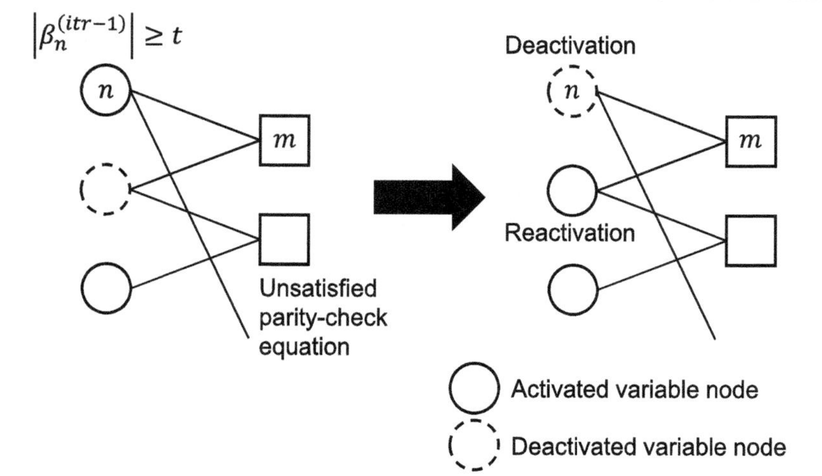

Fig. 1 Behavior of the FC algorithm, including deactivation and reactivation; variable nodes whose absolute value of APP-LLR is greater than a threshold t are deactivated. The deactivated variable nodes are reactivated when they are found in the unsatisfied parity-check equation

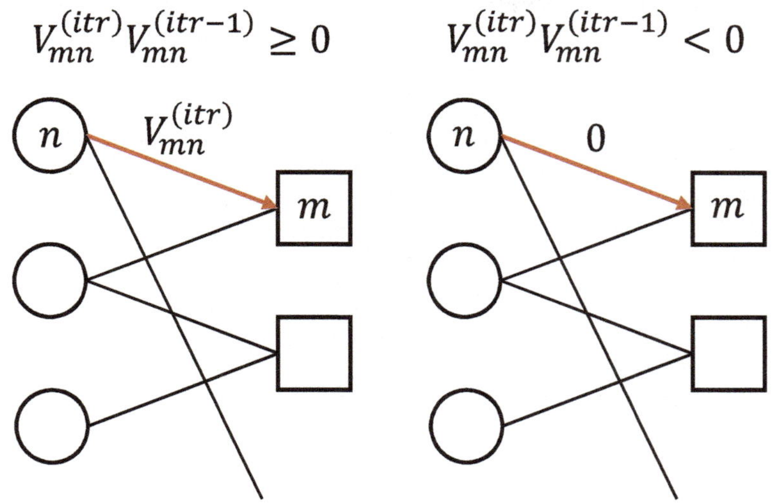

Fig. 2 Behavior of the SC scheme for the min-sum algorithm; variable node to check node (V2C) message that changes its sign between two consecutive iterations is set to zero

From the algorithm of the SC method, check node operation can be simplified according to the number of erased V2C messages. However, the main proposal of the SC method is an error correcting performance improvement. The number of erased V2C messages in the conventional SC method is not enough to obtain sufficient benefits in the viewpoint of complexity. Therefore, we proposed a threshold-based zero-forcing scheme for improving check node complexity reduction.

All thresholds used for the proposed scheme should be properly determined to deactivate nodes as much as possible but not to cause early error floor. We proposed a threshold decision method using an information theoretic analysis tool.

3.1 Adaptive node deactivation for reliable information

Assume that there is an LDPC code that is defined by an $M \times N$ parity check matrix H or an $M_p \times N_p$ base matrix B. Each element of base matrix $b_{(i,j)}$, $(1 \leq i \leq M_p, 1 \leq j \leq N_p)$ represents the number of connected edges between type-i check nodes and type-j variable nodes. As in condition (5), variable nodes satisfying the condition given in (7) are deactivated.

$$\left|\beta_n^{(itr-1)}\right| \geq t_v(j), n \in \text{set of type-}j \text{ variable nodes} \quad (7)$$

The parameter $t_v(j)$ is a threshold for type-j variable nodes. As mentioned previously, the thresholds for each type of variable node are increased during decoding process. The reason for a usage variable threshold instead of fixed threshold is a trade-off between error performance and complexity reduction. Although substantial complexity reduction can be obtained when the threshold is small, early error floor occurs. The reason is that the hastily deactivated variable nodes arising from the small threshold value degrade error performance. On the other hand, a large threshold value brings the opposite results.

In order to reactivate the hastily deactivated variable nodes, the threshold in the proposed scheme is increased when the rate of variable node deactivation exceeds a certain criterion. The increment of threshold value reactivates the deactivated variable nodes that do not satisfy condition (7). As a result, threshold values from the early phase to the last phase compensate each other by obtaining substantial complexity reduction with relatively less performance loss compared to when the fixed threshold value is used.

Initial thresholds $t_{v,ini}$ are determined using a protograph-based extrinsic information transfer (PEXIT) chart [27], which is an analysis tool for protograph-based and multi-edge type LDPC codes. In contrast to the extrinsic information transfer (EXIT) chart [28], which is based on the degree distribution of LDPC codes, a base matrix is used for the PEXIT chart. The PEXIT chart provides not only the threshold signal to noise (SNR) of LDPC codes but also the growth of reliability for each type of node in the iterative decoding process.

Let $J(\sigma)$ denote the mutual information between a binary random variable X with $Pr(X = \mu) = Pr(X = -\mu) = 1/2$, and a continuous Gaussian random variable Y with mean X and variance $\sigma^2 = 2\mu$. The function $J(\sigma)$ is calculated as follows:

$$J(\sigma) = 1 - \int_{-\infty}^{+\infty} \frac{1}{\sqrt{2\pi\sigma^2}} e^{-\frac{(y-\sigma^2/2)^2}{2\sigma^2}} \cdot \log_2(1+e^{-y})dy. \quad (8)$$

When a rate R LDPC code is used, the mutual information for the channel messages to type-j variable nodes is $I_{ch}(j) = J(\sqrt{8R \cdot SNR_{th}})$, where R is the code rate of LDPC codes, and SNR_{th} is the iterative decoding threshold SNR of LDPC codes. Then, transmitted mutual information between check nodes and variable nodes are calculated iteratively as iterative decoding method. The V2C mutual information between a type-i check node and a type-j variable node can be evaluated as

$$I_{V2C}(i,j) = J\left(\sqrt{\sum_{s\neq i}^{M_p} b_{(s,j)} \left[J^{-1}(I_{C2V}(s,j))\right]^2 + \left[J^{-1}(I_{ch}(j))\right]^2}\right), \quad (9)$$

Similarly, the C2V mutual information between a type-i check node and a type-j variable node can be represent as follows:

$$I_{C2V}(i,j) = 1 - J\left(\sqrt{\sum_{s\neq j}^{N_p} b_{(i,s)} \left[J^{-1}(1 - I_{V2C}(i,s))\right]^2}\right). \quad (10)$$

Using I_{ch} and the updated I_{C2V} at each iteration, the cumulative mutual information (CMI) for each type of variable node can be obtained as

$$I_{CMI}(j) = J\left(\sqrt{\sum_{s}^{M_p} b_{(s,j)} \left[J^{-1}(I_{C2V}(s,j))\right]^2 + \left[J^{-1}(I_{ch}(j))\right]^2}\right). \quad (11)$$

Since the mutual information of variable nodes increases at different rates, it is more reasonable to apply different threshold values for each type of variable node rather than an equal threshold value as in conventional FC schemes.

We decided the initial thresholds from the increment of CMI. Let $I^{l}_{CMI}(j)$ denote the CMI of a type-j variable node at the l-th iteration. The end point of minimum increment from the $(l-1)$-th iteration to the l-th iteration is calculated as

$$l^*(j) = \min_{l} \left(I^{l}_{CMI}(j) - I^{l-1}_{CMI}(j)\right). \quad (12)$$

The proposed technique uses APP-LLR at the $l^*(j)$-th iteration as the initial threshold value, $t_{v,ini}(j)$. APP-LLR can be obtained by applying the J^{-1} function to I_{CMI} as follows:

$$t_{v,ini}(j) = \frac{1}{2}\left[J^{-1}\left(I^{l^*(j)}_{CMI}(j)\right)\right]^2. \quad (13)$$

The deactivating process reduces the amount of variable node operations as the ratio of deactivated variable nodes gradually increases. However, the deactivation disturbs the decoding because some hastily deactivated variable nodes provide insufficient V2C LLR values whose magnitudes are not large enough. Therefore, we enabled the decoder to increase the threshold $t_v(j)$ several times in the middle of the decoding procedure as follows:

$$t_v(j) \leftarrow t_v(j) + \Delta t_v(j) \quad (14)$$

The threshold increasing operation is depicted in Fig. 3. However, the unlimited increasing threshold may reduce the ratio of deactivated nodes. Therefore, we set the threshold limit based on the PEXIT chart as the initial threshold. In this case, we found a maximum increment as follows:

$$l^{\dagger}(j) = \max_{l} \left(I^{l}_{\mathrm{CMI}}(j) - I^{l-1}_{\mathrm{CMI}}(j)\right), \quad (15)$$

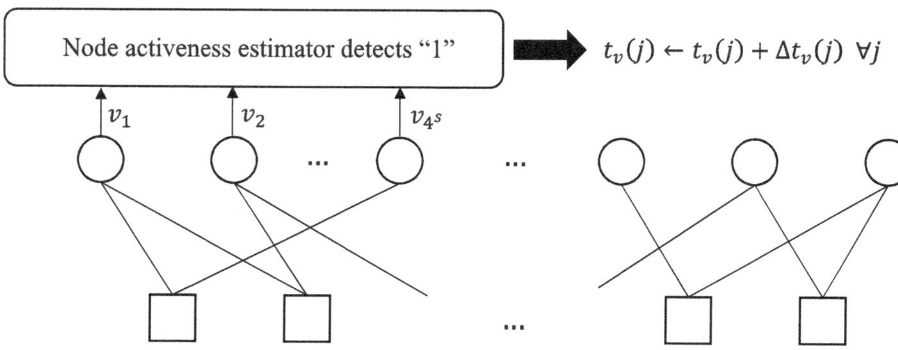

Fig. 3 Increasing threshold t_v when the node activeness estimator detects "1"

$$t_{v,max}(j) = \frac{1}{2}\left[J^{-1}\left(I_{\mathrm{CMI}}^{I^{\dagger}(j)}(j)\right)\right]^2. \quad (16)$$

Figure 4 shows an example of CMI and ranges for expected $t_{v,ini}$ and $t_{v,max}$ when a rate-1/2 LDPC code specified in the IEEE 802.11ad standard (WiGig)[29] is used. Because there are 16 types of variable nodes, several behaviors of CMI are depicted in Fig. 4.

When the value of threshold $t_v(j)$ is greater than $t_{v,max}(j)$, it is forced to $t_{v,max}(j)$. Increments of the threshold $\Delta t_v(j)$ are determined by the number of desired increment trials T and the limit of threshold $t_{v,max}(j)$.

$$\Delta t_v(j) = \frac{t_{v,max}(j) - t_{v,ini}(j)}{T} \quad (17)$$

The threshold t_v is increased when the output of a simple device called a node activeness estimator (NAE) is 1. The architecture of NAE is shown in Fig. 5. The estimator is constructed by concatenating AND gates and OR gates where it has 4^s inputs with a positive integer s, $v_1, v_2, \ldots v_{4^s}$, and one output. The parameter s is the number of stages that consist of AND gates and OR gates. First, choose 4^s out of n variable nodes. For the sake of simplicity, let the chosen variable nodes correspond to the first 4^s columns of H. Then determine the inputs of the estimator as follows:

$$v_i = \begin{cases} 1, & \text{if } i-\text{th variable node is deactivated} \\ 0, & \text{if } i-\text{th variable node is activated} \end{cases} \quad (18)$$

The NAE shown in Fig. 5 detects 1 with high probability when the ratio of 1s among the inputs is greater than 0.7. The probability that the NAE detects 1 can be calculated using the following equation recursively:

$$P_i = 1 - \left(1 - P_{i-1}^2\right)^2 \quad (19)$$

The parameter P_i is the probability that the output is 1 when the binary inputs are processed through a set of AND-OR logical circuits i times. The parameter P_0 is the proportion of the deactivated variable nodes in the 4^s samples. For instance, when the NAE consists of three stages, the probability of detection is $P_3 = 1 - (1-(1-(1-(1-(1-P_0^2)^2)^2)^2)^2)^2$. As a result, when the density of deactivated variable nodes is equal to 0.7, the probability of detection is 0.86 when $s = 3$ and increases to 0.94 when $s = 4$. In other words, the deactivating threshold t_v is increased with high probability

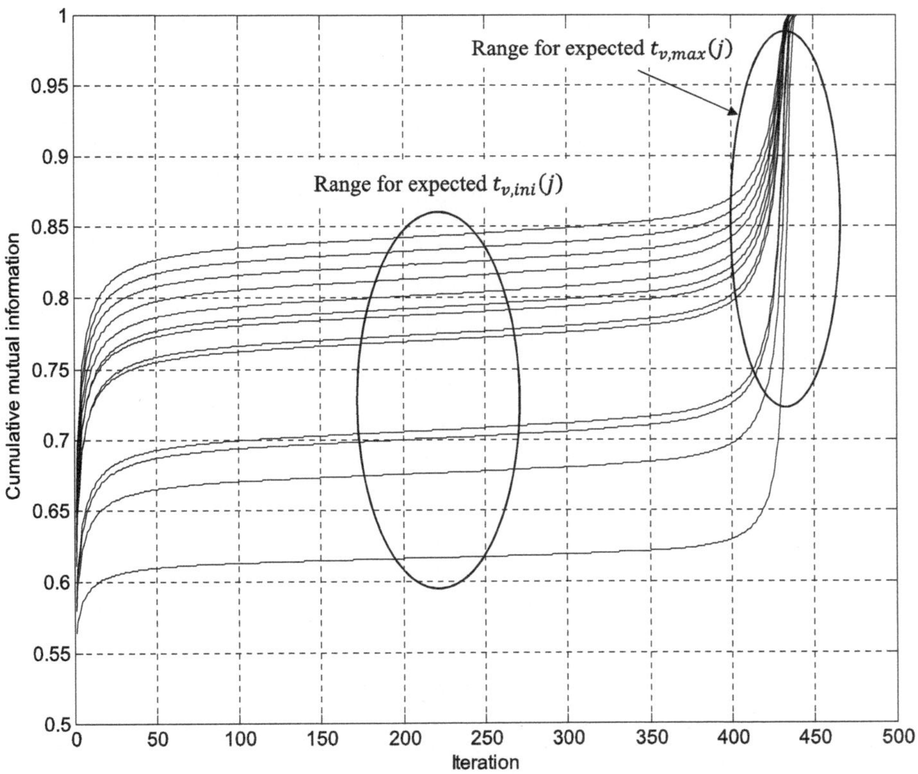

Fig. 4 Cumulative mutual information as a function of the number of iterations for each type of variable node when a rate-1/2 LDPC code specified in IEEE 802.11ad is used

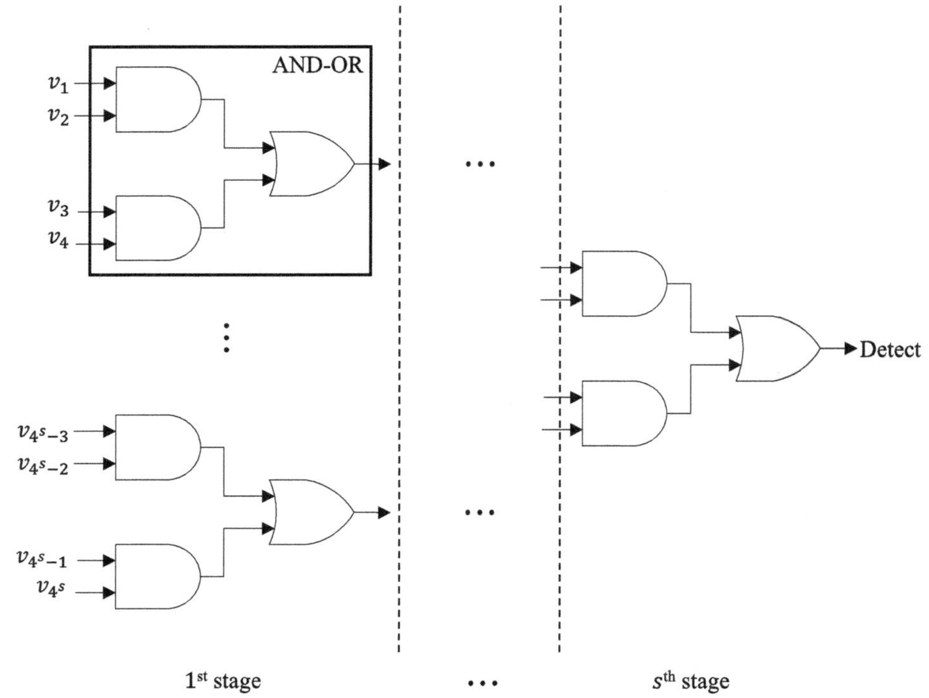

Fig. 5 An architecture of the NAE with 4^s inputs and 1 output

when more than 70% of the variable nodes are deactivated. Figure 6 shows the simulation result of probability that the NAE detects 1 as the density of binary inputs varies.

3.2 Zero-forcing scheme for unreliable information

In the check node operation for normalized MSA, it is required to find the minimum and second minimum values among the V2C messages that come from connected variable nodes. It can be simplified when some V2C messages are equal to 0. For example, let a check node m receive more than two zero-V2C messages. Then the check node does not have to calculate any C2V messages because all of the check node outputs are zero in this case. Consequently, it is known that zero-V2C messages enable check nodes to execute fewer operations. However, V2C messages are generally non-zero values in the conventional decoding algorithm. In order to generate zero-V2C messages, we proposed a threshold-based zero-forcing scheme. In the proposed method, V2C messages, which are regarded as unreliable information, are forced to be zero.

Then, check nodes can selectively calculate C2V LLR values from the received V2C LLRs. For a check node m, if more than two zero V2C LLR values are received, then every C2V LLR from the check node m, $U_{mn}^{(itr)}$ $\forall n$, is set to zero without any calculation. If only one zero V2C LLR, say $V_{mn}^{(itr-1)} = 0$, is received, the check node operation is executed only for $U_{mn}^{(itr)}$. The effects of zero-forced V2C LLRs on check node operations are depicted in Fig. 7. Because less calculation is required on the check-node side as more V2C LLRs are replaced with zero, a higher t_c is helpful for complexity. However, too high t_c results in performance degradation.

Thresholds for identifying unreliable information are obtained by using the PEXIT chart as thresholds for variable nodes. In this case, the mutual information I_{V2C}, which is transmitted from variable nodes to check nodes, is used. From the PEXIT chart, there are M_pN_p kinds of I_{V2C} that exist. Although we could achieve the thresholds for each type of V2C message, a large amount of memory is required for thresholds. Therefore, an equal threshold t_c is used for every V2C message from an equal variable node. The threshold t_c for each variable node is approximately calculated by averaging the values that are obtained by applying J^{-1} function to the corresponding I_{V2C}. Let $l^{\ddagger}(i,j)$ denote the number of iterations when the increment of I_{V2C} is a minimum. Then, the calculation of threshold for V2C messages from type-j variable nodes $t_c(j)$ can be represented as follows:

$$l^{\ddagger}(i,j) = \min_{l} \left(I_{V2C}^{l}(i,j) - I_{V2C}^{l-1}(i,j) \right), \qquad (20)$$

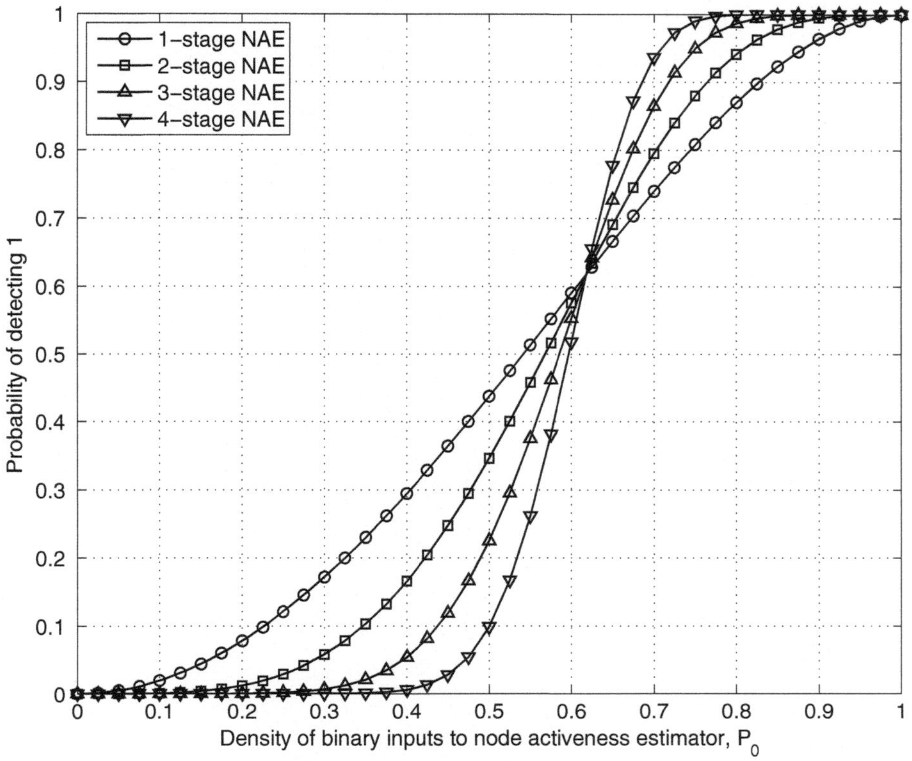

Fig. 6 Probability that the NAE detects 1 versus density of input vectors

$$t_c(j) = \frac{1}{d_{v,j}} \sum_{s=1}^{M_p} b_{(s,j)} \frac{1}{2} \left[J^{-1} \left(I_{V2C}^{\ddagger(s,j)}(s,j) \right) \right]^2, \quad (21)$$

where $d_{v,j}$ is the degree of type-j variable nodes. Figure 8 shows an example of V2C LLR and a range for expected t_c when the rate-1/2 LDPC code specified in IEEE 802.11ad is used. Similar to the case of CMI, because there are 16 types of variable nodes, several behaviors of V2C LLR are depicted in Fig. 8.

Since a V2C LLR value is said to be unreliable when its magnitude is small, our method forces V2C LLR values whose magnitudes are less than threshold to be zero as follows:

$$V_{mn}^{(itr)} = \begin{cases} 0, & \text{if } \left| V_{mn}^{(itr)} \right| \leq t_c(j), n \in \text{type-}j \text{ variable nodes} \\ V_{mn}^{(itr)}, & \text{otherwise} \end{cases} \quad (22)$$

For LDPC codes of length N and rate $R = \frac{N-M}{N}$, the proposed simultaneous node deactivation and zero-forcing scheme for highly reliable and unreliable information is described in Algorithm 1.

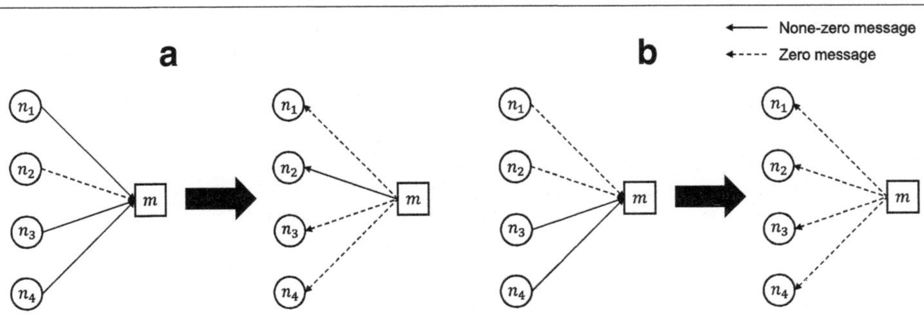

Fig. 7 Effects of zero V2C LLRs on check node operations. Dashed lines represent zero-LLR V2C or C2V messages. **a** The case where one zero V2C LLR is received. **b** The case where two zero V2C LLRs are received

Algorithm 1 Normalized Min-sum Decoding with the Proposed Algorithm

1: Initialization:
   $itr = 1$, set thresholds $t_v = t_{v,ini}$, t_c, and the maximum iteration I_{max}.
   $U_{mn}^{(0)} = 0$ and $\beta_n^{(0)} = V_{mn}^{(0)} = U_{n,ch}$ for all (m, n).
2: **for** every check node ($m = 1, \cdots, M$) **do**
3:   **if** ($V_{mn}^{(itr-1)} == 0$ and $V_{mn'}^{(itr-1)} \neq 0$ ($n' \neq n$)) **then**
4:     Calculate $U_{mn}^{(itr)}$ using (4) with scaling factor and $U_{mn'}^{(itr)} = 0$ ($n' \neq n$)
5:   **else if** (more than two V2C LLRs are zero) **then**
6:     $U_{mn}^{(itr)} = 0 \; \forall n$
7:   **end if**
8: **end for**
9: **for** every variable node ($n = 1, \cdots, N$) **do**
10:   **if** $\left(\left|\beta_n^{(itr-1)}\right| \geq t_v(j)\right)$ **then**
11:     Deactivate n-th variable node.
12:     $\beta_n^{(itr)} = \beta_n^{(itr-1)}$ and $V_{mn}^{(itr)} = V_{mn}^{(itr-1)}$
13:   **else**
14:     Calculate $V_{mn}^{(itr)}$ and $\beta_n^{(itr)}$ using (2) and (3)
15:     **if** $\left(\left|V_{mn}^{(itr)}\right| < t_c(j)\right)$ **then**
16:       $V_{mn}^{(itr)} = 0$.
17:     **end if**
18:   **end if**
19: **end for**
20: **if** stopping condition is not satisfied **then**
21:   **if** node activeness estimator detects 1 **then**
22:     Reactivate all variable nodes
23:     $t_v(j) \leftarrow t_v(j) + \Delta t_v(j) \; \forall j$
24:     **if** $t_v(j) > t_{v,max}(j)$ **then**
25:       $t_v(j) = t_{v,max}(j)$
26:     **end if**
27:   **end if**
28:   Go back to line 2
29: **end if**

4 Simulation results

In order to examine the performance of the proposed algorithm, simulations were performed over an additive white Gaussian noise (AWGN) channel with binary phase shift keying (BPSK) modulation. LDPC codes specified in the IEEE 802.11ad standard of length $N = 672$ and rate $R = 1/2$ and 3/4 were used to evaluate performance. In addition, the simulation results for LDPC codes specified in the IEEE 802.22 standard [30] of length $N = 384$ and IEEE 802.11n standard [31] of length $N = 1944$ are represented in the table at the end of this section. We performed all simulations using 7-bit fixed-point precision, with 4 bits for the integer part and 3 bits for the fractional part. The maximum number of iterations for all simulations was 20.

First, the frame error rates (FERs) of the proposed adaptive deactivation and zero-forcing (ADZF) scheme are compared to those of the normalized MSA without any additional manipulation, and the conventional FC schemes, as shown in Fig. 9. Performance of the proposed scheme was evaluated when the number of stages in NAE was equal to 4, and the number of increment trials was 10. The simulation results of the FC scheme applied to WiGig LDPC code with $R = 1/2$ show the influence of the threshold value on the error correcting performance of the FC scheme in a high SNR regime. Even though significant complexity reduction can be achieved by the FC scheme using a low threshold value, the performance loss in the target SNR region makes the scheme meaningless. Figure 10 shows the influence of threshold value on the required SNR for achieving FER $= 10^{-3}$. As represented in Fig. 10, in order to overcome performance loss, a sufficiently large threshold value should be used. In this section, a threshold value that show similar error performance at FER $= 10^{-3}$, such as $t = 9$ or 10, was used for the comparison of complexity.

Next, we evaluated the average number of iterations for each scheme. Figure 11 shows the average number of iterations where the maximum iteration $I_{max} = 20$. The ADZF and FC schemes require a similar average number of iterations to the normalized MSA. The reason for slight difference in iterations at low SNR region (FER $< 10^{-2}$) is that the initial thresholds $t_{v,ini}$ in the proposed ADZF scheme are smaller than the threshold value for the FC scheme. Figure 12 shows the influence of threshold value on the average number of iterations for achieving FER $= 10^{-2}$. When a smaller threshold value is used, the average number of iteration for maintaining error performance increases. In the proposed ADZF scheme, when the NAE detects 1, the threshold values increase from $t_{v,ini}$, which can be lower than 6, to $t_{v,max}$, which can be higher than 10. Since the NAE is operated at the end of each iteration, the opportunity for threshold increment increases at the low SNR region, where more iteration is required compared to the high SNR region. In other words, the thresholds for the proposed ADZF scheme can be higher than that of the FC schemes at the low SNR region. Nevertheless, since the difference in iterations decreases to zero at practical operation region (FER $\geq 10^{-2}$), it does not affect system's performance significantly.

Figure 13 shows the variable node complexity of each scheme when the complexity of the normalized MSA is evaluated as 1. The variable node complexity is defined as follows:

$$\sum_{itr=1}^{I_{max}} \frac{\sum_{i=1}^{N} \left(1 - v_i^{itr}\right) \cdot d_{v,i}}{\sum_{i=1}^{N} d_{v,i}} \quad (23)$$

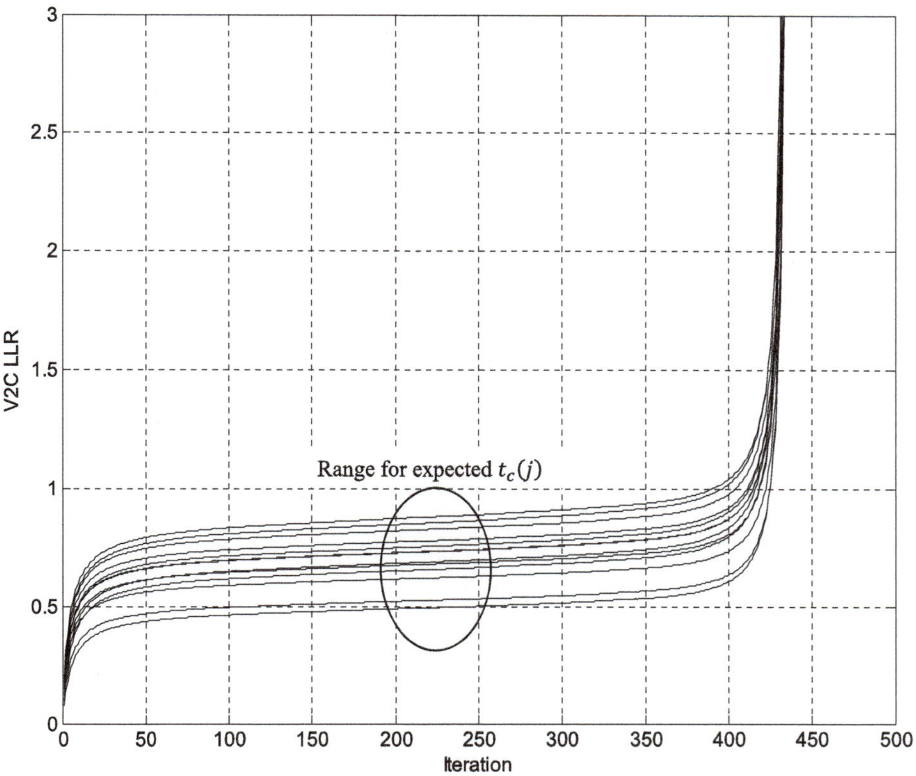

Fig. 8 Average V2C LLR curve for each type of variable node when a rate-1/2 LDPC code specified in IEEE 802.11ad is used

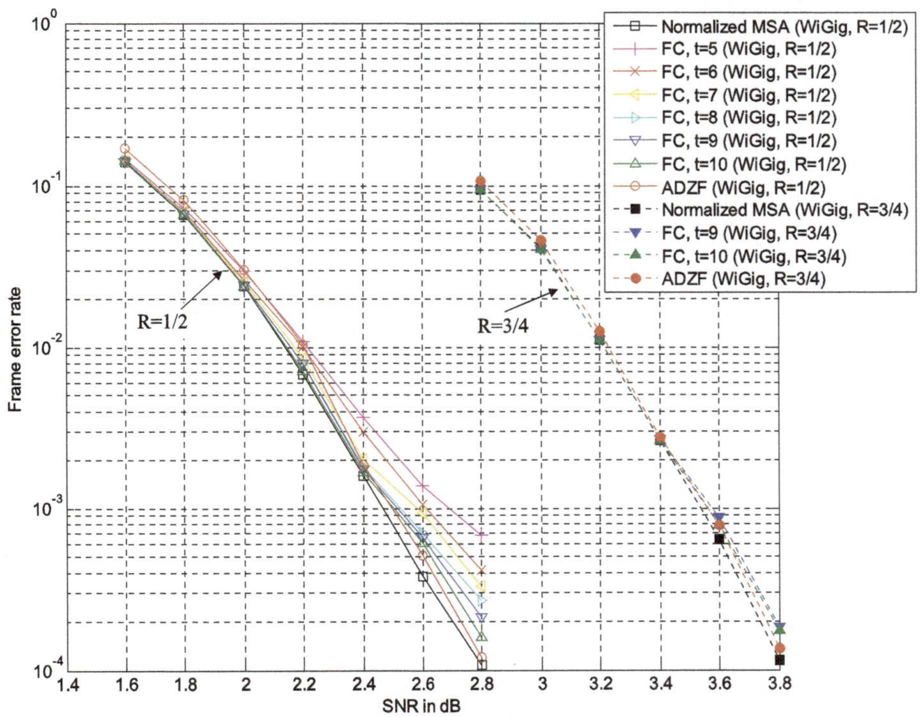

Fig. 9 Frame error rates of the normalized min-sum algorithm, conventional FC scheme, and proposed algorithm (ADZF)

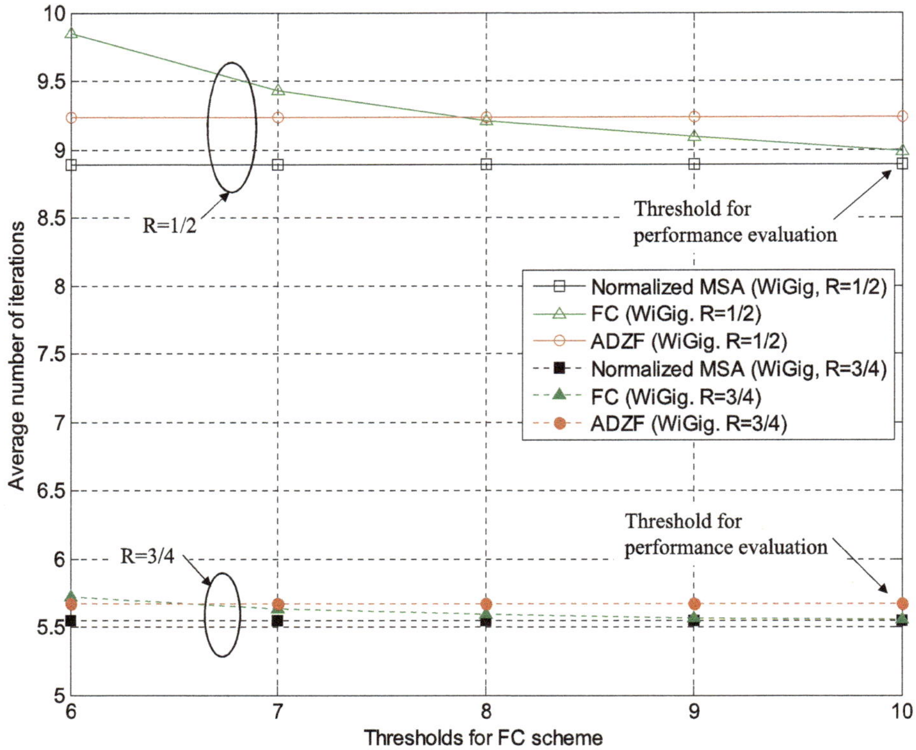

Fig. 10 Required SNR for achieving an FER = 10^{-3}

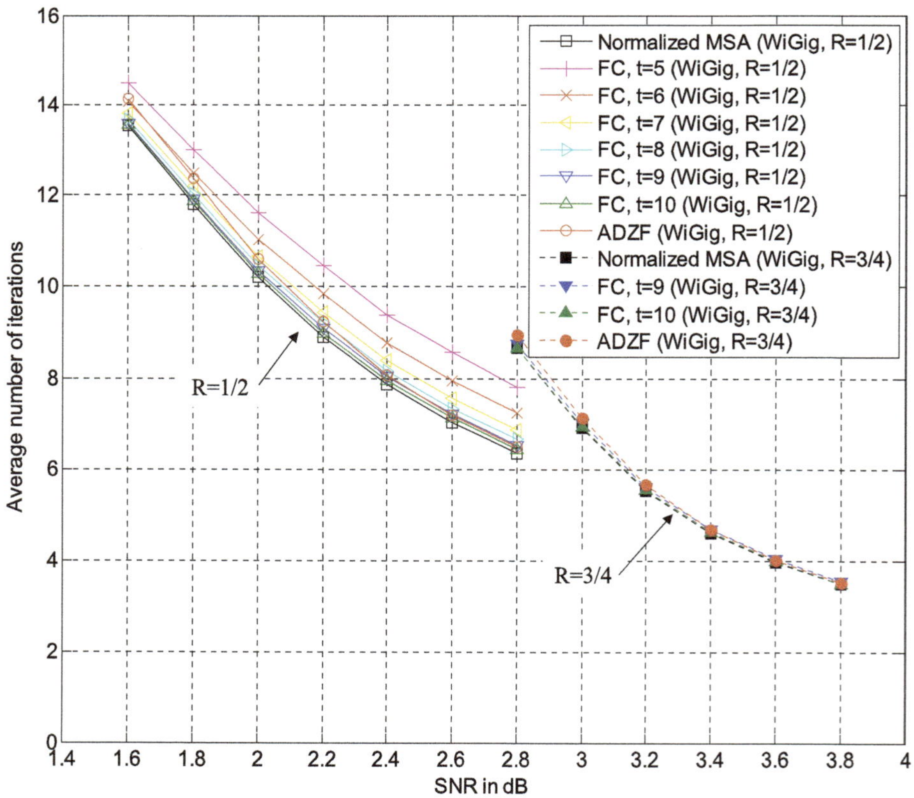

Fig. 11 Average number of iterations of the normalized min-sum algorithm, conventional FC scheme, and proposed algorithm (ADZF) where $I_{max} = 20$

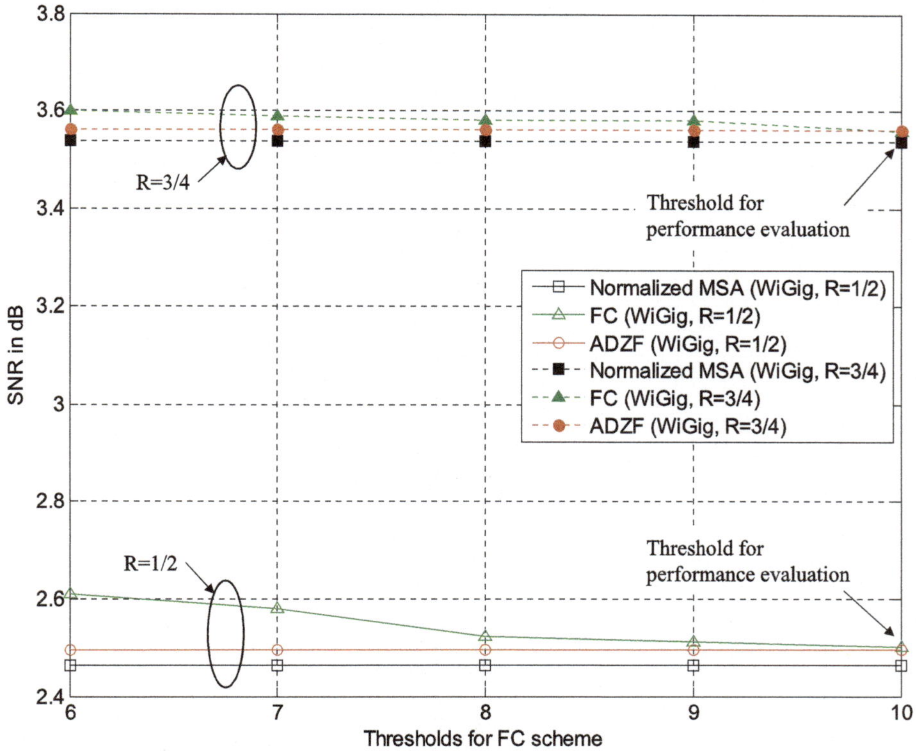

Fig. 12 Average number of iterations of the normalized min-sum algorithm, conventional FC scheme with various threshold values, and proposed algorithm (ADZF) where $I_{max} = 20$ and FER $= 10^{-2}$

The parameter v_i^{itr} means that the i-th variable node is activated(=0) or deactivated(=1) at the itr-th iteration. Because the effect of a deactivated high degree variable node on computation complexity is greater than that of a deactivated low degree variable node, we considered the degree of variable nodes as a weight. When the iteration is stopped before reaching the maximum number of iterations, all v_i^{itr} values for the rest of the iterations are equal to 1. In this way, a reasonable complexity considering early stopping can be obtained. For the proposed scheme and FC schemes, the complexity decreases as SNR increases, because each variable node is likely to have a higher APP-LLR at a higher SNR. As represented in Fig. 13, the variable node complexity of the proposed scheme applied to WiGig LDPC code of rate 1/2 is only 0.26, which is almost half of that of the FC scheme (0.53). When the threshold value for the FC scheme is smaller than 6, the complexity of the FC scheme is lower than that of the proposed scheme. However, as represented in Fig. 9, the FC scheme with a low threshold value is meaningless due to the error performance loss. For the proposed scheme, using an NAE with $s = 3$ shows less complexity than that with $s = 4$. The reason is that the probability of detecting 1 becomes higher when NAE has more stages, as shown in Fig. 6. Although more stages in NAE cause less complexity reduction compared to fewer stages, it has an advantage in the viewpoint of error floor. Similarly, the number of increment trials also influences error performance and decoding complexity. The large number of trials brings low decoding complexity and early error floor. On the contrary, lower complexity gain with less performance loss can be obtained when the smaller number of increment trials is used. Consequently, the number of stages in NAE and increment trials should be selected cautiously in accordance with the requirements of the communication system due to a trade-off between complexity and error performance.

Check node complexity is shown in Fig. 14, which is the effect of zero-forced V2C LLRs as explained in Section 3, and deactivated variable nodes. Check node complexity is obtained by evaluating the amount of check node calculations:

$$\sum_{itr=1}^{I_{max}} \frac{\sum_{i=1}^{M} \left(1 - c_i^{itr}\right) \cdot d_{c,i}}{\sum_{i=1}^{M} d_{c,i}} \tag{24}$$

The parameter $d_{c,i}$ is the degree of the i-th check node, and c_i^{itr} means that the operation of the i-th check node can be omitted (=1) or not (=0) at the itr-th iteration. Similar to the calculation of variable node complexity, the degree of check node is considered as a weight. In addition, when the iteration is stopped before reaching the

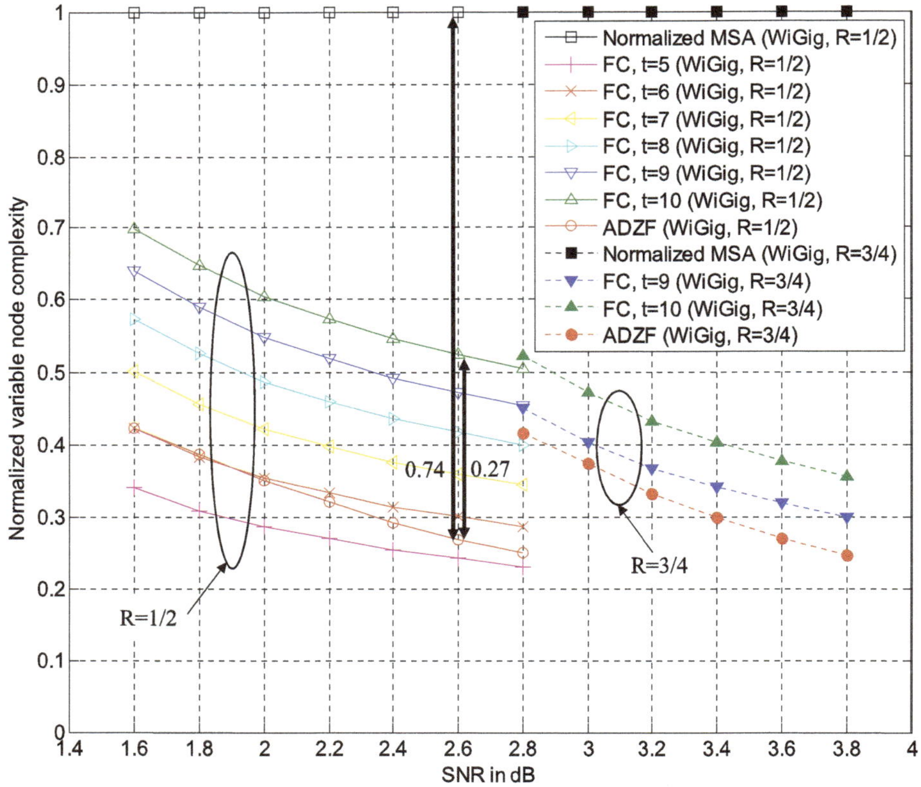

Fig. 13 Normalized variable node complexity of the normalized min-sum algorithm, conventional FC scheme, and proposed algorithm (ADZF)

maximum number of iterations, all c_i^{itr} values for the rest of the iterations are equal to 1. As mentioned previously, when a check node receives more than two zero-V2C messages, the corresponding check node operation also can be omitted, because all C2V messages from that check node are zero. In addition, because the operation of check nodes is meaningless when all related variable nodes are deactivated, we can omit the operation of corresponding check nodes. On the other hand, the FC scheme has only the effect of deactivated variable nodes. Moreover, as represented in Fig. 13, the number of deactivated variable nodes in the FC scheme is less than that in the proposed scheme. As a result, the check node complexity of the proposed scheme applied to WiGig LDPC code of rate 1/2 is 0.55, which is lower than that of the FC scheme (0.76). Complexity gain increases as SNR increases because the the portion of deactivated variable nodes is increased. Meanwhile, the influence of the SC scheme on the check node complexity reduction is decreased due to the increment of channel information reliability. Figure 15 shows the percentage of V2C messages for which the absolute value of LLR is lower than 0.5 and 1.0 as a function of SNR and iteration. As shown in Fig. 15, the percentage of unreliable V2C messages decreases significantly while SNR is increased.

Table 1 shows the effect of the proposed ADZF scheme when other LDPC codes are used. The average number of iterations, variable node complexity, and check node complexity of each scheme for obtaining FER $= 10^{-2}$ and 10^{-3} are represented. The represented variable node complexity and check node complexity are normalized by the complexity of the normalized MSA. Similar to the results for WiGig LDPC codes, the proposed scheme shows significantly lower node operation complexity compared to FC scheme. If the threshold value is smaller than the value represented in the Table 1, early error floor occurs. Figure 16 shows the error performance of each scheme when the smaller threshold value is used. To make it easier to see, we represent results only at the SNR regions that the FER of normalized MSA is approximately 10^{-2} and 10^{-3}. As represented in Fig. 16, when the smaller threshold is used, there is a noticeable performance gap at the SNR $= 2.8$ and 3.0 dB due to the early error floor.

For the sake of fair comparison, we should consider the additional complexity which is caused by variable node deactivation, zero-forcing message, and NAE. Then, the total complexity of ADZF scheme C_{ADZF} is calculated as

$$C_{\text{ADZF}} = C_{\text{variable}} + C_{\text{check}} + C_{\text{D}} + C_{\text{ZF}} + C_{\text{NAE}}, \quad (25)$$

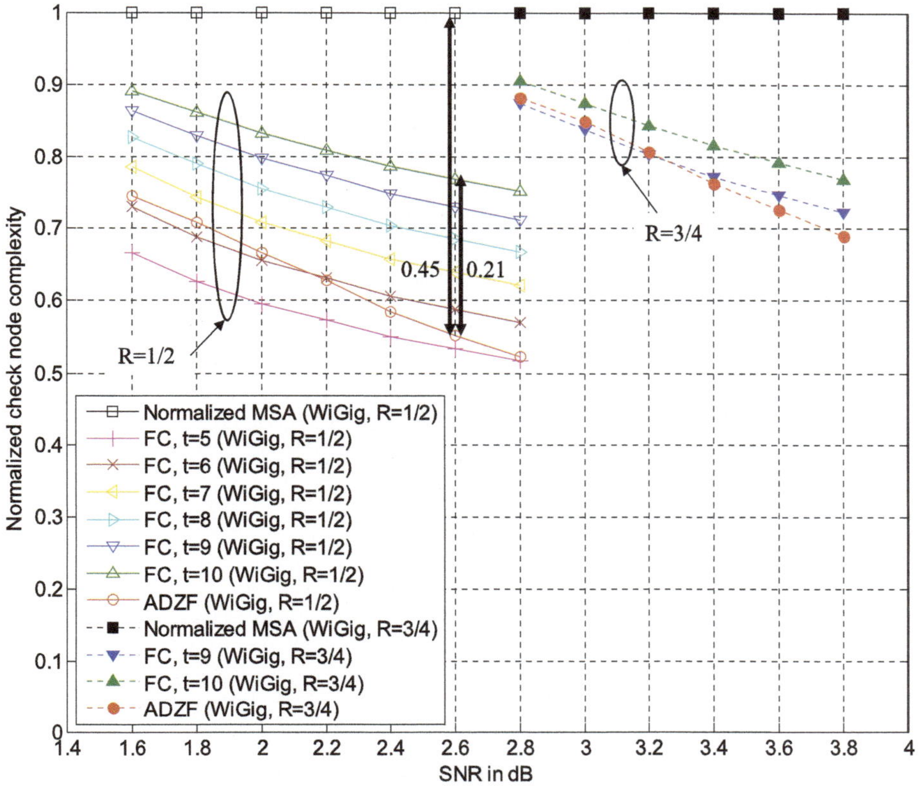

Fig. 14 Normalized check node complexity of the normalized min-sum algorithm, conventional FC scheme, and proposed algorithm (ADZF)

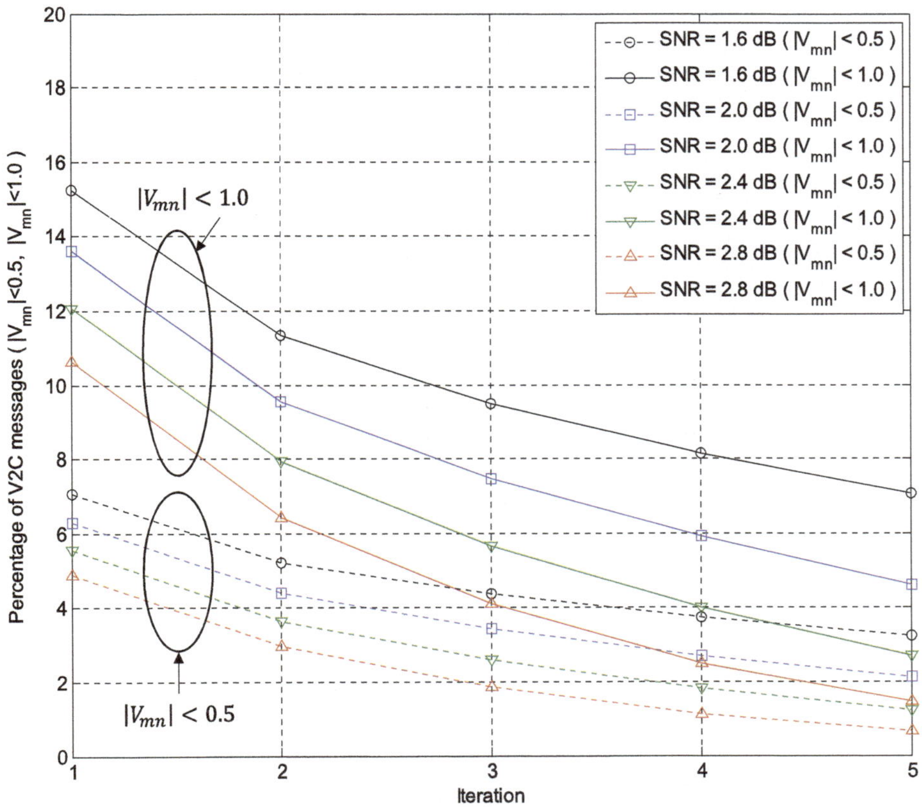

Fig. 15 Percentage of V2C messages for which magnitude is lower than 0.5 and 1.0

Table 1 Simulation results of the FC scheme and proposed ADZF scheme when 802.22 LDPC codes and 802.11n LDPC codes are used

Code	FER	Scheme	Average iteration	VN complexity	CN complexity
802.22 LDPC	10^{-2}	FC ($t = 10$)	6.637	0.631	0.894
($N = 384, R = 1/2$)		ADZF (proposed)	7.050	0.373	0.707
	10^{-3}	FC ($t = 10$)	5.186	0.586	0.861
		ADZF (proposed)	5.408	0.320	0.634
802.11n LDPC	10^{-2}	FC ($t = 12$)	9.091	0.531	0.854
($N = 1944, R = 1/2$)		ADZF (proposed)	9.390	0.225	0.507
	10^{-3}	FC ($t = 12$)	7.207	0.508	0.826
		ADZF (proposed)	7.377	0.202	0.458

where C_{variable} and C_{check} are the complexity of variable node operation and check node operation, respectively. The parameters C_{D}, C_{ZF} and C_{NAE} are the complexity of deactivation, zero-forcing, and NAE, respectively. Similarly, the total complexity of FC scheme C_{FC} is calculated as follows:

$$C_{\text{FC}} = C_{\text{variable}} + C_{\text{check}} + C_{\text{D}} \tag{26}$$

In order to combine the complexity of different operations, we first checked the number of additions, comparisons, and AND-OR logical circuits that are basic components of each operation. Variable node operation consists of addition. Comparison is used for check node operation, deactivation, and zero-forcing process. At last, node activeness estimator consists of AND-OR logical circuits as shown in Fig. 5. Then, we decided the weight of each component, since the computational complexity of each component are different. Using the number of components and their weight, the Eqs. (25) and (26) can be rewritten as

$$C_{\text{scheme}} = W_{\text{add}} \cdot N_{\text{add}} + W_{\text{comp}} \cdot N_{\text{comp}} + W_{\text{AND-OR}} \cdot N_{\text{AND-OR}}, \tag{27}$$

where N_{add}, N_{comp} and $N_{\text{AND-OR}}$ represent the number of additions, comparisons, and AND-OR logical circuits, respectively. The parameters W_{add}, W_{comp}, and $W_{\text{AND-OR}}$ mean the weight of addition, comparison, and AND-OR logical circuit, respectively. To quantify the computational complexity of each component, we used execution time. From the simulation, we obtained that the execution time for the one AND-OR logical circuit is approximately 1.5 times longer than that of addition and comparison. Therefore, the weight of the AND-OR logical circuit is 1.5 when that of the addition and comparison is 1. Figure 17 shows the normalized total complexity of each scheme when the total complexity of normalized-MSA is 1. Simulated error rate region is FER $= 10^{-3}$, and threshold of FC scheme is $t = 12$ for 802.11n code and $t = 10$ for the rest. As a result, even though ADZF scheme has extra operations

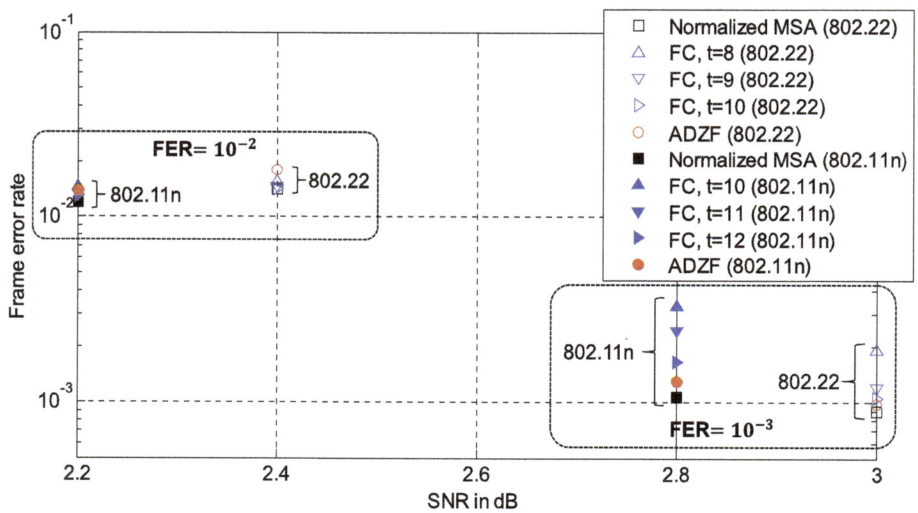

Fig. 16 Frame error rate of each scheme when 802.22 LDPC codes and 802.11n LDPC codes are used. The simulated error region is that the frame error rate of normalized min-sum is approximately 10^{-2} and 10^{-3}

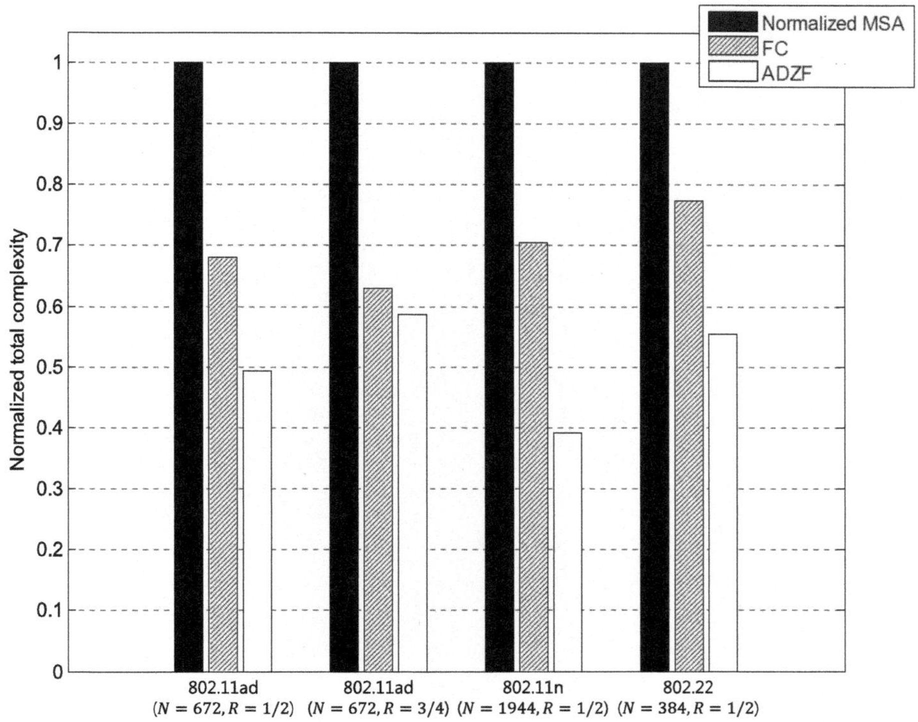

Fig. 17 Normalized total complexity of normalized min-sum algorithm, conventional FC scheme, and the proposed algorithm (ADZF) when FER= 10^{-3}

compared with FC scheme, it still have benefit in the viewpoint of complexity.

Finally, as a merit of the proposed scheme, ADZF scheme is compatible with the various existing schemes using LDPC codes, such as [32–34], in the viewpoint of the practical usage of their schemes. Decoding complexity of their schemes is reduced through simply adopting proposed scheme to LDPC decoder.

5 Conclusions

A node operation reducing technique, called the ADZF scheme, is proposed for a low-complexity decoder of LDPC codes. Variable nodes with high reliability are deactivated to omit their calculations. A simple NAE constructed with logical circuits is introduced to increase the threshold value when deactivated variable nodes are saturated. Hastily deactivated variable nodes are automatically reactivated by the increased threshold value. Messages with low reliability propagated from variable nodes are replaced with zeroes to reduce check node operations. Since the threshold value is decided only once in the offline mode, the proposed threshold value decision method does not incur additional computational complexity for the decoder. Simulation results show that the proposed scheme reduced decoding complexity more than the FC scheme. Moreover, the proposed scheme resolves the early error floor problem, which is a drawback of the FC scheme.

Acknowledgements

This research was supported by the MSIP (Ministry of Science, ICT and Future Planning), Korea, under the ITRC (Information Technology Research Center) support program (IITP-2017-2015-0-00385) supervised by the IITP (Institute for Information and communications Technology Promotion).

Authors' contributions

TK designed the main of the algorithm, analyzed the data, and wrote this paper. JB and ML designed a part of the algorithm and completed simulations. JH gave valuable suggestions on the idea of this paper. All authors read and approved the final manuscript.

Competing interests

The authors declare that they have no competing interests.

Author details

[1]The School of Electrical Engineering, Korea University, 145, Anam-ro, Seongbuk-gu, 02841 Seoul, Republic of Korea. [2]LG electronics, 19, Yangjae-daero 11-gil, Seocho-gu, 06772 Seoul, Republic of Korea. [3]Samsung electronics, 1-1, Samsungjeonja-ro, 18448 Hwaseong-si, Republic of Korea.

References

1. R Gallager, Low-density parity-check codes. IRE Trans. Inf. Theory. **8**(1), 21–28 (1962)
2. DJC Mackay, Good error-correcting codes based on very sparse matrices. IEEE Trans. Inf. Theory. **45**(2), 399–431 (1999)
3. C Berrou, A Glavieux, P Thitimajshima, in *IEEE International Conference on Communications (ICC 1993)*. Near Shannon limit error-correcting coding and decoding: Turbo-codes. 1 (IEEE, Geneva, 1993)

4. SK Chronopoulos, G Tatsis, V Raptis, P Kostarakis, in *Pan-Hellenic Conference on Electronics and Telecommunications (PACET 2012)*. A Parallel turbo encoder-decoder scheme (The Electronics and the Telecommunications laboratories of the Departments of Physics and Electrical and Computer Engineering, of the Aristotle University of Thessaloniki (AUTh), Thessaloniki, 2012)
5. SK Chronopoulos, T Giorgos, P Kostarakis, Turbo codes–a new PCCC design. Commun. Netw. **3**(4), 229–234 (2011)
6. SK Chronopoulos, T Giorgos, P Kostarakis, Turbo coded OFDM with large number of subcarriers. J. Signal Inf. Process. **3**(2), 161–168 (2012)
7. SK Chronopoulos, V Christofilakis, G Tatsis, P Kostarakis, Performance of turbo coded OFDM under the presence of various noise types. Wireless Pers. Commun. **87**(4), 1319–1336 (2016)
8. MPC Fossorier, M Mihaljevic, H Imai, Reduced complexity iterative decoding of low density parity check codes based on belief propagation. IEEE Trans. Commun. **47**(5), 673–680 (1999)
9. J Chen, MP Fossorier, Near optimum universal belief propagation based decoding of low density parity check codes. IEEE Trans. Commun. **50**(3), 406–414 (2002)
10. J Heo, Analysis of scaling soft information on low density parity check code. IEE Electron. Lett. **37**(25), 1530–1531 (2001)
11. V Savin, in *IEEE International Symposium on Information Theory (ISIT 2007)*. Iterative LDPC decoding using neighborhood reliabilities (IEEE, Nice, 2007)
12. V Savin, in *IEEE International Symposium on Information Theory (ISIT 2008)*. Self-corrected min-sum decoding of LDPC codes (IEEE, Toronto, 2008)
13. J Andrade, G Falcao, V Silva, JP Barreto, N Goncalves, V Savin, in *IEEE International Conference on Communications (ICC 2013)*. Near-LSPA performance at MSA complexity (IEEE, Budapest, 2013)
14. E Amador, V Rezard, R Pacalet, in *17th IFIP International Conference on Very Large Scale Integration (VLSI-SoC 2009)*. Energy efficiency of SISO algorithms for turbo-decoding message-passing LDPC decoders (IEEE, Florianopolis, 2009)
15. E Amador, R Knopp, V Rezard, R Pacalet, in *IEEE Computer Society Annual Symposium on VLSI (ISVLSI 2010)*. Dynamic power management on LDPC decoders (IEEE, Lixouri, 2010)
16. O Boncalo, A Amaricai, V Savin, in *21st IEEE International Conference on Electronics, Circuits and Systems (ICECS 2014)*. Memory efficient implementation of self-corrected min-sum LDPC decoder (IEEE, Marseille, 2014)
17. E Yeo, P Pakzad, B Nikolic, V Anantharam, in *IEEE Global Telecommunications Conference (GLOBECOM 2001)*. High throughput low-density parity-check decoder architectures (IEEE, San Antonio, 2001)
18. J Zhang, M Fossorier, Shuffled iterative decoding. IEEE Trans. Commun. **53**(2), 209–213 (2005)
19. CY Lin, MK Ku, in *IEEE International Symposium on Circuits and Systems (ISCAS 2009)*. Node operation reduced decoding for LDPC codes (IEEE, Taipei, 2009)
20. E Zimmermann, P Pattisapu, PK Bora, G Fettweis, in *7th International Symposium on Wireless Personal Multimedia Communications (WPMC 2004)*. Reduced complexity LDPC decoding using forced convergence (WPMC steering board, Abano Terme, 2004)
21. E Zimmermann, W Rave, G Fettweis, in *11th European Wireless Conference 2005-Next Generation Wireless and Mobile Communications and Services*. Forced convergence decoding of LDPC Codes - EXIT chart analysis and combination with node complexity reduction techniques (VDE, Nicosia, 2005)
22. M Sarajlc, L Liu, O Edfors, in *IEEE 25th Annual International Symposium on Personal, Indoor, and Mobile Radio Communication (PIMRC 2014)*. Reducing the complexity of LDPC decoding algorithms: an optimization-oriented approach (IEEE, Washington, 2014)
23. M Sarajlc, L Liu, O Edfors, in *IEEE 26th Annual International Symposium on Personal, Indoor, and Mobile Radio Communication (PIMRC 2015)*. Modified forced convergence decoding of LDPC codes with optimized decoder parameters (IEEE, Hong Kong, 2015)
24. J Fan, H Yang, in *IEEE International Conference on Communications Technology and Applications (ICCTA 2009)*. A new forced convergence decoding scheme for LDPC codes (IEEE, Beijing, 2009)
25. BJ Choi, MH Sunwoo, in *International SoC Design Conference (ISOCC 2013)*. Efficient forced convergence algorithm for low power LDPC decoders (IEEE, Busan, 2013)
26. BJ Choi, MH Sunwoo, in *IEEE Asia Pacific Conference on Circuits and Systems (APCCAS 2014)*. Simplified forced convergence decoding algorithm for low power LDPC decoders (IEEE, Ishigaki Island, 2014)
27. G Liva, M Chiani, in *IEEE Global Telecommunications Conference (GLOBECOM 2007)*. Protograph LDPC codes design based on EXIT analysis (IEEE, Washington, 2007)
28. S ten Brink, Convergence behavior of iteratively decoded parallel concatenated codes. IEEE Trans. Commun. **49**(10), 1727–1737 (2001)
29. (IEEE, New York, 2012). http://standards.ieee.org/getieee802/download/802.11ad-2012.pdf
30. (IEEE, New York, 2011). http://standards.ieee.org/getieee802/download/802.22-2011.pdf
31. (IEEE, New York, 2009). http://standards.ieee.org/getieee802/download/802.11n-2009.pdf
32. M Baldi, N Maturo, G Ricciutelli, F Chiaraluce, Security gap analysis of some LDPC coded transmission schemes over the flat and fast fading Gaussian wire-tap channels. EURASIP J. Wirel. Commun. Netw. **232**(1), 1–12 (2015)
33. M Baldi, N Maturo, E Paolini, F Chiaraluce, One the use of ordered statics decoders for low-density parity-check codes in space telecommand links. EURASIP J. Wirel. Commun. Netw. **272**(1), 1–15 (2016)
34. K Kwon, T Kim, J Heo, Pre-coded LDPC coding for physical layer security. EURASIP J. Wirel. Commun. Netw. **283**(1), 1–18 (2016)

A method for identifying faulty cells using a classification tree-based UE diagnosis in LTE

P. Muñoz[1*], R. Barco[1], E. Cruz[2], A. Gómez-Andrades[1], E. J. Khatib[1] and N. Faour[2]

Abstract

The latest advances in wireless technologies have led to a proliferation of data mobile devices and services. As a consequence, mobile networks have experienced a significant increase in data traffic, while voice traffic has shown nearly no growth. It is therefore essential for operators to understand the data traffic behavior at the user level in order to ensure a good customer experience. In the radio access networks (RANs), traditional solutions based on cell-level measurements are not adequate to analyze performance of individual users. Instead, novel alternatives such as the use of call traces and the definition of new user-centric indicators will provide detailed and valuable information for each connection. One of the key measurements related to data services is the user throughput. In this work, the user throughput is adopted as the main attribute to conduct diagnosis in the RAN, which has typically been the bottleneck for data services. To that end, a binary classification tree is proposed to determine the root cause of poor throughput in user-level data sessions. Then, this information is aggregated at the cell level in order to provide effective diagnosis of degraded cells. In particular, a correlation-based analysis of the cell status is proposed in order to identify abnormal cell behaviors in an automatic way. Evaluation has been carried out with datasets from live cellular networks. Results show that the proposed diagnosis approach is an effective means to identify the main factors that limit the user throughput in network cells.

Keywords: Self-healing, Fault diagnosis, Long-Term Evolution, Correlation, Self-Organizing Networks

1 Introduction

During the last years, the wireless data services have become the dominant traffic source in cellular networks. Behind this, there is an expansion of new mobile applications and a rapid growth in the number of subscribers, both motivated by the advances in cellular communication technologies and the development of user-friendly smartphones. According to a large network vendor [1], global mobile data traffic grew 69% in 2014 while the average smartphone usage grew 45% in the same year. This enormous increase in data traffic has forced operators not only to invest large amounts of money in new infrastructure but also to reduce operational expenditures (OPEX) in order to maintain the levels of user satisfaction.

To produce significant cost savings, one of the adopted solutions by standardization bodies was the creation of the Self-Organizing Networks (SONs) [2], which provide a new concept of network management where the maintenance and optimization tasks are carried out mostly in an automated way. Typically, technical experts in these fields have to deal with hundreds of traffic measurements and performance indicators every day [3, 4]. The vast diversity and quantity of these metrics makes the operational work very complex. Thus, the use of automated techniques for cellular traffic data analysis is essential to reduce human effort while expertise can be focused on new areas, bringing additional value to the operator [5].

Traditionally, mobile operators paid their attention in providing a good quality of the voice service, since it was the main offered service. To ensure this Quality-of-Service (QoS), troubleshooting experts mainly monitored the call blocking and dropping rates at the cell level to measure the levels of *accesibility* and *retainability*, respectively, in the network. However, with the explosion of Internet services, the QoS of multimedia and data applications is given by the data rates experienced by the users,

*Correspondence: pabloml@ic.uma.es
[1]Communications Engineering Dept., University of Málaga, Málaga, Spain
Full list of author information is available at the end of the article

where *integrity* metrics such as throughput and latency are essential traffic measurements [4]. The problem of throughput performance indicators is that they are often difficult to interpret because of their dependence on many factors. In particular, there are some aspects beyond the typical variables related to the radio environment (e.g., distance to base station, cell loading, user speed, etc.) that should be considered. First, unlike in traditional voice services, the mobile network is only one segment of the end-to-end connection in an IP world. For example, a router in the IP cloud that suffers congestion may influence the user perceived data rate. Second, the recent radio access technologies (RATs) such as Long-Term-Evolution (LTE) have included a class-based QoS model as a mechanism to differentiate between services, establishing various levels of service to the users. Third, the traffic pattern of new data services clearly impacts throughput measurements. Due to the increasing popularity of web navigation, streaming video, social networking, file sharing, online gaming, and other data services, there are significant differences in traffic patterns [6]. As a consequence, operators are investing a large amount of money to investigate traffic modeling and classification through packet inspection in order to better understand the characteristic of today's cellular data traffic. In addition, sophisticated traffic data filtering, processing, and correlation with other network metrics are also important features to identify root causes of any detected anomaly and increase the reliability of the network [7, 8].

The increasing complexity of network infrastructure and services has also led operators to be interested in managing performance at the user level, instead of the cell- or network level, with the aim of maintaining their competitiveness levels. Today's solutions based on per-cell performance counters are insufficient to perform adequate root-cause analysis. For this reason, the standardization bodies have proposed the use of user-centric indicators and call traces to support the optimization and troubleshooting processes [9]. With the Minimization of Drive Tests (MDT) described in [10], the collection of traffic measurements can be done in an autonomous manner. In other words, each device that is active in the network reports measurements and signaling messages (i.e., call events) to the base station. Unlike traditional drive tests, MDT avoids the use of expensive measurement equipment and it does not require human effort. The information provided by call traces and MDT is not aggregated and reflects the performance at the user level. However, operators can process this information and use it to identify problems with greater accuracy at higher levels (e.g., the cell level).

This paper presents a novel method for determining problems in cells using information at the user level. Specifically, the input of the method is given by the metrics contained in the call traces that provide specific information about data sessions. One of these metrics is the user throughput, which will act as the driver attribute to determine those data sessions experiencing bad QoS. This work focuses on the most common radio causes that may impact user throughput in a cellular network (i.e., congestion, lack of coverage, and interference). In case the user throughput is degraded, other radio measurements obtained from call traces will also be analyzed in order to identify the cause of bad QoS. To automate the analysis, a method based on a binary classification tree has been proposed. The adopted tree is generated from the analysis of real datasets and expert knowledge. Given the diagnosis of each data session, this information is aggregated for each network cell with the aim of creating a cell status or profile. Such a cell profile is then correlated to other cell profiles in order to identify faulty cells. The required calculations to classify the data and then compute the correlation values are carried out in an external server which communicates with the Operational Support System (OSS) in the network to obtain the network metrics. The study is carried out with different datasets from LTE networks where the proposed approach is applied to diagnose abnormal cells whose most users are experiencing poor user throughput.

Compared to other techniques, such as self-organizing maps [8], classification trees are an appropriate method to improve the effectiveness of diagnosis systems, especially when faults have a critical impact on network performance (i.e., various metrics are affected). Note that this kind of faults should be attended before any other. For example, a congestion situation is a critical problem that is reflected by a high number of connected users, but it may also be reflected by a high level of interference in the scenario. As a consequence, the diagnosis system could wrongly identify this situation as an interference problem. However, if the congestion situation is alleviated, the excessive interference level is also reduced. Thus, it is important for troubleshooting experts to have a clear map of which problems are prioritized by the diagnosis system. With classification trees, critical faults receiving higher priority should appear closer to the top of the tree. According to this, the classification trees enable fast visualization of the importance and prioritization of each fault.

The rest of the paper is organized as follows. The state-of-the-art is discussed in Section 2. The measurement setup and the real datasets to build and evaluate the proposed method are described in Section 3. The generation of the classification tree and the correlation-based method are covered in Section 4. The proposed method is evaluated using a real dataset in Section 5. Finally, Section 6 summarizes the main conclusions of this work.

2 Related work

The operational tasks in network management can be divided into three stages [11]: an initial measurement activity, a decision-making process, and lastly, a phase in which one or more actions are applied to the network. The first step means a continuous activity where a multitude of measurements are collected through different sources, including network counters and probes. Key Performance Indicators (KPIs) are continuously collected from network cells and then evaluated for optimization and troubleshooting purposes [12–14]. The main drawback of using those KPIs is that individual user performance may be lost if the data aggregation process at the cell level involves a considerable number of users. In addition, although they provide relevant information for managing the voice service, they are not enough for measuring the performance of data services. In this case, the use of user-centric KPIs would make the optimization and troubleshooting much more effective. For this reason, MDT is a feature introduced by the 3rd Generation Partnership Project (3GPP) allowing operators to utilize user devices to collect radio and traffic measurements in order to assess per-user level performance [15].

Some works addressing the MDT use case can be found in the literature. In [16, 17], MDT is utilized for coverage optimization, where the geo-localization of the user measurements enables powerful estimation and prediction of coverage holes. In [18], the location-aware radio measurements are employed for creating RF fingerprint databases which improves User Equipment (UE) positioning accuracy. In [19], a signal strength forecast method based on the classification and regression trees is proposed as another application of MDT. In [20], a KPI ranking system is proposed to significantly reduce the number of analyzed variables in MDT, while earlier work on QoS verification of MDT is described in [21]. In particular, the user experienced QoS in terms of throughput and its corresponding radio conditions are jointly analyzed. However, the evaluation is carried out by using a simulation tool instead of real traffic profiles and measurements.

The analysis of integrity performance (e.g., in terms of user throughput) in live networks has also gained attention in the research community. This kind of analysis can be carried out by means of field trials [22–24] or by using call traces (or MDT) [25]. However, in the case of field trials, the conclusions may not be representative of the real QoS experienced by the users. In the case of call traces, the correlation analysis presented in [25] is rather limited in terms of the number of radio measurements employed and no method for root-cause analysis was applied.

Due to the vast amount of data when per-user level information is collected, a new approach for network management is needed to address the requirements of the future fifth generation (5G). The main challenges in the current SON paradigm to make 5G technically feasible has also been investigated in the literature. In [26–28], empowering SONs with Big Data techniques is studied with the aim of transforming big data into a readily useable knowledge base. In the field of self-healing, several research efforts have been devoted to the development of usable automatic detection and diagnosis systems [29]. On the one hand, various mathematical approaches have been applied to analyze network measurements, such as Bayesian networks [30, 31], Neural Networks [5, 8], Fuzzy Logic combined with Genetic Algorithms [32], linear prediction [33], correlation [34], and statistical analysis [35, 36]. However, these data-driven algorithms have been exclusively evaluated with per-cell level measurements, which may not be sufficient to manage the new data services.

The work in [8] whose aim is to diagnose problems at the cell level from traditional KPIs has been further investigated in [37] by employing call traces (as opposed to traditional KPIs) and applying a rule-based system to them. That work has been extended in [38], where a method based on Neural Networks (similar to that in [8]), is applied to diagnose the users. Then, from such a user diagnosis, a threshold-based method is applied to diagnose faulty cells. Due to its benefits, the present work also employs call traces as in [37, 38], instead of traditional KPIs. However, the proposed method improves the results presented in [38], whose method for cell diagnosis shows a clear dependence on thresholds settings. Unlike [38], the proposed method in this paper for cell diagnosis is based on the correlation to specific cell profiles so that the use of thresholds is avoided. In addition, the work in [37, 38] focuses on connections whose release has been abnormal, which can be considered a limitation in the case of data traffic, where the QoS evaluation during the session in both uplink and downlink is essential for diagnosis purposes.

On the other hand, there are some works in the literature that investigate the problem of cell outage by employing user-centric measurements [39–41]. Nevertheless, this information is commonly related to the radio environment (e.g., signal strength) derived from MDT functionality, while other measurements related to integrity performance such as user throughput are ignored.

Thus, there is a large fragmentation in references related to the abovementioned topics. For this reason, this paper attempts to unify such a fragmentation and overcome some limitations that have been found in previous works. First, the user throughput is used as a key indicator to estimate QoS of data traffic in mobile networks. Second, a method for troubleshooting network cells based on analysis of call traces is proposed. Third, rather than using traffic measurements from simulation tools or field tests, the evaluation is carried out with large-scale datasets of real subscribers provided by operators.

3 Measurement setup and datasets

Figure 1 depicts the architectural model of 3GPP Radio Access Networks (RANs), together with the extensions to support the proposed data analysis framework. Since the data services are the dominant traffic source, this study focuses on LTE networks. The OSS is in charge of collecting events and measurements. These recordings are stored in files every report period, typically of 15 min duration. On the one hand, the connection between the OSS and the base station (e-NodeB) supports the collection of radio events. On the other hand, the connection between the OSS and the Mobility Management Entity (MME) enables the collection of user session details that can be used for mapping of radio events with measurements from another location of the network. Other network nodes (e.g., S-GW and P-GW) could also be employed to measure the quality of the services given to the user.

The tracing system provides detailed information at the session level for the UE. This information is contained in events which include measurements reported by the UE and signaling messages exchanged between the network elements. There are two different approaches for recording this information: UE traces and cell traces. The UE traces record events from those UEs which have been selected by the operator. This allows operators to monitor a certain user if it is not getting the appropriate QoS. The cell traces record information from all UEs or a subset of UEs (provided some filters) in a selected cell. This approach can be used by operators for network optimization and troubleshooting purposes, since it provides larger datasets of per-user level statistics than UE traces. For this reason, cell traces have been used in this work.

After trace collection, the information has to be processed (trace processing in Fig. 1) in order to obtain relevant statistics at the user session level and KPIs with adequate granularity. For example, the data provided by an event that periodically reports a certain measurement can be aggregated in time to produce values in a longer time scale (e.g., at the user session level). A priori, the information could also be aggregated in the space domain (e.g., cells, cell clusters, or network). However, this may be equivalent to traditional approaches based on cell counters, where the measurements are typically aggregated at the cell level. Instead, the user-level statistics are directly analyzed by means of the trace analysis module. At this stage, the analysis of traces can be extended to higher levels, such as the cell level, keeping in mind that individual user performance should always be reflected to avoid hiding problems in the context of root-cause analysis.

The present work is carried out with large-scale measurements coming from performance recording applications which are executed in four different commercial

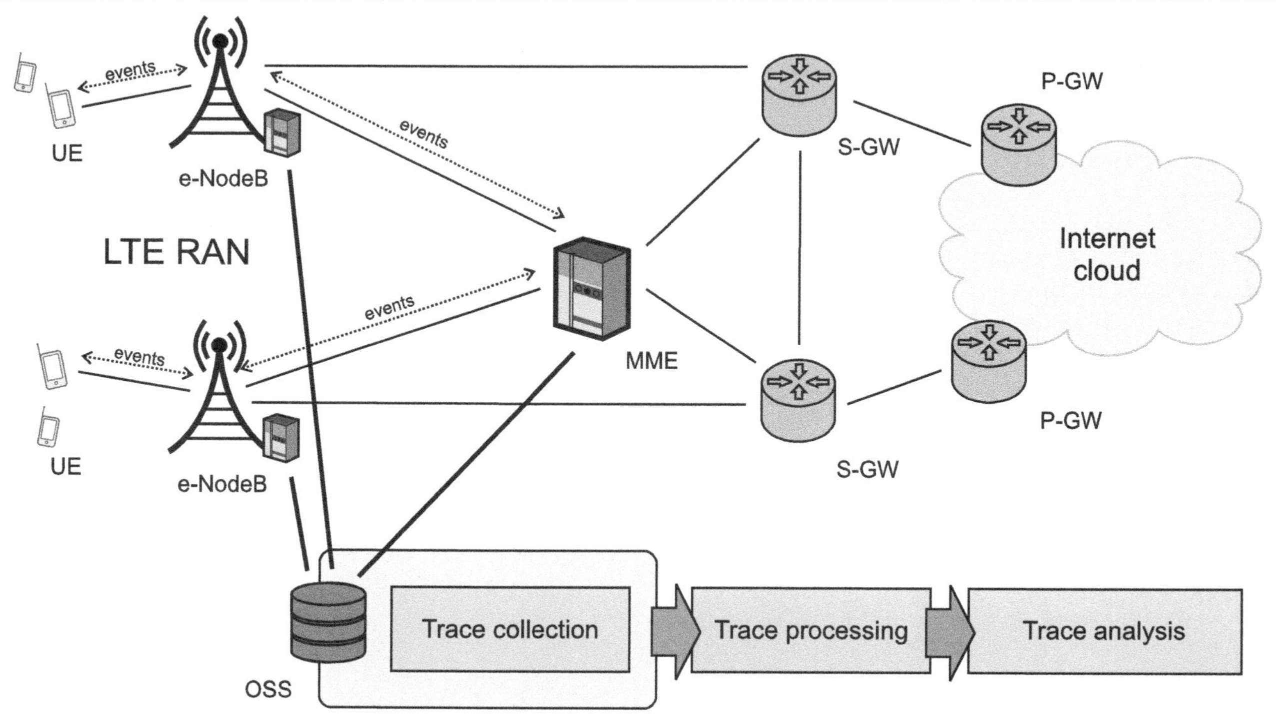

Fig. 1 LTE architecture and tracing system

LTE network deployments. They are owned by different operators and located in metropolitan areas. The main characteristics of the networks are described in Table 1. Note that these values are calculated after filtering the data. Thus, the actual number of data sessions per cell is presumably higher; however, only the data sessions that provide reliable statistics have been considered in the study. The data filtering process will be explained in the next section. Another remarkable observation is that the cell bandwidth varies depending on the dataset. This may have an impact on user throughput. For example, it is observed that the average session throughput is higher in dataset 4 since the cell bandwidth in this network is greater.

4 Cell diagnosis based on per-user level traffic measurements

Cell traces provide very detailed information at the session level on every UE served by a cell. In particular, they give instantaneous values for a specific event. The events related to radio measurements play an important role in activities such as determining the root cause of malfunctioning UEs, analyzing dropped calls and optimizing resource usage and quality. In this work, the information contained in the traces is used to diagnose UEs whose performance in terms of throughput is degraded. This is carried out by means of a binary classification tree. The diagnosis made for each UE will serve to identify problems at the cell level. The following sections explain in detail how the classification tree is built and the subsequent cell diagnosis based on a correlation analysis.

4.1 Data input definition and filtering

In the context of data mining, the data pre-processing is essential to carry out troubleshooting and optimization tasks in an effective manner. Due to this, the first step in the construction of the classification tree is the definition of the input metrics from the information contained in the cell traces, followed by the filtering and cleaning processes. The metrics that have been defined in this work are shown in Table 2. As observed, they represent the main aspects in a cellular network such as coverage, quality, and capacity in both directions of the radio link. By analyzing the distribution of these metrics over the network, most problems in the RAN can be diagnosed. For this reason, the number of selected metrics in this paper has been limited to 7. However, after building the classification tree, new metrics could be included in the tree in order to identify a larger number of problems, such as those related to mobility issues. To do this, the classification tree must comprise at least one leaf (i.e., a class) that represents the data sessions whose root cause remains unknown. From this leaf, the classification tree would be grown by introducing the conditions related to the new metrics. It is worth noting that the priority of the problems is affected by their position in the tree. The following paragraphs explain in detail each of these metrics as well as some aspects related to their filtering and cleaning processes.

To evaluate network performance, the 3GPP defines integrity as one of the basic categories for KPIs [4]. It attempts to measure how the RAN impacts the service quality provided to the users. Within this category, two different types of metrics are commonly defined: latency and throughput. In the downlink, the former is related to the delay experienced by the users, measured as the time from the reception of data in the e-NodeB to the transmission of the first packet over the radio interface. The latter is the data rate experienced by the users, measured as the data volume per elapsed time unit on the radio interface. In this work, the session throughput (*Ses_Thp*) is computed to identify users whose performance in terms of QoS is degraded. This requires the activation of the trace event that provides the corresponding data to compute the throughput for each UE in the downlink radio interface. In particular, this event periodically (every 2 s) reports the data rate for each radio bearer on the Downlink Shared Channel (DL-SCH) and Uplink Shared Channel (UL-SCH), which are the main transport channels for downlink and uplink data transfer, respectively. Let r_k be the kth

Table 1 Characteristics of the datasets and network-related information

	Dataset 1	Dataset 2	Dataset 3	Dataset 4
Technology	LTE	LTE	LTE	LTE
Area	Urban	Urban	Urban	Urban
Cell type	Macrocells	Macrocells	Macrocells	Macrocells
Measurement period [min]	15	15	15	15
Cell bandwidth [MHz]	5 and 10	10	10	15
Number of cells	6421	238	318	655
Average number of data sessions per cell	4.07	4.07	4.38	5.19
Average session duration [s]	73.7	75.3	91.2	92.2
Average session throughput [Mbps]	5.1	5.3	6.7	11.6

Table 2 Selected metrics from cell traces

Metric	Link direction	Related magnitude
Ses_Thp	Downlink	QoS
Num_UE_MHz_DL	Downlink	Cell load per MHz
RSRP	Downlink	Signal strength
RSRQ	Downlink	Signal quality
Num_UE_MHz_UL	Uplink	Cell load per MHz
Restr_Pwr_Ratio	Uplink	Signal strength
HARQ_Fail_Ratio	Uplink	Signal quality

sample of data rate in the downlink collected during a session. The data rate r_k is internally computed by measuring the data volume in the so-called Time Transmission Intervals (TTIs).

Once the data rate samples at each measurement period k are gathered, *Ses_Thp* is calculated as an average of these samples. However, some of them should be excluded from the computation of this indicator in order to have a more accurate estimation of the throughput. The reason for this is that, with the new data services such as the online instant messaging, the level of bursty traffic has significantly increased. This obviously affects the evaluation of throughput performance indicators. On the one hand, there are a lot of TTIs in the traces where there are no data to transmit by the base station. These time intervals should be ignored in the computation of these indicators to make them independent of the file size. On the other hand, there are TTIs where the measured throughput is not representative of the data rate experienced by the UE. More specifically, given a traffic burst, the data volume transmitted in the TTI that empties the buffer (i.e., the last TTI in the burst) can negatively impact the average user throughput, especially if the amount of data in the last TTI is relatively much smaller than in the previous TTIs. The data volume in the last TTI also depends on the size of the packets at the user plane. As stated in [25], in the LTE downlink, almost 14% of data are transmitted in the last TTI, while around 40% of TTIs are last TTIs. To avoid the effect of the last TTI, in this paper, the data rate samples below a certain level, r_{min}, have been removed from the calculation of *Ses_Thp*. Assuming that N is the total number of samples in a data session, *Ses_Thp* can be formally expressed as:

$$Ses\_Thp = \frac{\sum_{k=1}^{N} f(r_k > r_{\min}) \cdot r_k}{\sum_{k=1}^{N} f(r_k > r_{\min})}, \tag{1}$$

where $f(\cdot)$ is a function that returns "1" if the condition within the brackets is true, otherwise it returns "0." In this paper, $r_{\min}$ is set to 250 Kbps.

The metric *Ses_Thp* is also impacted by other factors, so that data filtering will be required at the session level. For example, the Outer Loop Link Adaptation (OLLA) function [42] leads data sessions of short duration to abnormally low values of *Ses_Thp* [43]. The reason is that this mechanism adapts the modulation and coding scheme to provide certain block error rate in the connection. In particular, the OLLA starts with a conservative parameter setting and, after some iterations, the algorithm must converge. However, the target value is reached or not depending on the session duration. For short transmissions, the steady state of OLLA may not be reached and the modulation ramp-up is not long enough to compensate the initial setting of the algorithm. As a consequence, *Ses_Thp* will be negatively impacted. To avoid this problem, short data sessions should be removed from the dataset. Specifically, the event parameter available in the tracing system that measures the time to transmit a data burst excluding the last TTI can be used for this purpose [44]. Thus, those data sessions with this parameter below a certain threshold (set to 1500 ms in this work) are removed from the considered datasets.

While *Ses_Thp* is a measurement of the per-user QoS, the rest of metrics presented in Table 2 are measurements that describe the channel conditions in the radio environment for each UE. An estimation of the cell load in the downlink and uplink is given by *Num_UE_MHz_DL* and *Num_UE_MHz_UL*, respectively. They represent the mean number of UEs considered active in the corresponding link per TTI, calculated over one measurement period (2 s) and normalized by the cell bandwidth. Formally, it can be formulated as:

$$Num\_UE\_MHz\_xL = \frac{Num\_UE\_xL}{Cell\_BW}, \tag{2}$$

where x is "D" or "U" depending whether it is related to the downlink or the uplink, *Num_UE_xL* indicates the mean number of active UEs in the cell and *Cell_BW* expresses the cell bandwidth in MHz. The denominator can be obtained from the configuration management functional area, and the numerator is taken from an internal trace event in the cells. This event reports information about the cell traffic. It can be expected that a congested area would simultaneously lead to a high value of both downlink and uplink metrics (*Num_UE_DL* and *Num_UE_UL*). This effect is represented by the squared Pearson correlation coefficient (r^2) calculated for these two metrics, as shown in Fig. 2a. However, there are also frequent situations where only one link is congested, as illustrated in Fig. 2a, where the samples of dataset 1 (considering only cells with 10 MHz bandwidth) are represented. This means that the two metrics are meaningful from the diagnosis perspective. The metrics have also been normalized by the cell bandwidth since today's LTE network deployments may have cells with different bandwidths, meaning that cells with higher bandwidth will provide higher

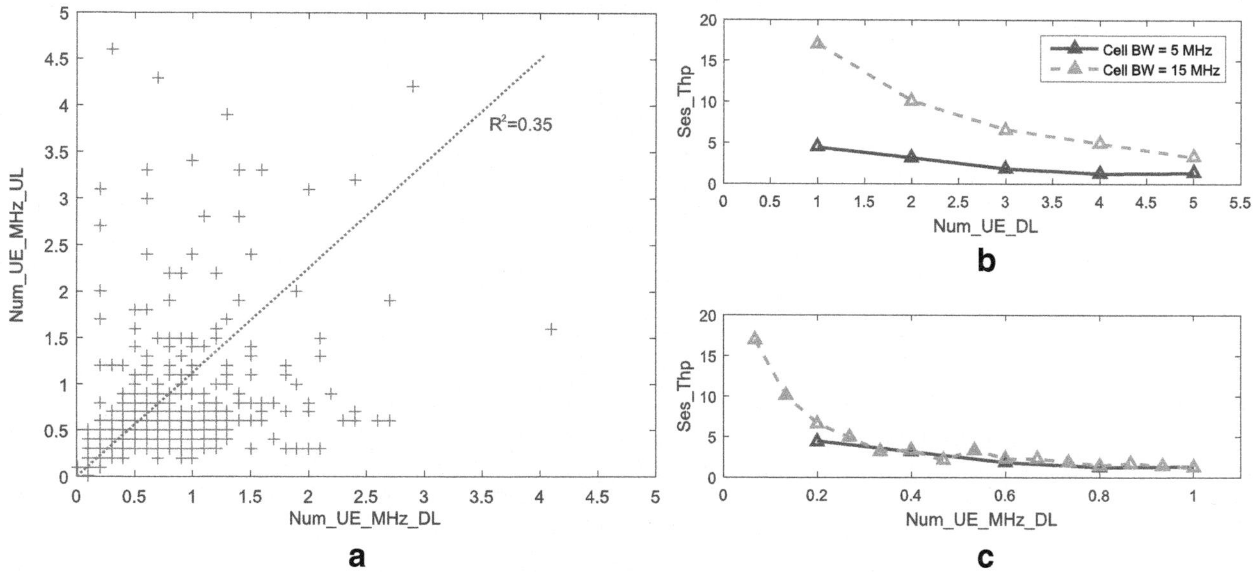

Fig. 2 Analysis of the mean number of active UEs in a cell: **a** Correlation between the downlink and uplink. **b**, **c** Impact of the cell bandwidth

user throughput and support higher number of users. In Fig. 2b, c, the mean number of active UEs against the session throughput is represented in two cases: when the cell bandwidth is omitted, Fig. 2b, and when it is included in the definition, Fig. 2c. In the first case, given a number of active UEs, there is a big difference in *Ses_Thp* between cells with different bandwidth. Conversely, in the latter case, the values are very similar. For this reason, the cell bandwidth has been considered in the definition of cell load according to (2).

Another kind of metrics is related to the measurement of the signal strength. In the downlink, it is given by the Reference Signal Received Power (RSRP), defined as the average power of resource elements that carry cell specific reference signals over the entire bandwidth. RSRP levels typically range from around −75 dBm (close to an e-NodeB) to −120 dBm (at the cell edge). In the uplink, the information is taken from a trace event that reports the number of transport blocks on the Medium Access Control (MAC) level that are scheduled in the uplink, distinguishing between two cases. In particular, one is when the UE was considered to be power limited (counted by *Sched_TP_Restr_Pwr*) and another is when the UE was not limited in terms of power (counted by *Sched_TP_Unrestr_Pwr*). Based on these event parameters, the *Restr_Pwr_Ratio* is defined as follows:

$$Restr\_Pwr\_Ratio\,[\%] = 100 \times \frac{Sched\_TP\_Restr\_Pwr}{Sched\_TP\_Unrestr\_Pwr + Sched\_TP\_Restr\_Pwr}. \tag{3}$$

A high value of this metric may represent a situation where the UE suffers from a lack of coverage since the UE is transmitting with the maximum power.

Finally, the measurement of the signal quality provides additional information of the radio environment, since it includes the interference component. In the downlink, it is given by the Reference Signal Recieved Quality (RSRQ), defined as:

$$RSRQ = L \times \frac{RSRP}{RSSI}, \tag{4}$$

where *RSSI* is the average total received power including the intra-cell power, interference, and noise, and *L* is the number of resource blocks over which the RSSI is measured (typically equal to the cell bandwidth). While the RSRP and RSRQ are reported by the UEs, the RSSI can simply be computed from RSRP and RSRQ. The range of RSRQ is normally from −19.5 to −3 dB.

In the uplink, an estimation of the signal quality is obtained from a trace event that reports the number of successful and unsuccessful transmissions at the hybrid automatic repeat request (HARQ) level in the uplink direction. The failed transmissions at the HARQ level are detected by means of a cyclic redundancy check (CRC). The provided information is also collected for each modulation format that is chosen by the UE. In this work, the statistics related to the quadrature phase-shift keying (QPSK) modulation have been considered since this modulation is typically used under the worst radio conditions.

Thus, from these event parameters, the following metric has been defined:

$$HARQ\_Fail\_Ratio\,[\%] = 100 \times \frac{HARQ\_Fail\_QPSK}{HARQ\_Succ\_QPSK + HARQ\_Fail\_QPSK}. \quad (5)$$

where *HARQ_Succ_QPSK* and *HARQ_Fail_QPSK* are the number of successful and failed transmissions, respectively, at the HARQ level in the uplink direction using the QPSK modulation.

To avoid certain dependence of performance evaluation on the datasets and, thus, facilitate the application of the proposed method to any LTE network, it is observed that the used metrics are not influenced by the network configuration. For example, the indicator to measure the traffic load is given per unit of bandwidth to avoid dependence on the system bandwidth. In addition, it is important that the number of considered cells (i.e., the network's size) in the datasets is large to include a significant number of faulty cells. Finally, note that the lack of data for any of the above described metrics in a few cells of a large network is a common situation, e.g., because the corresponding trace events have been erroneously deactivated. For this reason, as part of the data pre-processing, the cleaning function copes with incomplete and incorrect samples in the dataset. In this work, to avoid having inconsistent data, the entries (sessions) in the dataset with missed values are removed.

4.2 Construction of the classification tree

Traditionally, the RAN has been the bottleneck for data-hungry services. From the perspective of the QoS, it is important to identify if any degradation in the user throughput is due to the radio environment. The proposed classification tree aims at determining the root cause of abnormally low throughput values in user data sessions. This problem is addressed in the context of unsupervised learning since the training dataset is composed of unlabeled data, i.e., the root cause is a priori unknown. Considering a sufficient number of varied samples in a cellular network, the main problems related to radio aspects, such as congestion, lack of coverage and interference in either downlink or uplink, can be found in the given dataset. These root causes are determined by establishing adequate thresholds to radio KPIs. In the present section, the structure of the binary classification tree is proposed, while in the next section the adjustment of the thresholds is covered.

To better understand the binary classification problem, let Ω be the population of a training dataset, $X_1, X_2, \ldots, X_q$ be the set of q selected KPIs and C be the target attribute that indicates the root cause and takes only a finite number of different values, i.e., $C = \{c_1, c_2, \ldots, c_{q+1}\}$, where c_1 represents the normal behavior and q is the number of potential faults. Note that $q + 1$ categories can be distinguished using a binary classification tree with q KPIs. The goal of the tree is to generate a model $\phi(X_1, X_2, \ldots, X_q)$ in the form of a decision tree for predicting the value of C from the values taken by the predictive variables. The solution is obtained by partitioning the Ω space into $q + 1$ disjoint sets, $A_1, A_2, \ldots, A_{q+1}$. Thus, given a member of the space Ω denoted by ω, the predicted value of Y is c_j if ω belongs to A_j. The construction of the tree is based on iteratively splitting the dataset at each step by using one of the predictive attributes, $X_1, X_2, \ldots, X_q$. In this work, a binary partition of the attributes is assumed since each attribute can be represented by two states, corresponding to normal and anomalous values. In most cases, one of the two children of the node will be a leaf node representing a specific problem. In addition, the nodes of the tree (referred to as s_j) obtained at each step t of the growing process define a partition that becomes finer with t. To make a binary partition, the logical operator "<" is used to compare each metric to a given threshold.

Figure 3 shows the logic of the proposed classification tree. Internal nodes and leaf nodes are colored in gray and white, respectively. The first partition of the tree, L_1, acts as a detection step where data sessions with poor values of throughput are separated from those whose values are good. Thus, as illustrated in Fig. 3, the first leaf-node s_1 represents the group of data sessions with acceptable QoS, while the other branch of the split (i.e., node s_2) is subject to further analysis. As observed, the position of each variable in the tree will determine the priority of each problem in the diagnosis process. This establishes the way that troubleshooting experts operate when simultaneous problems happen in a network cell. To decide which variables are placed at higher levels in the tree, the statistical significance of the variables with respect to the primary variable, *Ses_Thp*, is first analyzed. More specifically, the predictive variables are used to fit a linear regression model where a set of coefficients are derived to describe the statistical relationship between *Ses_Thp* and the rest of KPIs. Then, the p values of the t statistics are computed for each coefficient in order to test the null hypothesis that the coefficient is equal to zero. In other words, a predictive variable X_j that has low p value is suitable to be included in the regression model since the changes in the variable are expected to be related to changes in *Ses_Thp*. On the contrary, a larger p value indicates that the changes in the predictive variables are not associated with changes in the primary variable. Table 3 shows the p values for each coefficient of the predictive variables in the linear regression model when 2000 samples from dataset 1 are used

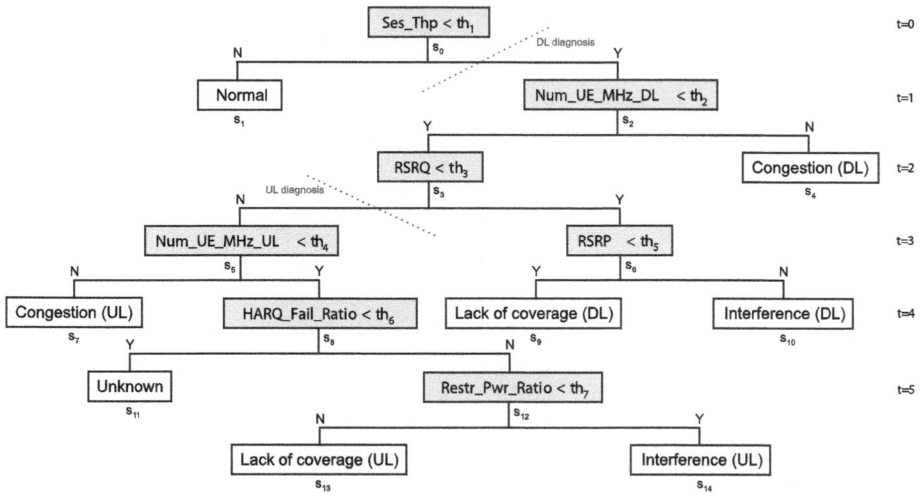

Fig. 3 Structure of the proposed binary classification tree

as input. In the downlink, two metrics result in p values equal to zero, meaning that there exists a strong relation between the user throughput and such metrics. For this reason, in order to analyze the downlink user throughput (*Ses_Thp*), the radio conditions in the downlink are evaluated first in the proposed tree. This is represented by "DL diagnosis" in Fig. 3. After checking that the conditions in the downlink are good, the radio environment is then analyzed in the uplink (see "UL diagnosis" in the figure). For each link direction, the decision flow in the tree is started by checking the traffic load (nodes s_2 and s_5), because the p value obtained for *Num_UE_MHz_xL* is zero in both link directions. The next conditions to be evaluated are those related to coverage and quality issues. In the downlink, it is observed that *RSRQ* is notably more correlated to *Ses_Thp* than *RSRP* since there is a big difference between their p values. As a consequence, the *RSRQ* is used in the node s_3 to separate UEs with coverage and quality issues from those with good radio conditions in the downlink. The first group of UEs, related to node s_5, are further diagnosed by analyzing the uplink through a decision flow that, for simplicity, is similar to the one applied in the downlink. The second group of UEs, related to node s_6, are separated by using the *RSRP* with the aim of distinguishing UEs having a lack of coverage from those suffering from interference. Finally, in Fig. 3, there is one leaf node (s_{11}) representing the group of UEs that, having low values of *Ses_Thp*, are not associated to any of the considered root causes. This kind of UEs has been labeled as "Unknown" meaning that the root cause of their problem is not known.

Table 3 The p-values of the predictive variables

Metric	p value	Related root cause	Link direction
Num_UE_MHz_DL	0.0	Congestion	Downlink
RSRP	$7.5e-3$	Lack of coverage	Downlink
RSRQ	0.0	Interference	Downlink
Num_UE_MHz_UL	0.0	Congestion	Uplink
Restr_Pwr_Ratio	$1.7e-12$	Lack of coverage	Uplink
HARQ_Fail_Ratio	$3.9e-7$	Interference	Uplink

Perturbation from external systems may produce false or very biased values of the metrics. However, this normally happens in a small number of UEs/cells compared to the size of the analyzed cell cluster/network. Thus, the "Normal" UE class and cell profile can be properly determined by the proposed method. If the external perturbation degrades the throughput but it does not produce any effect on the selected metrics, then the cause of this perturbation will be identified as "Unknown" by the system. On the contrary, if one or more selected metrics are degraded, then the system will probably produce diagnostic errors. Lastly, in case of a global impact of the perturbation, the expert team should be responsible for monitoring and troubleshooting this kind of problems.

The adjustment of each threshold th_j present in the classification tree is addressed in the next section. In particular, the thresholds from th_2 to th_7 are configured by calculating a certain percentile of data, given that degradation is associated to extreme values of the indicators. After this, the data distribution over the classes is analyzed for different values of th_1. Based on this analysis, th_1 is configured.

4.3 Adjustment of the thresholds

The binary classification tree performs test on numeric features to divide the data into two groups: those whose values for the variable are less than a threshold and those whose values are greater than or equal to the given threshold. Thus, the configuration of the thresholds determine the data distribution over the different rootcauses. The metrics placed in higher levels of the tree will have greater impact on the overall data distribution. In particular, the metric *Ses_Thp*, located at the top of the tree and used for detection purposes, affects the distribution of all the considered root causes. This variable, which is not related to a single problem, requires some practical experience to define the corresponding threshold. The problem of defining low/abnormal values of throughput is not straightforward, as it depends on many aspects such as the radio access technology or the user perception. Thus, a sensitivity analysis is needed to evaluate the overall impact on the diagnosis process. Conversely, the rest of variables in the tree are related to specific problems. To calculate the thresholds for such variables, an automatic technique referred to as Percentile-Based Discretization (PBD) [45] is used. In particular, each threshold is set to the Xth percentile of all the values of the corresponding metric in the dataset. Such a percentile represents an estimation of the percentage of data samples considered to be degraded over the total, assuming that the worst values of each variable most likely correspond to UEs having the associated problem. For example, UEs with bad RSRP are assumed to have a lack of coverage. The PBD technique is applicable when the training dataset is large enough to be statistically meaningful (as it is the case of the datasets used in this work), including not only the well-performing UEs but also the problematic UEs. Thus, the thresholds from th_2 to th_7 are adjusted by computing a certain percentile of the data.

Table 4 shows the corresponding percentiles for each metric of the datasets used in this work. Typically, if the metric is degraded by reaching high values, the 80th percentile has been selected. On the contrary, if degradation is given by the lower part of the range, the 20th percentile is taken. Note that the selected percentiles are not extreme values since there can be different levels of the magnitude of the problems. It is observed that the calculated percentiles, especially for the measurements RSRP and RSRQ, are similar to those used by troubleshooting experts in practice. In addition, the obtained values are similar over the different datasets. Only dataset 4 reflects a notable deviation in the RSRQ values since its greater cell bandwidth has resulted in lower inter-cell interference conditions. Finally, note that the thresholds from th_2 to th_7 in the classification tree will be those derived from the PBD method, depending on the specific dataset. Thus, each dataset has its own set of thresholds to be used in the classification tree.

Once the mentioned thresholds have been adjusted, the sensitivity analysis of th_1 can be performed. Figure 4 shows the distribution in percentage of the data sessions over the root causes for each dataset with cell bandwidth equal to 10 MHz. The distributions are represented for three different values of th_1. As observed, results between datasets 1 and 2 are quite similar in general. Compared to them, dataset 3 presents some differences in specific classes. In particular, the percentage of data sessions for the Normal class in dataset 3 is higher than in datasets 1 and 2, while the percentage of data sessions experiencing congestion in the downlink in dataset 3 is lower than in datasets 1 and 2. This observation reveals that the network in dataset 3 is less loaded and, as a consequence, better QoS is provided to the UEs. However, when th_1=10 Mbps is used, the differences between datasets in the percentage of Normal UEs are significantly reduced. This is mainly because of an increase of the percentage of UEs suffering from interference and UL congestion in dataset 3. Thus, setting th_1 to 10 Mbps would lead to similar distributions between normal and abnormal cases for the given datasets. The main disadvantage of using this setting is that the percentages of normal cases are significantly low, specifically below 20%, while the percentage of data sessions with unknown root cause is above 25%. Obviously, this represents a pessimistic view of the network status and may block the detection of those problems with major impact on QoS. On the contrary, the setting with the lowest tested value, th_1=2 Mbps, provides distributions where some fault classes are empty. This is also a bad choice provided that all the considered problems should be present in large datasets even in a small proportion. For these reasons, the setting th_1=5 Mbps has been selected in this work to provide more effective diagnosis of

Table 4 Defined thresholds based on the Xth percentile of the metrics for each dataset

Metric	X^{th} -ile	Dataset 1	Dataset 2	Dataset 3	Dataset 4	Units
Num_UE_MHz_DL	80	0.4	0.7	0.2	0.2	[UE/MHz]
RSRP	20	−118.0	−111.5	−112.1	−113.4	[dBm]
RSRQ	20	−18.0	−16.5	−17.5	−13.7	[dB]
Num_UE_MHz_UL	80	0.4	0.7	0.3	0.2	[UE/MHz]
Restr_Pwr_Ratio	80	99.0	98.7	99.9	99.8	[%]
HARQ_Fail_Ratio	80	10.4	9.2	16.0	13.4	[%]

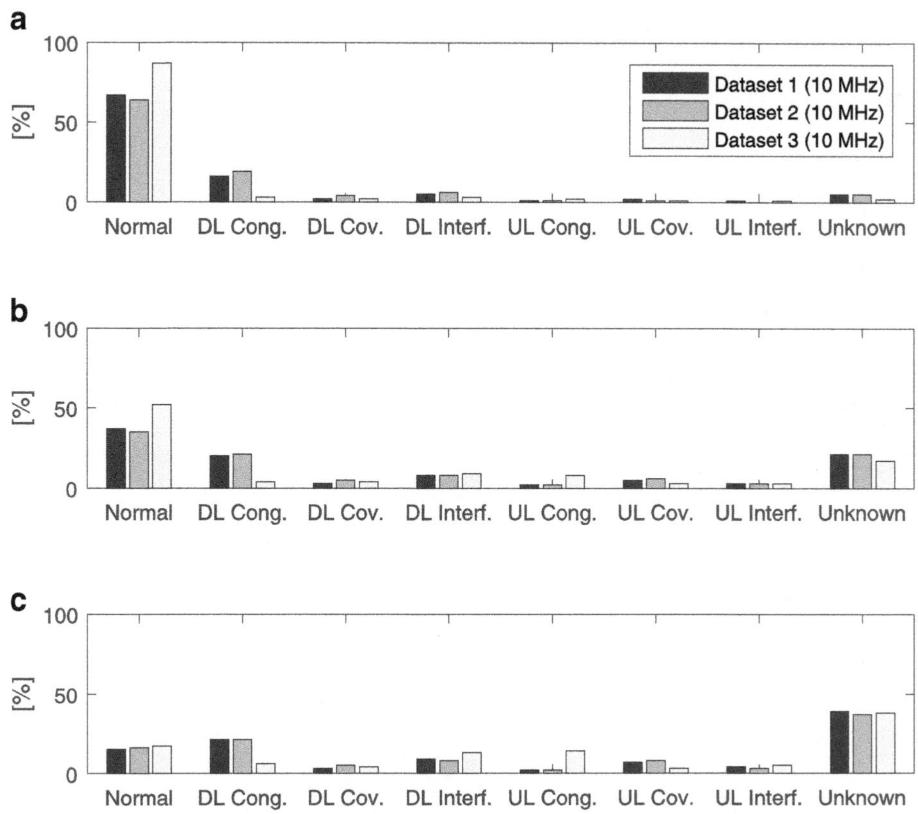

Fig. 4 Sensitivity analysis of th_1 for datasets with the same (10 MHz) cell bandwidth. **a** $th_1 = 2$ Mbps. **b** $th_1 = 5$ Mbps. **c** $th_1 = 10$ Mbps

the radio problems. In this sense, note that the Unknown class includes the UEs with bad QoS due to non-radio issues. Since most UEs are accessing to Internet services, the contribution to this class should be important. With the proposed threshold, the percentage of data sessions whose root cause is unknown remains relatively high but in a lesser extent than the Normal class, as expected.

Finally, Fig. 5 shows a comparison of the data distribution over the classes between the datasets with different values of cell bandwidth. The analysis is also made for each value of th_1 previously analyzed. As expected, there is a clear trade-off between the cell bandwidth (related to capacity) and the QoS of the data sessions. In particular, the higher the cell bandwidth, the higher the percentage of UEs belonging to the 'Normal' class is expected to be. On the one hand, it is noted that the main affected root cause due to a change in the cell bandwidth is congestion, while the data distribution over the rest of root causes is not significantly impacted. Thus, the proposed classification tree effectively establishes a relation between the cell bandwidth and congestion issues: a high contribution of the classes related to congestion suggests that there is a lack of extra bandwidth. On the other hand, the same tendency is observed regardless of the specific value of the analyzed threshold, th_1. Thus, the optimal setting for th_1 previously selected can be applied to scenarios with different cell bandwidth.

4.4 Cell-level correlation-based diagnosis

The final cause of the problem in a specific UE can be very diverse, especially when it connects to Internet. Abnormally low values of throughput can be registered, for example, when a router in the core network is malfunctioning, the server is overloaded or when an application in the UE is crashing. In this paper, the previous sections have proposed a practical system with the aim of determining the root cause of problematic data sessions when their radio conditions are not favorable. This means that isolated problems (such as the previously mentioned) are not considered in this work. On the contrary, the final goal of the proposed method is the diagnosis of problems in the radio environment that affect a significant number of UEs. More specifically, the diagnosis made for every UE is used to find localized problems in network cells. This is achieved by defining cell profiles (explained later), which represents the distribution of the diagnosed UEs over the predefined fault classes. Then, the comparison between different cell profiles using correlation techniques allows

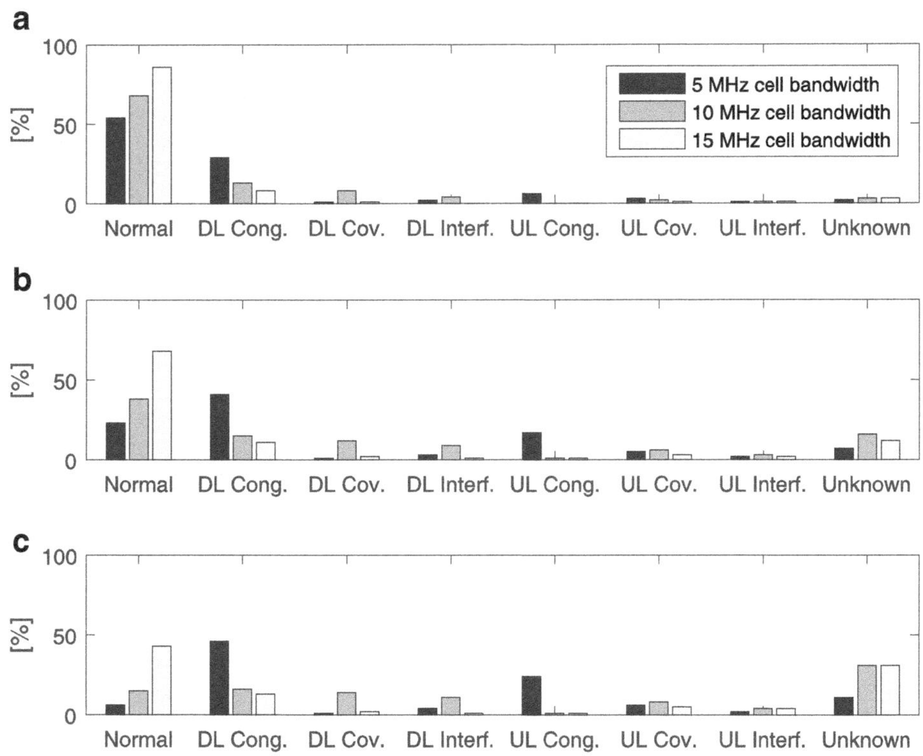

Fig. 5 Sensitivity analysis of th_1 for datasets with different cell bandwidth. **a** $th_1 = 2$ Mbps. **b** $th_1 = 5$ Mbps. **c** $th_1 = 10$ Mbps

identifying problematic cells. This differs from traditional approaches where individual measurement samples are aggregated in a non-selective way at the cell-level to provide the diagnosis. The main benefit of the proposed method is that the performance of the problematic UEs is not biased by the performance of the dominant UEs in a cell, since the diagnosis is first made at the user level. Essentially, the method is composed of five steps represented in Fig. 6.

The first step of the proposed method is to apply the binary classification tree to every UE in a certain cell in order to obtain the diagnosis at the user level. After this, the percentage of UEs having a specific issue (i.e., those UEs belonging to a certain class in the tree) is computed. In particular, the following cell-level indicator, denoted as ξ_j^k, is calculated for each class j in every cell k of the scenario:

$$\xi_j^k\,[\%] = 100 \cdot \frac{|A_j^k|}{\sum_{i=1}^{q+1} |A_i^k|}, \qquad \forall j = 1, 2, \ldots, q+1. \quad (6)$$

where A_j^k is the set of data sessions associated to class j and served by cell k.

Once the above indicator is computed for the $q + 1$ classes, the profile of a cell k is determined by the tuple $\boldsymbol{\xi}^k = \left\{\xi_1^k, \xi_2^k, \ldots, \xi_{q+1}^k\right\}$. A visual representation of the cell profile provides an effective means of diagnosing cells. For example, in the case of cells without problems, the value of ξ is expected to be high for the Normal class and low for the rest of (fault) classes. However, this can be an overly simplistic view of the diagnosis process. Specifically, the profile of a healthy cell depends on many factors and it may vary from one network to another. To explain this, Fig. 4b can be seen as a representation of the profile of a network, in a similar way to cell profiles. Then, a close examination of this figure reveals that, although the UEs without problems are the dominant class in the network, there are still significant differences in the profiles between the different datasets. For example, the levels of interference in healthy cells of dataset 3 are expected to be higher than in the other two datasets. This does not necessarily mean that most cells of dataset 3 suffer from interference issues, rather it is merely an indication of the overall radio conditions in the network. For this reason, cell profiles should be adapted to every network in order to provide a better diagnosis.

Based on the above explanation, the next step of the method (Fig. 6) is the creation of the Reference Cell Profiles (RCPs). There exists one RCP that represents the faultless or normal behavior of a cell in the considered network, referred to as $\boldsymbol{\xi}^{\text{normal}}$. Additionally, other RCPs

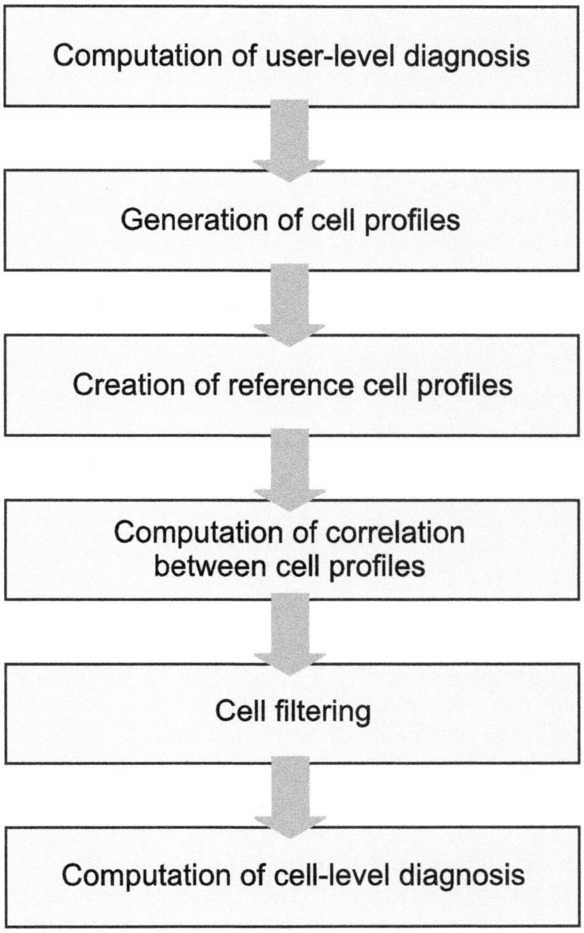

Fig. 6 Block diagram of the proposed cell-level correlation-based algorithm

for faulty situations $\left(\text{e.g.}, \xi^{DL\_Cong}, \xi^{DL\_Cov}\right)$ can be generated. The main problem is that the RCPs must be created in an unsupervised manner because the actual performance of each cell (normal or faulty) is unknown. Clustering methods are useful to discover the most frequent cell profiles existing in the datasets. In this paper, data clustering is carried out by using the well-known k-means algorithm, which is the most popular clustering tool used in scientific and industrial applications. However, the RCPs can also be obtained by using other clustering methods, such as Self-Organizing Maps (explained later).

The k-means algorithm partitions the dataset into k clusters. Such a partition minimizes the sum of discrepancies between a point and its centroid expressed through appropriate distance (in this case, Squared Euclidean distance is used). Each cluster is represented by the mean (or weighted average) of its points, i.e., the so-called *centroid*. In this work, each centroid represents a RCP. The mechanism to choose the initial positions of the cluster centroids is based on randomly selecting k observations from the dataset. Then, the algorithm iteratively executes three steps until convergence is reached: first, the distance of one data point to the centroids is calculated; second, the class label of the cluster centroid having the minimum distance is set to the given point; third, the centroids are updated based on the assignment. The convergence is reached when the partitioning error is not reduced by the relocation of the centroids or the maximum number of iterations allowed is reached. The number of clusters k establishes how many RCPs are identified. This number must be at least equal to the number of categories that have been defined in the classification tree for the user diagnosis (i.e., $q + 1$). A greater value of k can be used to identify new RCPs that would be associated by troubleshooting experts to one of the given categories. Note that an excessively high value might lead to wrong associations between the RCPs and the fault categories. For this reason, the value of k has been set to 12.

Table 5 shows the results of applying the k-means algorithm to dataset 1. Each row represents one obtained RCP or centroid. The first column is the identifier of the RCP, the second column shows the category assigned to the RCP, which can be the normal state or a fault, and the last columns provide the percentage of UEs associated to a specific issue for the given RCP. It is observed that RCPs from 1 to 8 are associated to categories where the dominant diagnosis at the user level corresponds to the labeled category at the cell-level. More specifically, RCP 1 is linked to the normal state, since not only the percentage of well-performing UEs is the highest, but also the percentage of UEs for the rest of (faulty) categories is very low. In line with the network profile represented in Fig. 4b, the second dominant category in RCP 1 is Unknown. This means that a cell normally has a tolerable percentage of abnormal UEs, labeled as Unknown, under normal cell conditions. RCPs from 2 to 8 are mainly related to specific problems, having a large percentage of UEs experiencing the dominant problem, but also a small percentage of UEs belonging to Normal and Unknown categories. Such particular distributions between categories, which varies among datasets, is an interesting reason to create specific RCPs to every network. Another reason comes from analyzing RCPs from 9 to 12, where it is observed that the Normal (and in many cases Unknown) category achieves similar levels to a certain fault category. Recall that the parameter k has been configured in order to find a reasonable number of these situations. This set of RCPs represents the most difficult faulty situations to be identified by the diagnosis system since the problematic UEs are not a clear dominant group within the coverage area of the affected cell. Moreover, in these cases, using traditional KPIs (i.e., cell-level measurements) may not be enough to identify the problem.

Table 5 Set of RCPs obtained by applying *k*-means with *k*=12

		ξ_i [%]							
No. of RCP	Labeled categ. c_i	Normal (i=1)	DL Cong (i=2)	DL Cov (i=3)	DL Int (i=4)	UL Cong (i=5)	UL Cov (i=6)	UL Int (i=7)	Unknown (i=8)
1	Normal	98.41	0.07	0.16	0.28	0.03	0.15	0.11	0.78
2	DL Cong	4.38	92.65	0.13	0.51	0.85	0.13	0.18	1.17
3	DL Cov	11.24	0.10	65.25	6.12	1.84	5.55	1.90	8.00
4	DL Int	0.80	0.00	0.85	93.38	0.66	0.16	0.71	3.44
5	UL Cong	3.59	3.24	1.21	3.71	85.23	0.98	0.51	1.54
6	UL Cov	3.66	0.00	0.78	2.03	1.06	86.83	1.22	4.43
7	UL Int	2.87	0.00	0.99	4.39	0.00	5.90	83.13	2.73
8	Unknown	1.77	0.03	0.25	0.67	0.08	0.35	0.32	96.54
9	DL Int	40.44	0.58	2.35	39.28	1.58	1.95	1.84	11.98
10	DL Int	7.96	2.48	4.51	17.30	1.71	11.13	8.35	46.57
11	UL Cov	53.69	6.05	2.30	1.73	8.19	15.40	7.77	4.87
12	Unknown	51.31	0.37	1.45	1.38	0.38	2.25	1.14	41.72

The next step of the proposed method is the calculation of a correlation coefficient that measures the similarity between a given cell profile and a reference cell profile. Specifically, the Pearson's correlation coefficient has been used in this work. Its possible values are in the range -1 to $+1$, indicating high linear correlation when it is close to any of those values. The plus sign represents positive correlation between the variables, while the minus sign indicates negative correlation. The Pearson's correlation between the cell profiles $\boldsymbol{\xi}^k$ and $\boldsymbol{\xi}^{ref}$, referred to as $\rho_{k,ref}$, can be calculated according to this formula:

$$\rho_{k,ref} = \frac{\sum_{j=1}^{q+1} \xi_j^k \xi_j^{ref} - (q+1)\,\overline{\xi^k}\,\overline{\xi^{ref}}}{\sqrt{\left(\sum_{j=1}^{q+1}\left(\xi_j^k\right)^2 - (q+1)\left(\overline{\xi^k}\right)^2\right)}\sqrt{\left(\sum_{j=1}^{q+1}\left(\xi_j^{ref}\right)^2 - (q+1)\left(\overline{\xi^{ref}}\right)^2\right)}} \quad (7)$$

where $\overline{\xi^k}$ (and similarly $\overline{\xi^{ref}}$) is computed as:

$$\overline{\xi^k} = \frac{1}{q+1}\sum_{j=1}^{q+1} \xi_j^k. \quad (8)$$

To improve the accuracy of the proposed system, the cells are filtered before providing the final diagnosis. The filtering module is composed of two filters. First, the cells are filtered by the number of data sessions that have been used to compute their cell profile. Note that the cells with a low number of data sessions have been used to build the RCPs. Since this group of cells can be large in the dataset, they can provide (on average) meaningful information to discover frequent patterns. However, these cells are not adequate to be individually diagnosed since their cell profiles may not be statistically significant. For example, several UEs in a cell could provide an abnormal cell profile if these UEs are all located at the cell edge. Thus, cell k is considered for diagnosis only if the condition $D_k > D_{th}$ is fulfilled, where D_k is the number of data sessions in cell k and D_{th} is a configurable threshold of the filter. Second, the cells are also filtered by their correlation value obtained in the previous step of the method. This filter would act as a detection stage in order to discard the cells whose profile is not similar to any of the RCPs and, as a consequence, they cannot be correctly diagnosed. It might happen that a cell profile is given by a high contribution of two or more fault classes. In such cases, the proposed system is not capable of providing a diagnosis and they should be analyzed by troubleshooting experts in order to determine whether the source of the problem is the same or not and, if not, to solve the problems independently. Thus, cell k is considered for diagnosis only if the condition $\rho_{k,ref} < \rho_{th}$ is satisfied, where ρ_{th} is a configurable threshold.

Lastly, once the cells have been filtered, the diagnosis at the cell level is computed. For cell k, the diagnosis is given by the RCP having the highest correlation to the cell profile of cell k. More specifically, the root cause c_i associated to that RCP is the diagnosis for the given cell.

5 Performance evaluation

In this section, the proposed method is evaluated and compared to other reference methods. To this end, dataset 1 is used due to its larger number of data sessions compared to the other datasets. In addition, the filters of the proposed system have the following configuration: $D_{th} = 8$ data sessions and $\rho_{th} = 0.70$.

The methods are evaluated by using a key metric in diagnosis referred to as Diagnosis Success Rate (DSR). It is estimated as the ratio between the number of well-diagnosed cells and the total number of cells. Note that, to calculate the proportion of observations correctly diagnosed, the real root cause, or "label" associated to the observed cell, must be known. This labeling has been carried out in this paper by applying the rules that troubleshooting experts use to diagnose faults in their networks. Such rules, manually derived from experience, are verified by means of the inspection personnel that is in charge of checking the status of the affected network elements, either remotely or on-site. Thus, these labeled data have been used as reference to calculate the diagnose success rate of the evaluated methods.

To benchmark the proposed method, the results are compared to baseline methods based on previous work. On the one hand, regarding the user diagnosis, the proposed method (i.e., a classification tree) is compared to other approaches that are based on the following machine learning techniques:

- The k-means clustering: as explained in the previous section, this technique is used to partition the data into a number of clusters. In this case, the data sessions (regardless of the serving cell) are the input samples to be divided into various clusters. Then, each cluster is analyzed by experts and labeled with the normal state or one of the considered faults as in the classification tree. In this way, users belonging to a certain cluster share the same diagnosis. The result of this process (i.e. the user diagnosis), which is equivalent to that obtained from applying the classification tree, can be used for cell diagnosis. The k-means clustering has previously been applied to fault detection in the context of wireless networks [46].
- The self-organizing map (SOM): it is a type of Neural Network used for clustering. The methodology is similar to the k-means algorithm, since both are example methods of unsupervised learning. The approach followed in this paper is explained in [8], with the difference that the diagnosis in [8] is carried out by using traditional KPIs, while in this case call traces are used.

On the other hand, regarding the cell diagnosis, the proposed method (that uses k-means to generate RCPs and Pearson's correlation coefficient to compute cell diagnosis) is compared to the method presented in [38]. Such a method requires manual adjustment of several thresholds and configurable parameters. Its operation is divided into two stages, as follows. In the first stage, problematic cells are detected by comparing the percentage of well-performing UEs to a certain threshold. Then, in the second stage, the cells detected as faulty cells in the previous stage are diagnosed only if the percentage of UEs related to at least one fault is above a threshold. In that case, the selected fault is the one with the largest percentage of UEs having the same fault in the cell. Although the method is mainly based on selecting the most dominant fault (i.e., the one having the maximum percentage of UEs), note that the particular distribution of UEs (e.g., in Normal and Unknown categories) is key in the diagnosis process. Hereafter, the state-of-the-art method in [38] is referred to as 'Max,' as opposed to the correlation-based approach.

Note that the methods for user diagnosis can be interchangeably combined with the methods for cell diagnosis. All these combinations have been implemented for evaluation. Lastly, a reference method for cell diagnosis is based on applying the classification tree to traditional (cell level) KPIs. To do this, the average per cell of the selected metrics is firstly calculated. Then, the binary classification tree is applied to the obtained samples with the aim of determining the diagnosis. Note that the population Ω of the input space is composed of the set of cells included in the dataset rather than the carried data sessions. Thus, a difference with the proposed method is that only one fault class is associated to a given cell. On the contrary, with the proposed method, a distribution of data sessions over the fault classes is provided for each cell, namely the cell profile. A configuration parameter that both methods have in common is D_{th} in order to focus only on cells with a significant number of data sessions.

The first analysis is devoted to the user diagnosis. Figure 7 shows the DSR obtained in each category for the proposed (tree) and reference (k-means and SOM) methods. The 'Normal' category is not included in this figure since all data sessions with a throughput below th_1 are labeled (as real cause) with a faulty state, thus in the same way that the evaluated approaches. The last bars provides the global DSR, which is obtained by aggregating the UEs from each category (rather than averaging the DSRs). The values in brackets below each category represent the number of data sessions whose real diagnosis corresponds to such a category. According to these values, the dataset appears to be unbalanced, i.e. the number of samples varies greatly from one class to another. This is a consequence of using real data where some faults are more likely than others. It is observed that, from a perspective of the clustering techniques, the smallest class ("UL Int") is more difficult to be identified than others. However, the accuracy of the evaluation should not be affected since the number of samples per class is in the order of hundreds. A closer inspection of Fig. 7 reveals that the proposed method provides good DSR for "DL Cong" since this problem is placed at the top of the tree. This means that, in situations where DL Cong

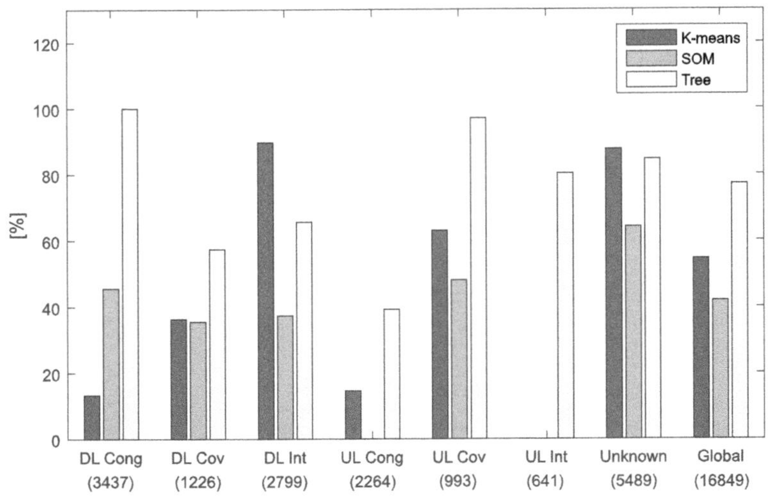

Fig. 7 Diagnosis success rate per class for user diagnosis methods

might be confused with other fault classes, the classification tree will select DL Cong as diagnosis. Another class with very high DSR for the proposed method is "UL Cov." In this case, the good correlation between the metric and the associated fault has led to a higher DSR with respect to other classes, as occurred in the reference methods. It is also important that the methods provide high DSR for the Unknown class, since the addition of new faults in the diagnosis system will rely on the accuracy of this class. Lastly, regarding the last columns, it is noted that the proposed classification tree outperforms the reference methods in the global DSR. The main differences between the *k*-means and SOM methods are related to DL Cong and "DL Int" classes. Note that, since these two faults are closely related (i.e. a congestion problem usually entails higher interference levels), the bad DSR in DL Cong achieved by the *k*-means method is counteracted with a good DSR in DL Int. This issue is less pronounced in the case of the SOM method.

The existence of noise into the metrics may affect the performance of the user diagnosis. The PBD method mitigates the impact of the bias into the selected metrics in the same way as the purging methods. To analyze the impact of the noise variance, a synthetic noise has been added to the metrics. In particular, the noise is generated following the Additive white Gaussian noise (AWGN) model. The average of the AWGN is set to zero and the standard deviation is set to $0.2 \times \sigma$, where σ is the standard deviation of the metric data. Figure 8 shows the global DSR obtained for each method in presence/absence of AWGN. As observed, the greater the DSR is, the higher the impact of the noise is on the performance. Consequently, the most affected method is the proposed classification tree, experiencing a decrease of about 5% with respect to the case without AWGN. However, the obtained value (71.6%) is still much greater than the value obtained by other methods.

With respect to intermittent perturbation, e.g., due to user mobility, the potential impact on the system performance is very limited due to temporal and spatial diversity factors. Specifically, in the time domain, the call traces are collected during a large period (typically, 15 min). This means that all the trace events and samples gathered during this period are considered for the elaboration of the metrics. Hence, this period is large enough to filter most propagation effects. In the spatial domain, all the users located in the service area of a cell contribute to determine its diagnosis. Thus, if one user experiences an eventual perturbation due to propagation effects, the users in other locations of the cell will counteract it.

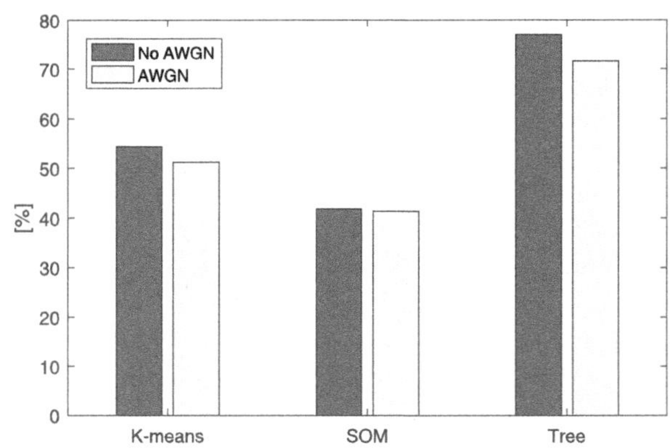

Fig. 8 Impact of noise on the global diagnosis success rate

The following analysis is related to the cell diagnosis, which is conducted based on the user diagnosis obtained in the previous evaluation. To do this, the methods "Max" and "Correlation" are combined with the previous methods, *k*-means, SOM and classification tree. Note that, in the case of solutions that comprise the proposed 'Correlation' method, a *k*-means clustering has been applied in order to find adequate RCPs for correlation. Figure 9 represents the DSR for such combined approaches. The method that applies the classification tree to cell-level KPIs is also included as a baseline. It is observed that the solutions based on reference clustering methods (i.e., *k*-means and SOM) provide worse performance than the classification tree with cell-level KPIs. This is mainly due to the bad user diagnosis that is obtained in the first stage of the solution. Moreover, the benefits of the proposed Correlation method are not evident using these solutions. If, on the contrary, the user diagnosis is carried out by means of the classification tree, the obtained DSR is better than applying the classification tree to cell-level KPIs. However, it is essential to use the proposed Correlation method in order to obtain a clear advantage over the reference method.

To further explain this, Fig. 10 shows the DSR obtained in each category for each method. As before, the values in brackets represent the actual number of data sessions in each category. It is observed that the amount of samples in some categories is scarce, so that the accuracy of the evaluation can be affected, especially for UL Int. The reason why Normal is not the dominant category is that cells with a low number of carried data sessions have been filtered. Due to the low number of served UEs, this group of cells are not critical from a troubleshooting perspective. Moreover, most of these cells belong to the Normal category. Regarding the methods, it is noted that using the classification tree with cell-level KPIs leads to a worse diagnosis, especially for coverage and interference issues. One reason is that the performance of the classification tree is more sensible to threshold settings when applied to cell-level KPIs (instead of call traces), since only one sample per cell is used. Another reason is that, if the proportion of problematic UEs in the cell is small, the abnormal values are hidden when the average over all the UEs in the cells is calculated. This idea can also be applied to the 'Max' method, since in some cases the identification of faults is hindered by the dominance of 'Normal' class, having the maximum percentage of UEs. Therefore, it can be concluded that the proposed system, based on a classification tree and a correlation-based approach, provides more accurate diagnosis to troubleshooting experts than the baselines. In addition, under the SON framework, an improved DSR will increase the effectiveness of the subsequent stages in the process, such as the fault compensation.

To evaluate the capability of detection of the proposed method, the false positive and false negative rates have been calculated. In particular, Fig. 11 shows this information together with the number (in brackets) of data sessions that are classified as false positives or false negatives. As observed, the method based on a classification tree and a correlation-based approach provides a better trade-off between false positives and false negatives, because the number of cases is similar in both cases (i.e., 17 and 11). In addition, the number of false negatives is much lower than in other methods.

The gain in accuracy achieved by the proposed method is at the cost of increasing the operational complexity due to the management of a larger amount of information,

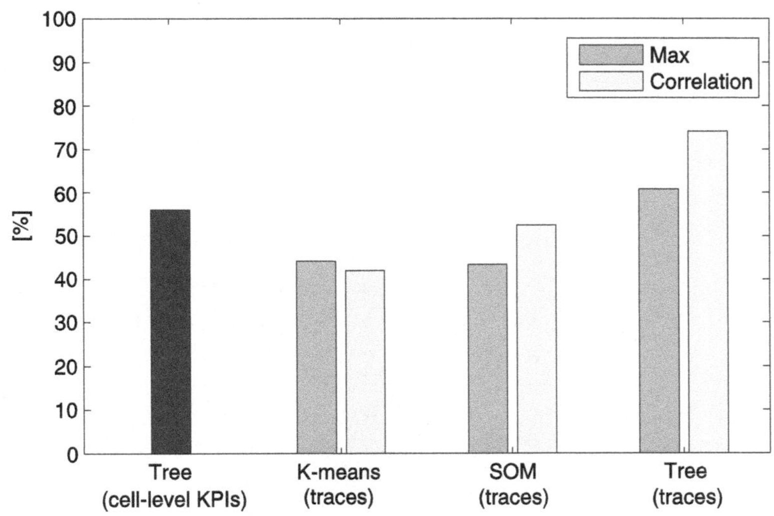

Fig. 9 Global diagnosis success rate for cell diagnosis methods

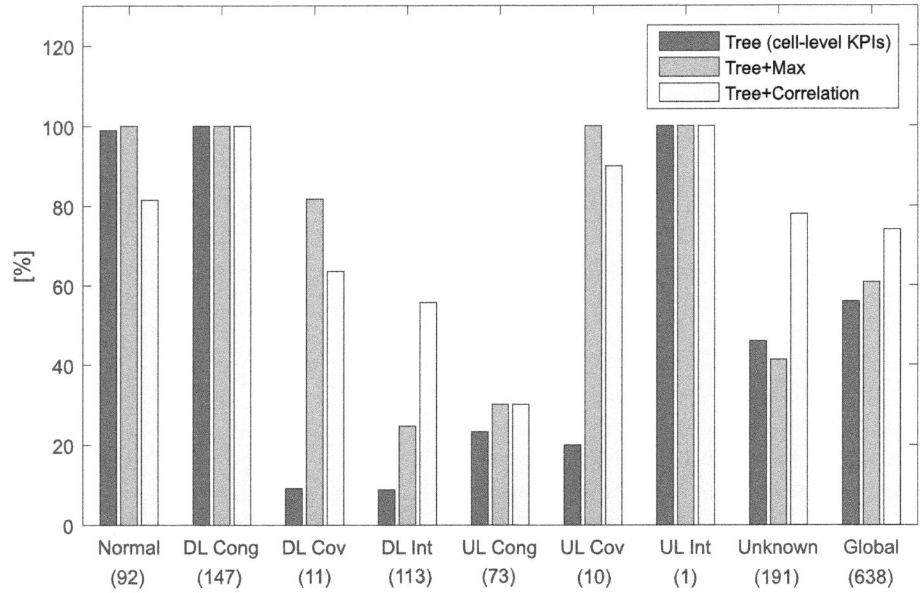

Fig. 10 Diagnosis success rate per class for cell diagnosis methods

given by the call traces. To estimate the computational cost, let C_{tree} be the cost of executing the classification tree. Its computation only requires elementary arithmetical operations, thus the complexity is $O(n)$. Let C_{profile} be the cost of generating the cell profile, whose complexity is also $O(n)$, since it mainly comprises data aggregation. The RCPs are not needed to be generated every time the cells are evaluated for diagnosis, e.g. they can be obtained from past evaluations in the same network. Thus, the computational cost of applying a clustering technique to find RCPs is not considered in this estimation. Let C_{corr} be the cost of calculating the Pearson's correlation coefficient for two data arrays. In this case, the computational cost is expected to be higher than C_{tree} due to the presence of the square root, whose complexity is $O(M(n))$ if the Newton's method is used. Lastly, let N_{cell} be the number of cells in the dataset and N_{session} be the average number of data sessions per cell. Then, the complexity of the proposed method is estimated as $N_{\text{cell}} \cdot (N_{\text{session}} \times C_{\text{tree}} + C_{\text{profile}} + C_{\text{corr}})$, while in the case of using traditional KPIs (i.e. cell-level measurements), the estimated complexity is $N_{\text{cell}} \times C_{\text{tree}}$. Although there are substantial differences in the computational cost, however, with the current solutions of Big Data and Edge Computing, the required complexity is not an issue. In addition, the temporal restrictions associated to cell diagnosis are not the same as those applied to other network management functions such as packet scheduling at the link layer, which typically are more critical. In particular, the time to collect cell traces, which is defined by the Report Output Period (ROP), in general is much larger than the required time to execute the diagnosis method. The ROP is typically 15 min, while the proposed classification tree is executed in the order of milliseconds to seconds, depending on the number of cells. Thus, the rate at which the diagnosis system is fed with measurements is much slower than the required time to compute the diagnosis.

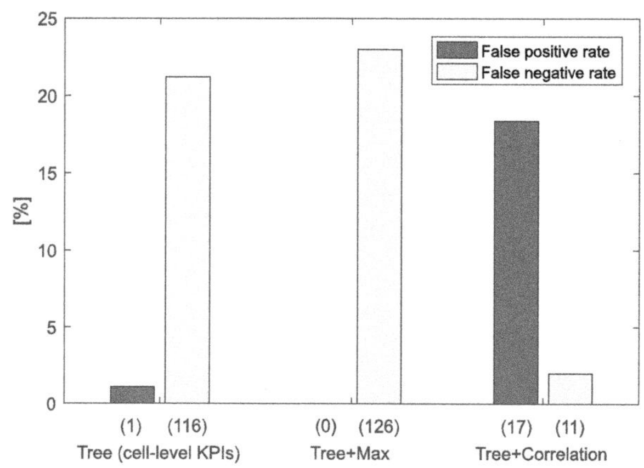

Fig. 11 False-positive and false-negative rates

6 Conclusions

Mobile operators have focused their attention in improving the user satisfaction for data services in their networks. In an IP world, a user connection experiencing low throughput can be given by many factors. The RAN is commonly the critical segment of the end-to-end path that affects the service integrity due to the limited amount of resources available in the radio interface. The correlation between the user throughput and the related radio conditions has been investigated in this paper with the aim of root-cause analysis. The study has been carried out through the use of large datasets collected from mature real networks.

More specifically, a diagnosis system has been developed to identify the most common radio causes that may impact user throughput. It takes advantage of the measurements that are collected per UE-basis as opposed to traditional counters and KPIs given on a cell-basis. The proposed system is based on a binary classification tree that is used to diagnose every data session in a cell. This tree allows the identification of data sessions suffering problems related to congestion, coverage or interference in either the downlink or the uplink from a set of metrics that have been carefully obtained considering some aspects such as the cell bandwidth and the data session duration. The diagnosis for each data session is then aggregated at the cell-level to provide a cell status (profile). After this, the set of cell profiles is then compared to several reference cell profiles in order to find anomalous cells. Such reference cell profiles has been previously generated by applying clustering techniques, e.g., the k-means. The comparison between cell profiles is made by calculating a correlation coefficient which is then used to determine the diagnosis of the cell.

Results show that the proposed system does not require a thorough adjustment of thresholds such as in other approaches, since the reference cell profiles automatically includes the particular characteristics of each network. Regarding the user diagnosis, the proposed method has been compared to other common clustering methods. It has been shown that the priority given by the proposed classification tree to the faults provides better accuracy in the diagnosis. With respect to the cell diagnosis, it has been shown that the performance of a small group of UEs experiencing poor radio conditions may be hidden by the dominant user performance in a cell. Due to this, the accuracy of the baselines is negatively affected. Conversely, the proposed method, by means of the generated RCPs, provides a better diagnosis for this kind of situations. Such an improvement leads troubleshooting experts to take appropriate recovery actions to solve the faults.

In addition, the proposed method can be applied to other radio access technologies since the thresholds for the metrics in the classification tree are calculated using percentiles. Considering that degradation is associated to extreme values of these metrics, the thresholds are automatically adapted to the radio access technology of the dataset. However, beyond the particular range of the metrics, it is important that the metrics are equivalent to those utilized in this paper in order to detect the considered faults. For example, to identify the problem of lack of coverage, a metric related to the signal strength should be used.

Further work is required to identify, in an automated manner, what values of throughput are considered to be abnormally low in a certain network based on other metrics. In particular, to calculate the threshold for the throughput metric, automatic learning techniques can be applied to the datasets, replacing the sensitivity analysis.

Acknowledgements
This work has been partially funded by Optimi-Ericsson, Junta de Andalucía (Agencia IDEA, Consejería de Ciencia, Innovación y Empresa, ref. 59288, and Proyecto de Investigación de Excelencia P12-TIC-2905) and ERDF.

Authors' contributions
All authors contributed extensively to the work presented in this paper. PM designed the proposed algorithm, performed the experiments, wrote the paper. RB supervised the design of the algorithm and the performance evaluation, wrote the paper. EC was in charge of the data collection and development of the supporting tools to process the data; he revised the paper. AG designed the reference algorithms, analyzed the results, wrote the paper. EJK designed the reference algorithms, analyzed the results, wrote the paper. NF formulated the problem, supervised the design of the algorithm, revised the paper. All authors read and approved the final manuscript.

Competing interests
The authors declare that they have no competing interests.

Author details
[1]Communications Engineering Dept., University of Málaga, Málaga, Spain.
[2]Ericsson, 29590 Campanillas, Málaga, Spain.

References
1. Cisco, Visual networking index: global mobile data traffic forecast update. Tech. Rep. Cisco, 2014–2019 (2015)
2. 3GPP, Self-organizing networks (SON): concepts and requirements, version 12.1.0, TS 32.500 (2014–12)
3. 3GPP, Evolved Universal Terrestrial Radio Access (E-UTRA); Physical layer; Measurements, version 12.2.0, TS 36.214 (2015–03)
4. 3GPP, Key Performance Indicators (KPI) for Evolved Universal Terrestrial Radio Access Network (E-UTRAN): definitions, version 12.0.0, TS 32.450 (2014–10)
5. J Laiho, K Raivio, P Lehtimaki, K Hatonen, O Simula, Advanced analysis methods for 3G cellular networks. IEEE Trans. Wirel. Commun. **4**(3), 930–942 (2005)
6. Y Zhang, A Årvidsson, in *Proc.of 2012 ACM SIGCOMM Workshop on Cellular Networks: Operations, Challenges, and Future Design*. Understanding the characteristics of cellular data traffic (ACM, Helsinki, 2012), pp. 13–18
7. EJ Khatib, R Barco, I Serrano, P Muñoz, in *Proc. of IEEE GLOBECOM 2014*. LTE performance data reduction for knowledge acquisition (IEEE, Austin, 2014), pp. 270–274
8. A Gómez-Andrades, P Muñoz, I Serrano, R Barco, Automatic root cause analysis for LTE networks based on unsupervised techniques. IEEE Trans. Veh. Technol. **65**(4), 2369–2386 (2015)
9. 3GPP, Telecommunication management; Subscriber and equipment trace; Trace concepts and requirements, version 13.0.0, TS 32.421. (2015-06)
10. 3GPP, Universal terrestrial radio access (UTRA) and evolved universal terrestrial radio access (E-UTRA); Radio measurement collection for minimization of drive tests (MDT); Overall description; Stage 2, version 12.2.0, TS 37.320 (2014-09)
11. LC Schmelz, et al., in *Wireless World Research Forum Meeting 20*. Self-configuration, -optimisation and -healing in wireless networks (WWRF, Stockholm, 2008)

12. R Barco, P Lázaro, V Wille, L Díez, S Patel, Knowledge acquisition for diagnosis model in wireless networks. Expert Syst. Appl. **36**(3), 4745–4752 (2009)
13. R Barco, P Lázaro, P Muñoz, A unified framework for self-healing in wireless networks. IEEE Commun. Mag. **50**(12), 134–142 (2012)
14. GF Ciocarlie, CC Cheng, C Connolly, U Lindqvist, S Nováczki, H Sanneck, M Naseer-ul-Islam, in *Proc. of 11th International Symposium on Wireless Communications Systems (ISWCS)*. Managing scope changes for cellular network-level anomaly detection (IEEE, Barcelona, 2014), pp. 375–379
15. J Johansson, WA Hapsari, S Kelley, G Bodog, Minimization of drive tests in 3GPP release 11. IEEE Commun. Mag. **50**(11), 36–43 (2012)
16. J Puttonen, J Turkka, O Alanen, J Kurjenniemi, in *Proc.of IEEE 21st International Symposium on Personal Indoor and Mobile Radio Communications (PIMRC)*. Coverage optimization for minimization of drive tests in LTE with extended RLF reporting (IEEE, Instanbul, 2010), pp. 1764–1768
17. A Galindo-Serrano, B Sayrac, S Ben Jemaa, J Riihijarvi, P Mahonen, in *Proc. of 8th International Conference on Cognitive Radio Oriented Wireless Networks (CROWNCOM)*. Harvesting MDT data: Radio environment maps for coverage analysis in cellular networks (IEEE, Washington, 2013), pp. 37–42
18. RU Mondal, T Ristaniemi, J Turkka, in *Proc. of International Conference on Localization and GNSS (ICL-GNSS)*. Genetic algorithm optimized grid-based RF fingerprint positioning in heterogeneous small cell networks (IEEE, Gothenburg, 2015)
19. PC Lin, in *Proc. of IEEE 3rd Global Conference on Consumer Electronics (GCCE)*. Minimization of drive tests using measurement reports from user equipment (IEEE, Tokyo, 2014), pp. 84–85
20. F Chernogorov, J Puttonen, in *Proc.of IEEE 24th International Symposium on Personal Indoor and Mobile Radio Communications (PIMRC)*. User satisfaction classification for Minimization of Drive Tests QoS verification (IEEE, London, 2013), pp. 2165–2169
21. F Chernogorov, T Nihtilä, in *Proc.of IEEE 75th Vehicular Technology Conference (VTC Spring)*. QoS verification for minimization of drive tests in LTE networks (IEEE, Yokohama, 2012)
22. MP Wylie-Green, T Svensson, in *Proc. of IEEE Global Telecommunications Conference (GLOBECOM 2010)*. Throughput, capacity, handover and latency performance in a 3GPP LTE FDD field trial (IEEE, Miami, 2010)
23. B McWilliams, Y Le Pezennec, G Collins, in *Proc.of XVth International Telecommunications Network Strategy and Planning Symposium (NETWORKS)*. HSPA+ (2100 MHz) vs LTE (2600 MHz) spectral efficiency and latency comparison (IEEE, Rome, 2012)
24. L Zhang, T Okamawari, T Fujii, in *Proc.of IEEE 75th Vehicular Technology Conference (VTC Spring)*. Performance evaluation of end-to-end communication quality of LTE (IEEE, Yokohama, 2012)
25. V Buenestado, JM Ruiz-Avilés, M Toril, S Luna-Ramírez, A Mendo, Analysis of throughput performance statistics for benchmarking LTE networks. IEEE Commun. Lett. **18**(9), 1607–1610 (2014)
26. A Imran, A Zoha, Challenges in 5G: how to empower SON with big data for enabling 5G. IEEE Netw. **28**(6), 27–33 (2014)
27. N Baldo, L Giupponi, J Mangues-Bafalluy, in *Proc.of 20th European Wireless Conference*. Big data empowered self organized networks (VDE, Barcelona, 2014)
28. EJ Khatib, R Barco, P Muñoz, I de la Bandera, I Serrano, Self-healing in mobile networks with big data. IEEE Commun. Mag. **54**(1), 114–120 (2016)
29. P Szilagy, S Novaczki, An automatic detection and diagnosis framework for mobile communication systems. IEEE Trans. Netw. Serv. Manag. **9**(2), 184–197 (2012)
30. R Barco, V Wille, L Díez, M Toril, Learning of model parameters for fault diagnosis in wireless networks. Wirel. Netw. **16**(1), 255–271 (2010)
31. L Bennacer, L Ciavaglia, A Chibani, Y Amirat, A Mellouk, in *Proc.of IEEE Network Operations and Management Symposium (NOMS)*. Optimization of fault diagnosis based on the combination of Bayesian Networks and Case-Based Reasoning (IEEE, Maui, 2012), pp. 619–622
32. EJ Khatib, R Barco, A Gómez-Andrades, I Serrano, Diagnosis based on genetic fuzzy algorithms for LTE self-healing. IEEE Trans. Veh. Technol. **65**(3), 1639–1651 (2016)
33. Y Zhang, N Liu, Z Pan, T Deng, X You, in *Proc.of IEEE/CIC International Conference on Communications in China (ICCC)*. A fault detection model for mobile communication systems based on linear prediction (IEEE, Shanghai, 2014), pp. 703–708
34. P Muñoz, R Barco, I Serrano, A Gómez-Andrades, Correlation-based time-series analysis for cell degradation detection in SON. IEEE Commun. Lett. **20**(2), 396–399 (2016)
35. B Cheung, G Kumar, S Rao, Statistical algorithms in fault detection and prediction: toward a healthier network. Bell Labs Tech. J. **9**(4), 171–185 (2005)
36. N Samaan, A Karmouch, Network anomaly diagnosis via statistical analysis and evidential reasoning. IEEE Transactions on Network and Service Management. **5**(2), 65–77 (2008)
37. A Gómez-Andrades, R Barco, I Serrano, P Delgado, P Caro-Oliver, P Muñoz, Automatic root cause analysis based on traces for LTE self-organising networks. IEEE Wirel. Commun. Mag. **23**(3), 20–28 (2016)
38. A Gómez-Andrades, R Barco, P Muñoz, I Serrano, Data analytics for diagnosing the RF condition in Self-Organising Networks. IEEE Trans. Mob. Comput. In press, (2016)
39. J Turkka, F Chernogorov, K Brigatti, T Ristaniemi, J Lempiäinen, An approach for network outage detection from drive-testing databases. J. Comput. Netw. Commun. **2012**, 1–13 (2012)
40. A Zoha, A Saeed, A Imran, M Imran, A Abu-Dayya, in *Proc.of 11th International Conference on the Design of Reliable Communication Networks (DRCN)*. Data-driven analytics for automated cell outage detection in Self-Organizing Networks (IEEE, Kansas City, 2015), pp. 203–210
41. O Onireti, A Zoha, J Moysen, A Imran, L Giupponi, M Imran, A Abu Dayya, A cell outage management framework for dense heterogeneous networks. IEEE Trans. Veh. Technol. **65**(4), 2097–2113 (2015)
42. H Holma, A Toskala, *LTE for UMTS: Evolution to LTE-Advanced*. (John Wiley & Sons Ltd, The Atrium, Southern Gate, Chichester, 2011)
43. K Aho, O Alanen, J Kaikkonen, in *Proc.of 10th International Conference on Networks (ICN)*. CQI reporting imperfections and their consequences in LTE networks (IARIA, St. Maarten, 2011), pp. 241–245
44. 3GPP, Evolved Universal Terrestrial Radio Access (E-UTRA); Layer 2 - Measurements, version 12.0.0, TS 36.314 (2014–09)
45. R Khanafer, B Solana, J Triola, R Barco, L Moltsen, Z Altman, P Lazaro, Automated diagnosis for UMTS networks using Bayesian network approach. IEEE Trans. Veh. Technol. **57**(4), 2451–2461 (2008)
46. M Wazid, A Das, An efficient hybrid anomaly detection scheme using K-means clustering for wireless sensor networks. Wirel. Pers. Commun. **90**(4), 1971–2000 (2016)

11

Performance analysis for an enhanced architecture of IoV via Content-Centric Networking

Zhuo Li[1,2,3*], Yutong Chen[4], Deliang Liu[5] and Xiang Li[6]

Abstract

TCP/IP protocol gradually exposes many shortcomings such as poor scalability and mobility. Content-Centric Networking is a new architecture which cares about the content itself rather than its source. Therefore, this paper proposes a novel IoV architecture which based on Content-Centric Networking and tests its transmission interference time, transmission delay, and throughout in network layer. The experimental results show that the novel architecture is superior to the current IoV in the communication performance.

Keywords: Content-Centric Networking, Internet of Things, Internet of Vehicles, Performance

1 Introduction

As a key member of the Internet of Things (IoT), Internet of Vehicles (IoV) will greatly change the future life [1]. IoV turns each participating vehicle into a mobile node and make vehicles connect to each other; hence, it creates a network with a wide range. The basic types of IoV communications could be divided into Vehicle to Vehicle (V2V) communications and Vehicle to Road (V2R) communications [2]. With On-board Unit (OBU) and Roadside Unit (RSU) as network nodes, IoV enables vehicles to transmit data in a single hop or multi-hop [3]. OBU is used as the network terminal node to transmit information toward other OBU and RSU. RSU which deploys along the road is the fixed node in V2R communications [4]. OBU links with local Internet service provider through RSU, so it can visit remote Internet server. Therefore, RSU plays a role of gateway in IoV.

To achieve many outstanding functions such as intelligent traffic management, intelligent vehicles control, and dynamic information service, IoV needs to meet following basic communication requirements [5]: (1) relatively low communication delay, (2) sufficient transmission distance, and (3) high reliability under high-speed mobile environment. Meanwhile, the network topology of IoV is complicated and changes rapidly, the time of duration of communications link is short, and the moving range of nodes is broad [6]. The current IoV which mainly adopts IEEE WAVE protocol stack, deploys TCP/IP protocol on transport layer and network layer to be compatible with IP network [7, 8]. Compared with other wireless communication technique, WAVE already has excellent communication performances such as low transmission delay (0.0002 s), long transmission distance (1000 m), and high transmission rate (27 Mbit/s) [9]. But because of basing on IP address, WAVE could not support the mobility well. Continuously, changes in the network topology will cause frequent reallocation of IP address and make current IoV architecture not applicable to communication scenarios which focus on the content itself rather than its provenance [10, 11]. So, it is urgent to find a revolutionary solution to build a new architecture of IoV which can response to the development better [12].

Therefore, a novel architecture of IoV named CCN-IoV was firstly proposed. Compared with IP network, CCN-IoV located the content chunks of CCN in network layer, fundamentally changes the traditional end-to-end transmission mode, and improves mobile communication performance of IoV. Then, the paper develops experimental measurements for the network layer to find transmission interference time, transmission delay, and throughout under different application

* Correspondence: zli@tju.edu.cn
[1]Tianjin Key Laboratory of Wireless Mobile Communications and Power Transmission, Tianjin Normal University, Tianjin 300387, China
[2]College of Electronic and Communication Engineering, Tianjin Normal University, Tianjin 300387, China
Full list of author information is available at the end of the article

scenarios. Experimental results show that CCN-IoV has better communication performance.

The other parts of the paper are as follows: the second section introduces CCN, the third section presents CCN-IoV in detail, the forth section evaluates the performance of CCN-IoV, and the last section is the conclusion.

2 Content-Centric Networking

Aiming at the deficiency of IP network, Van Jacobson put forward Content-Centric Networking (CCN) in 2009. Different from the IP network, CCN is content-oriented, no longer focuses on interface address or host location information [13]. So, CCN fundamentally changes the IP package structure as well as addressing mode. What is more, CCN has characteristics of short transmission delay, low power consumption, and high reliability, so it can fully respond to the future development of mobile Internet [14].

Compared with IP network, CCN exchanges Interest and Data for communication which is driven by consumers [15]. Interest from customers and Data from producer are the two types of CCN packets, both of them identify the content by name which is opaque to the network. Inspired by the structure of URLs, CCN uses a hierarchical data structure for name [16]. And the name of content is divided into a number of components which represent the context and relationships of data elements. Compared with the IP address, name has stronger expressive ability due to the semantic meaning and improves the efficiency of routing and forwarding.

Furthermore, CCN uses the adaptive forwarding plane to guide the routing and forwarding of packets [17]. An adaptive forwarding strategy can send Interests along optimum paths to avoid congestion and failures, balance payload, detect, and react to attacks such as prefix hijacking and Distributed Denial of Service (DDoS). On the adaptive forwarding plane, each NDN router maintains three data structures: a Pending Interest Table (PIT), a Forwarding Information Base (FIB) [18, 19], and a Content Store (CS). PIT stores all the Interests that a router has forwarded but not satisfied yet. FIB is similar to the routing table in IP network, which provides the path information for forwarding. However, CCN allows the simultaneous use of multiple interfaces to forward the same Interest and the IP network can only use one interface. What is more, CS is a temporary cache of Data which received recently, so it is convenient to satisfy future Interests and save the bandwidth of network.

CCN has advantages of balancing network payload, reducing network traffic, and improving communication reliability. Applying CCN to IoV can meet needs of IoV development for big data, high data rate, and low latency communication. At present, there are few researches focus on combining CCN with IoV, and the research on the performance test of the novel architecture has not been reported.

3 CCN-IoV

In this section, combined with IoV characteristics and demands, we introduce a novel architecture of IoV which based on CCN in detail and describe the communication process of CCN-IoV.

The architecture of CCN-IoV is divided into five parts: physical layer, data link layer, network layer, perception layer, and application layer, among which perception layer is composed with various kinds of sensors and gateway, so it could collect information of vehicles, traffic state, and road environment. Network layer seems to be the neural center and brain of IoV, it uses content chunks of CCN to transmit and handle information that acquired from perception layer. Application layer will support different applications which pay attention to what the content is, such as accident handling and dynamic traffic guidance.

Network layer of IoV based on CCN exchanges Interest and Data to communicate according to names. Compared with traditional IP network architecture, CCN-IoV could support in-network caching and multi-cast natively without any additional protocols or mechanisms. The communication process is shown as Fig. 1: vehicle A packs name into Interest and broadcasts it toward potential producers. In the course of broadcasting,

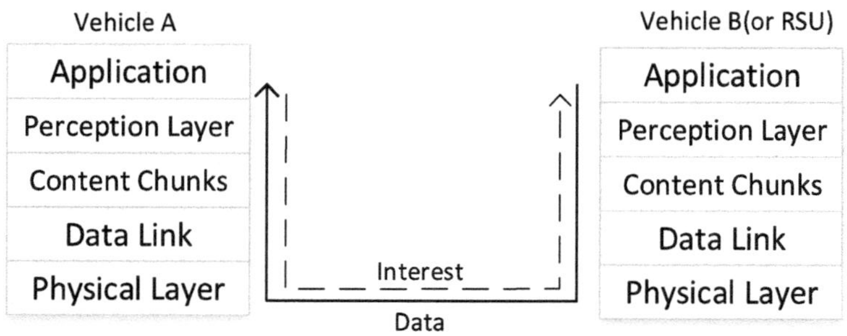

Fig. 1 The communication of CCN-IoV

when a middle node of the route receives the Interest and has data matching it, the Data will be returned to vehicle A using the Interest path in reverse. Otherwise, vehicle A needs to get response from the server of CCN-IoV. Then, the Data will be cached in the middle nodes on the path according to the name so that they can reply following Interests which request the same data. Therefore, CCN-IoV will greatly reduce the post-back pressure of network and improve the efficiency of flow management. So, it is quite fit for the situation which topology changes frequently and supports content-oriented applications.

In conclusion, CCN-IoV can realize the high reliability and low delay of data transmission. Compared with the current IoV, CCN-IoV has more simplified structure and more efficient communication process.

4 Evaluation and discussion

CCN-IoV uses content chucks of CCN in network layer, so, in this section, we mainly evaluated the performance of network layer by measuring transmission interference time, transmission delay, and throughout. In particular, there are few researches on the transmission interference time of IoV, so we refer to the parameter of high-speed railway mobile communication system GSM-R [20]. Results show that CCN-IoV has more excellent communication performance than the current IoV.

4.1 Experimental setup

Experiments are implemented in C++ and Python languages, tested on a PC with an Inter Corei5-3470 CPU of 3.20 GHz and DDR3 SDRAM of 4GB. We use the CCN simulator called ndnSIM 2.0 [21] to simulate and experiment. To make the same underlying transport environment as IP network, we configure and modify the interface module of physical layer and data link layer in ndnSIM 2.0 [22], so the experimental results are more comparable and persuasive.

By using ndn::AppDelayTracer of ndnSIM 2.0, we can record the transmission status of packets in a text file, so it is easy to analyze the experimental results in various communication environments. The experimental topology consists of 10 nodes is shown in Fig. 2.

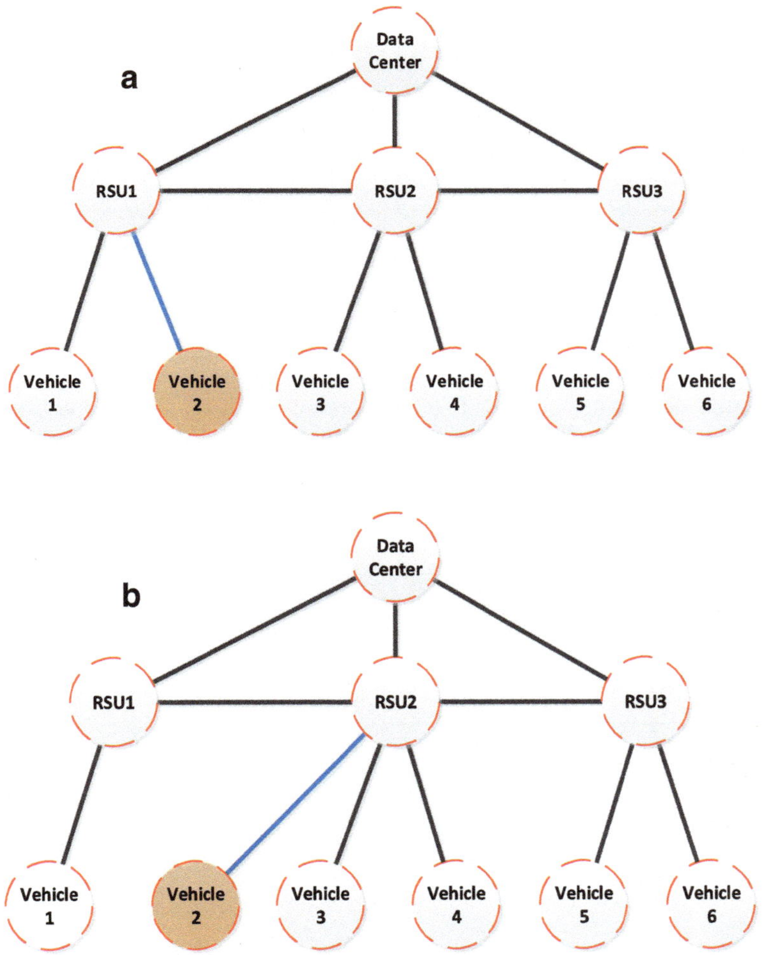

Fig. 2 Topology of simulations: **a** topology before communication link interrupt and **b** topology after communication link reconstruction

According to the current demands for communication, we test transmission interference time, transmission delay, and throughout under different application scenarios. Packets are routed and forwarded based on the best route strategy. The network bandwidth is set to be 10 MHz, costumer sends 200 Interests per second. By simulating the communication process that vehicles exchange 100 bytes Data, the communication performance of V2V are tested. In the same way, by simulating the process that RSU releasing 340 bytes Data to vehicles, we can estimate the communication performance of V2R. Meanwhile, in order to evaluate the mobility of CCN-IoV, we let vehicles move rapidly and then respectively simulate conditions that vehicle interrupts connection with surrounding vehicle and RSU.

4.2 Transmission interference time

The set maximum moving speed of vehicles is 180 Km/h, the distance between two base stations is 500 m, and the initial distance between vehicles is 250 m. Vehicles and RSUs have the same communication range which is 250 m. Because of referring to the transmission interference time of GSM-R, the switching time between two base stations is set to be 300 m. To test the mobility, vehicle 5 is set to be disconnected with RSU 1 at 1.000000 s and connected with RSU 2 at 1.300000 s. Besides, the transmission interference time of CCN is related to the network bandwidth as well as the amount of Interests sent by customer per second, but it has nothing to do with the packet payload. Therefore, whether in V2V or V2R communications, CCN-IoV has the same transmission interference time. In this paper, we only analyze the transmission interference time in the V2R communications, and vehicle 5 requests data to RSU 2.

Before the communication link broke, vehicle 5 broadcast Interests toward RSU 2, then Interests passed through RSU 1 and arrived at RSU 2 finally. When RSU 2 received Interests, it packed corresponding content in Data and sent it to vehicle 5. The experiment results show in Fig. 3. The serial numbers that vehicle 5 received changed linearly before 1.000000 s, which proves that the data transmission is normal. Vehicle 5 received no.199 Data at 0.995135 s and continued sending Interests to RSU 2 in order from 0.995135 to 1.000000 s. Because moving rapidly, vehicle 5 disconnected with RSU 1 at 1.000000 s, Data and Interest cannot be transferred through the communication link. But because of in-network caching, Data that responded by RSU 2 would be cached in the RSU 1 and RSU 2. When vehicle 5 connected with RSU 2 at 1.300000 s, it started to receive no.259 Data. Later, the serial numbers of Data that vehicle 5 received increases linearly until no.268 Data is received at 1.345140 s. Owing to not receiving Data that vehicle 5 requested during 0.995135s~1.000000 s, vehicle 5 would retransmit expired Interests. Then, vehicle 5 may receive Data that both cached in the network routers, and RSU 2 responded latest during 1.300000~1.690140 s.

In conclusion, after the reconstruction of the communication link, the Data transmission process (from 1.300000 to 1.690140 s) is disturbed. At last, the communication process recovered at 1.690140 s. Therefore, the transmission interference time of CCN-IoV is 390 ms and can meet future demands of mobility well.

4.3 Transmission delay

4.3.1 Transmission delay in the V2V communications

Vehicle 5 is set to request data from vehicle 6. Before the communication link broke, vehicle 5 broadcasts

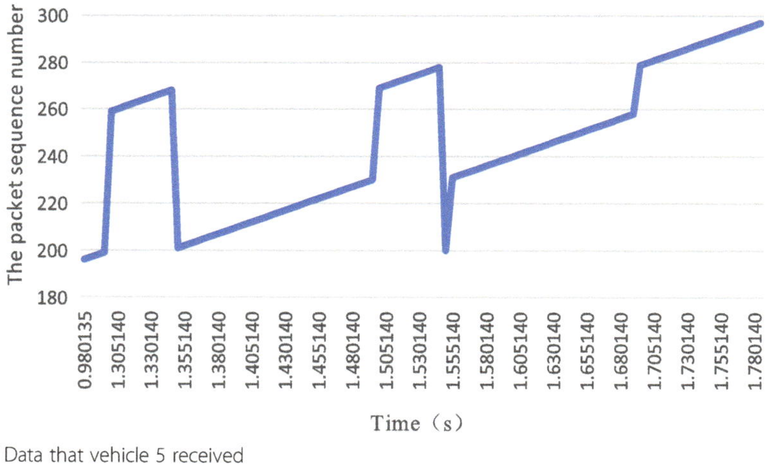

Fig. 3 Sequence number of Data that vehicle 5 received

Table 1 Transmission delay in the V2V communications

Time	Node	AppId	Seq.	Type	Delay (ms)	RetxCount	HopCount
1.48008	5	257	227	LastDelay	0.076264	1	4
1.48008	5	257	227	FullDelay	345.076	2	4
1.48508	5	257	228	LastDelay	0.076264	1	4
1.48508	5	257	228	FullDelay	345.076	2	4
1.49008	5	257	229	LastDelay	0.076264	1	4
1.49008	5	257	229	FullDelay	345.076	2	4
1.49508	5	257	230	LastDelay	0.076264	1	4
1.49508	5	257	230	FullDelay	345.076	2	4
1.50008	5	257	269	LastDelay	0.077116	1	4
1.50008	5	257	269	FullDelay	0.077116	1	4

Interests toward vehicle 6, Interests passed through RSU 1、RSU 2 then arrived at vehicle 6. Then, vehicle 6 returned Data which payload is 100 bytes to vehicle 5. The experimental results are shown in Table 1, when the communication link was not interrupted, the transmission delay of network layer is less than 0. 076264 ms. But vehicle 5 disconnected with RSU 1 at 1.000000 s, and the Data that vehicle 6 responded subsequently would be cached in RSU 1 and RSU 2 due to in-network caching. So, when vehicle 5 connected to RSU 2 at 1.300000 s, the transmission delay of Data no.200 to no.258 requested before needs less than 0.071160 ms. Hence, in the V2V, the transmission delay of CCN-IoV is less than 0.076264 ms.

4.3.2 Transmission delay in the V2R communications

Vehicle 5 is set to request data from RSU 2. Before the communication link broke, vehicle 5 broadcasts Interests toward RSU 2, Interests passed through RSU 1, then arrived at RSU 2 finally. After that, RSU 2 returned Data which payload is 340 bytes to vehicle 5. The experimental results are shown in Table 2, when the communication link was not interrupted, the transmission delay of network layer is less than 0.135918 ms. But vehicle 5 disconnected with RSU 2 at 1.000000 s, and the Data that RSU 2 responded subsequently would be cached in RSU 1 and RSU 2 due to the in-network caching. So, when vehicle 5 connected to RSU 2 at 1.300000 s, the transmission delay of Data no.200 to no.258 which requested before needs less than 0.135276 ms. Hence, in the V2V, the transmission delay of CCN-IoV is less than 0.135918 ms.

In conclusion, whether in the V2V or V2R communications, the transmission delay of network layer is less than 0.200000 ms which WAVE requires. From this point of view, CCN-IoV is better than the current IoV.

Table 2 Transmission delay in the V2R communications

Time(s)	Node	AppId	Seq.	Type	Delay (ms)	RetxCount	HopCount
1.48014	5	257	227	LastDelay	0.135276	1	3
1.48014	5	257	227	FullDelay	345.135	2	3
1.48514	5	257	228	LastDelay	0.135276	1	3
1.48514	5	257	228	FullDelay	345.135	2	3
1.49014	5	257	229	LastDelay	0.135276	1	3
1.49014	5	257	229	FullDelay	345.135	2	3
1.49514	5	257	230	LastDelay	0.135276	1	3
1.49514	5	257	230	FullDelay	345.135	2	3
1.50014	5	257	269	LastDelay	0.135918	1	3
1.50014	5	257	269	FullDelay	0.135918	1	3
1.50514	5	257	270	LastDelay	0.135918	1	3
1.50514	5	257	270	FullDelay	0.135918	1	3
1.51014	5	257	271	LastDelay	0.135918	1	3
1.51014	5	257	271	FullDelay	0.135918	1	3

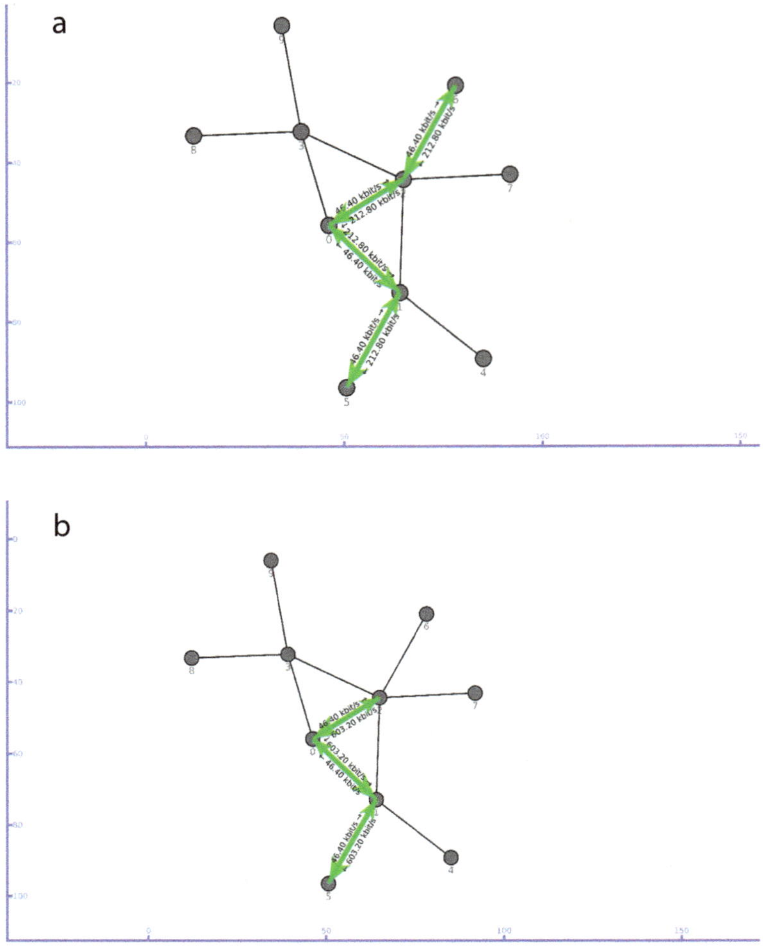

Fig. 4 Throughout of CCN-IoV: **a** throughout in the V2V communications and **b** throughout in the V2R communications

4.4 Throughout

Throughout is defined as the number of data that nodes successfully transmit in unit time. Experimental results are shown in Fig. 4, throughout is 212.8 kbit/s in the V2V communications and 603.2 kbit/s in the V2R communications. Throughout is related to the payload of Data. Compared with 100~300 kbps which is demanded for IoV based on WAVE, the network layer of CCN-IoV has more reliable communication performance.

5 Conclusions

In view of the rapid development of IoV and disadvantages of IP network, this paper proposed a novel IoV architecture named CCN-IoV which deploys content chunks of CCN in the network layer. The paper also tested and evaluated the performance of CCN-IoV, experimental results show that the communication performance is better than the current IoV.

Acknowledgements

This research was supported by the National Natural Science Foundation of China (61602346).

Authors' contributions

ZL conceived this study, performed the simulations, and wrote the manuscript. YC, DL, and XL participated in the co-ordination of the study, data monitoring, and drafted the manuscript. All authors read and approved the final manuscript.

Competing interests

The authors declare that they have no competing interests.

Author details

[1]Tianjin Key Laboratory of Wireless Mobile Communications and Power Transmission, Tianjin Normal University, Tianjin 300387, China. [2]College of Electronic and Communication Engineering, Tianjin Normal University, Tianjin 300387, China. [3]Department of Computer Science, University of Arizona, Tucson, AZ 85721, USA. [4]School of Microelectronics, Tianjin University, Tianjin 300072, China. [5]Shijiazhuang Mechanical Engineering College, Shijiazhuang 050003, China. [6]College of computer and Information Engineering, Tianjin Normal University, Tianjin 300387, China.

References

1. F Yang, S Wang, J Li et al., Wireless communication over ZigBee for automotive inclination measurement. China Commun. **11**(10), 1–15 (2014)
2. KM Alam, M Saini, AE Saddik, Toward social internet of vehicles: concept, architecture, and applications. Access IEEE **3**, 343–357 (2015)
3. BC Cheng, CT Chou, CH Chang et al., Survey on networking for internet of vehicles. J. Internet Technol. **14**(7), 1007–1019 (2013)
4. A Abdrabou, W Zhuang, Probabilistic delay control and road side unit placement for vehicular ad hoc networks with disrupted connectivity. IEEE J. Sel. Areas Commun. **29**(1), 113–120 (2011)
5. C Englund, L Chen, A Vinel et al., Future applications of VANETs, in *Vehicular ad hoc Networks*, 2015, pp. 525–544
6. HJ Tang, Research on system architecture and key technologies of IOV. Netw. Secur. Technol. Appl. (2013)
7. VD Khairnar, K Kotecha, Performance of vehicle-to-vehicle communication using IEEE 802.11p in vehicular ad-hoc network environment. Int. J. Netw. Secur. Appl. **5**(2), 78–83 (2013)
8. YL Morgan, Notes on DSRC & WAVE standards suite: its architecture, design, and characteristics. IEEE Commun. Surv. Tutorials **12**(4), 504–518 (2010)
9. S Eichler, Performance evaluation of the IEEE 802.11 p WAVE communications standard, in *IEEE 66th Vehicular Technology Conference*, 2007
10. W Lucas, Z Lixia, W Ryuji et al., Data naming in vehicle-to-vehicle communications, in *IEEE INFOCOM 2012 Workshop on Emerging Design Choices in Name-Oriented Networking*, 2012, pp. 382–333
11. C Jaeyoung et al., A survey on content-oriented networking for efficient content, in *IEEE Communications Magazine*, 2011, pp. 121–126
12. Q Karagiannis, O Altintas, E Ekici et al., Vehicular networking: a survey and tutorialon requirements, architectures, challenges, standards and solutions. IEEE Commun. Surv. Tutorials **13**(4), 584–616 (2011)
13. V Jacobson et al., *Networking Named Content. CoNEXT'09, New York*, 2009, pp. 1–12
14. L Zhang, Deborah Estrin, Jeffrey Burke, et al. Named data networking (NDN) project, *Technical Report NDN-0001*, 2009. http://www.named-data.net/
15. D Saxena, V Raychoudhury, N Suri, Named data networking: a survey. Comput. Sci. Rev. **19**(1), 15–55 (2016)
16. L Zhang, A Afanasyev, J Burke, Named data networking[J]. ACM SIGCOMM Comput. Commun. Rev. **44**(3), 66–73 (2014)
17. Y Kim, Y Kim, J Bi et al., Differentiated forwarding and caching in named-data networking. J. Netw. Comput. Appl. **60**, 155–169 (2015)
18. T Song et al., Scalable name-based packet forwarding: from millions to billions, in *Proceedings of the 2nd International Conference on Information-Centric Networking* (ACM, 2015), pp. 19–28. http://dl.acm.org/citation.cfm?id=2810166.
19. Z Li et al., Hybrid wireless networks with FIB-based named data networking. EURASIP J. Wirel. Commun. Netw. **1**(2017), 54 (2017)
20. SM Siour, MM Mousa, Analysis of modified high-speed GSM-R path loss model and handoff hysteresis margin of viaduct scenario, in *International Conference on Computational Intelligence, Communication Systems and Networks* (IEEE, 2015), pp. 219–223. http://ieeexplore.ieee.org/document/7311159/.
21. S Mastorakis, A Afanasyev, I Moiseenko, ndnSIM 2.0: a new version of the NDN simulator for NS-3. Technical Report NDN-0028 **1**(1), 1–8 (2015)
22. JA Fernandez, K Borries, L Cheng, V Kumar, D Stancil, F Bai, Performance of the 802.11p physical layer in vehicle-to-vehicle environments. IEEE Trans. Veh. Technol. **61**(1), 3–14 (2012)

12

Design of adaptive constellations and error protection coding for wireless network coding in 5-node butterfly networks

Pavel Prochazka[1], Tomas Uricar[1], David Halls[2*] and Jan Sykora[1]

Abstract

In this paper, we focus on the design of a reliable communication scheme for generic 5-node wireless butterfly networks (WBN). Since the WBN represents a natural extension of the basic two-way relay communication channel, we follow the paradigm of *Wireless Physical Layer Network Coding* and we show that the design of novel adaptive constellations is desirable to exploit the promising performance of wireless network coding in WBN. We introduce a *systematic constellation design* algorithm, and we show that the proposed constellations are capable of outperforming conventional point-to-point modulations in WBN over the whole range of channel signal to noise ratios. To further support the applicability of the proposed constellations in a practical system, we integrate a simple binary channel coding scheme, providing a robust *adaptive modulation and coding scheme* for relaying in symmetric WBNs with arbitrary channel conditions.

Keywords: Wireless network coding, Butterfly network, Relaying, Superposition coding, Superposition modulation

1 Introduction

The invention of Wireless Network Coding (WNC) has provided a means for a significant enhancement in wireless system performance [1–3]. However, WNC processing has also revealed several non-trivial research problems which do not have counterparts in conventional point-to-point or multi-user systems, like the sensitivity to channel parameterization [4–7] or challenging multi-source transmission synchronization [8–10]. Furthermore, the specific characteristics of multi-node WNC systems can call also for novel constellation design approaches (see e.g., [7, 11]).

In WNC, the transmitted information symbols are allowed to interact directly in the constellation space, inducing a specific set of requirements on the constellations' properties and characteristics (see e.g., [7, 11]). Generally, we can declare [11] that any constellation suitable for the specific nature of WNC processing should be able to forma *superimposed constellation* with an appropriate structure at all receiving nodes (allowing direct decoding of specific WNC functions of user data from the received non-orthogonal signals [12, 13]) and simultaneously it should be able to exploit the *broadcast nature* of wireless channels (allowing a reduction in the consumption of channel resources in the system [1]). Since both of these properties hinge on the actual SNR conditions in the system, the constellation design itself can become a relatively challenging task.

One of the first attempts to design a suitable multi-source constellation for communication in a multi-node WNC system was introduced in our previous work [11], where the *adaptive 2-source constellations* for WNC-based communication in a 5-node *wireless butterfly network* (WBN) were developed. However, the results presented in [11] are limited to an uncoded system, which significantly limits its practical applicability.

In this paper, we extend the communication scheme from [11] by appending appropriate *error protection coding* to the system, resulting in a practical channel-coded adaptive communication scheme for WBN. We provide the following results and contributions:

1. We summarize the *systematic constellation design algorithm* from [11] and we provide additional

*Correspondence: david.halls@toshiba-trel.com
[2]Toshiba—Telecommunications Research Laboratory, 32 Queen Square, Bristol, BS1 4ND, UK
Full list of author information is available at the end of the article

numerical results analyzing its promising performance.

2. We revise and extend the results discussing the *adaptive capabilities of the constellations,* including *the constellation mapping regions* for adaptive WBN relaying.
3. We validate the proposed design in a *real-world hardware (HW) setup,* including the HW evaluation of adaptive system performance.
4. We show how the *channel coding* can be integrated into the proposed constellation design, including the analysis of proper source transmission rates.
5. We evaluate numerically the performance of the resulting *adaptive modulation and coding* scheme for a wide range of channel conditions in the WBN.
6. We show that the proposed design is viable even if the real-world conditions induce some deviation from the assumed system model.
7. We compare the proposed design with state of the art reference scenarios to highlight its superior performance.

The rest of the paper is organized as follows. The system model, including all necessary definitions and details about WNC processing in WBNs, is summarized in Section 2. A systematic *constellation design algorithm* for WBNs is revised in Section 3. The performance analysis of the proposed constellations (including the HW evaluation) in an uncoded system is available in Section 4. Section 5 discusses an adaptive constellation design, and Section 6 introduces the channel coding extension to the system. The performance of the novel adaptive modulation and coding scheme is analyzed in Section 7, including a simple robustness analysis. Concluding remarks are presented in Section 8.

2 WNC relaying in wireless butterfly network

We assume a *symmetric* WBN model [14, 15] with a half-duplex constraint (Fig. 1). An example practical scenario complying with the WBN model is a communication among vehicles and road-side infrastructure nodes for traffic information exchange on a bidirectional highway within two time slots [16]. The direct channels between the intended source → destination pairs are not available in WBN and thus the help of the intermediate relay node is necessary to enable end-to-end communication. Note that some partial source material from the implementation of this model can be found in [17].

2.1 Relaying strategy

An information-theoretic relaying strategy (based on superposition coding (SC) [18–21]) for WBNs was proposed in [14, 15]. The fundamental idea of the SC approach is to split each data symbol $\mathbf{d}_i$ at source $i \in$

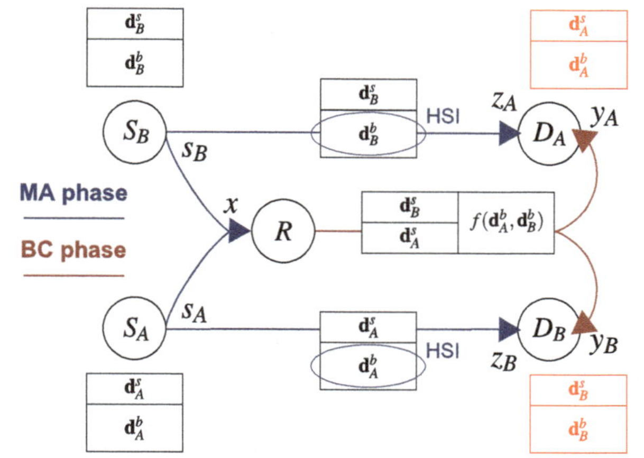

Fig. 1 Symmetric WBN model with half-duplex constraint and two-phase communication

$\{A, B\}$ into the *basic part* $\mathbf{d}_i^b = \left[d_{i,0}^b, \ldots, d_{i,N_b-1}^b\right]$ and the *superposed part* $\mathbf{d}_i^s = \left[d_{i,0}^s, \ldots, d_{i,N_s-1}^s\right]$, $d_{i,n}^b, d_{i,n}^s \in \{0,1\}$, and then process each of the resulting data streams separately throughout the system (see Fig. 1). While the basic part represents a "true WNC" information stream (limited by the amount of available HSI) in the system, the superposed part defines the "routed" information which is appended on top of the WNC stream to exploit any potential spare capacity of related channels in the system (see [14, 15] for more details). A simplified description of SC-based relaying in an uncoded WBN system is summarized in Table 1 and depicted in Fig. 2.

2.2 System model

For the sake of simplicity of the constellation design analysis, we provisionally assume that the system is uncoded[1]. The signal space representations of the transmitted channel symbols are $s_A = \mathscr{A}^A(\mathbf{d}_A)$, $s_B = \mathscr{A}^B(\mathbf{d}_B)$, where $\mathscr{A}^i(.)$ is the memoryless constellation mapper and $\mathbf{d}_i$ is the source $i \in \{A, B\}$ data symbol. Each communication round can be divided into two phases (Fig. 1). In the first phase (Multiple Access—MA phase) both sources S_A, S_B simultaneously transmit their signal to the relay, which receives a noisy superimposed constellation:

$$x = h_{AR}s_A + h_{BR}s_B + w_R. \tag{5}$$

Additionally, the source S_i ($i \in \{A, B\}$) transmission is overheard (at no cost) by its "unintended" destination D_j ($j \in \{A, B\}$, $j \neq i$) as:

$$z_j = h_{ij}s_i + w_j \tag{6}$$

Even though the overheard information (6) does not carry the desired data for the respective destination, it can be efficiently exploited to enable WNC-based processing in WBNs [14]. We denote the specific overheard information as the *Hierarchical Side Information* (HSI) [15].

Table 1 SC-based relaying scheme in uncoded WBN

Step 1: Source transmission • Source $S_i, i \in \{A,B\}$ maps its data symbol $\mathbf{d}_i = [\mathbf{d}_i^b; \mathbf{d}_i^s]$ into a $(N_b + N_s)$-bit superposition modulation symbol $$s_i = \mathscr{A}^i\left(\left[\mathbf{d}_i^b; \mathbf{d}_i^s\right]\right), \quad (1)$$ • S_A, S_B simultaneously transmit their signals to the relay and unintended destinations (Fig. 1). • Relay receives $$x = h_{AR}\mathscr{A}^A\left(\left[\mathbf{d}_A^b; \mathbf{d}_A^s\right]\right) + h_{BR}\mathscr{A}^B\left(\left[\mathbf{d}_B^b; \mathbf{d}_B^s\right]\right) + w_R \quad (2)$$ and decodes $[\mathbf{d}_A^s; \mathbf{d}_B^s; f(\mathbf{d}_A^b, \mathbf{d}_B^b)]$, where f is a hierarchical WNC function [13]. • D_j receives $$z_j = h_{ij}\mathscr{A}^i\left(\left[\mathbf{d}_i^b; \mathbf{d}_i^s\right]\right) + w_j, \quad (3)$$ where $i, j \in \{A,B\}, i \neq j$ and stores the signal for further processing.
Step 2: Relay broadcast • Relay sends $N_R = 2N_s + N_b$-bit modulation symbol to both destinations: $$s_R = \mathscr{A}^R\left(\left[\mathbf{d}_A^s; \mathbf{d}_B^s; f(\mathbf{d}_A^b, \mathbf{d}_B^b)\right]\right). \quad (4)$$ • D_j $(i, j \in \{A,B\}, i \neq j)$ decodes: – $[\mathbf{d}_A^s; \mathbf{d}_B^s; f(\mathbf{d}_A^b, \mathbf{d}_B^b)]$ (from the relay signal (4)) – $\mathbf{d}_i^b$ (from the stored signal z_j (3), after interference cancellation (IC) of known $\mathbf{d}_i^s$) – $\mathbf{d}_j^b$ (from $\mathbf{d}_i^b$ and $f(\mathbf{d}_A^b, \mathbf{d}_B^b)$, using a standard WNC decoding [13]) • D_j merges $\mathbf{d}_j^b$ with $\mathbf{d}_j^s$ to obtain the desired $\left[\mathbf{d}_j^b; \mathbf{d}_j^s\right]$.

The relay performs WNC-based decoding [13] of the received signal and broadcasts the decoded WNC (hierarchical) information as s_R to D_A, D_B in the second phase (Broadcast—BC phase). Both destinations are able to decode the desired data using the relay's (hierarchical) signal (5) and the overheard HSI (6) if a proper WNC processing is employed in the system.

Scalar complex channel coefficients h_{ij}, $i,j \in \{A,B,R\}$, $i \neq j$ are assumed to be constant during the communication round and known at the respective receiving node. Zero mean i.i.d. complex Gaussian noise sample w_i, $i \in \{A,B,R\}$ has variance σ_i^2. Signal to noise ratios (SNRs) in the system are given by $(\gamma_{\text{MAC}}, \gamma_{\text{HSI}}, \gamma_{\text{BC}})$, where γ_{MAC} is the SNR on the source $\rightarrow$ relay multiple-access channel—MAC $((S_A, S_B) \rightarrow R)$, γ_{HSI} is the SNR on the unintended source $\rightarrow$ destination (HSI) channels $(S_A \rightarrow D_B, S_B \rightarrow D_A)$ and γ_{BC} is the SNR on the relay $\rightarrow$ sources broadcast channel $(R \rightarrow (D_A, D_B))$.

3 Superposition modulation design

In this section, we summarize the design of constellation mappers $\mathscr{A}^A$, $\mathscr{A}^B$ as presented in [11]. The output constellation is created as

$$s = \mathscr{A}\left(\left[\mathbf{d}^b; \mathbf{d}^s\right]\right) = \sum_{n_s=0}^{N_s-1} L_{n_s}^s \left(2d_{n_s}^s - 1\right) + \sum_{n_b=0}^{N_b-1} L_{n_b}^b \left(2d_{n_b}^b - 1\right), \quad (7)$$

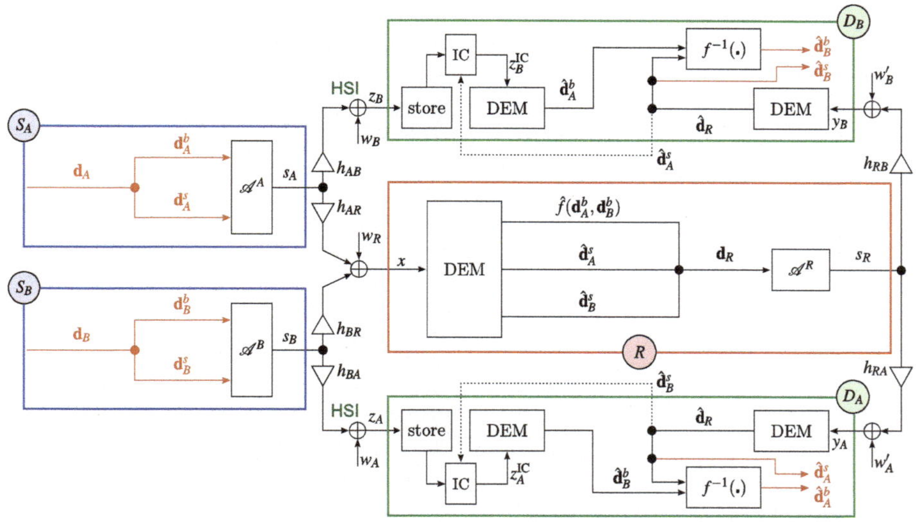

Fig. 2 Relaying scheme for the uncoded WBN system. DEM stands for a hard decision demodulator, IC is the interference canceler

where L_n^b, L_n^s are *Superposition Modulation* [22] *scaling coefficients.*

The coefficients L_n^b, L_n^s (7) are generally taken from the set of complex numbers. However, for the sake of simplicity, we limit the scaling coefficients L_n^b, L_n^s to purely imaginary or real numbers. Even though this simplified approach can potentially result in a sub-optimal performance in fading channels (complex channel coefficients h_{iR}, $i \in \{A, B\}$ can rotate the component superposition constellations), we follow this simplification to show the basic ideas of the proposed superposition modulation design. The extension to general complex L_n^b, L_n^s can naturally embrace the source phase pre-rotation [23–25], which virtually reverts the fading channel to the conventional additive white Gaussian noise (AWGN) channel with $\sphericalangle h_{ij} = 0$.

As shown in Table 1, the basic and superposed parts of the source information are processed in a different way through the system and so we will discuss the constellation design for these two specific information streams separately. We briefly introduce the incentives and main ideas of the design for the superposed and basic parts, and then we summarize the complete multi-source constellation design in a systematic algorithm.

3.1 Constellation design: basic part

A specific WNC function $f\left(\mathbf{d}_A^b, \mathbf{d}_B^b\right)$ has to be decoded by the relay (Table 1) and hence some overlapping of symbols in the superimposed constellation (observed by the relay node) can be allowed, resulting in improved Euclidean distance properties. More precisely, two (or more) pairs of basic information symbols $\left(\mathbf{d}_A^b, \mathbf{d}_B^b\right)$, $\left(\mathbf{d}'^b_A, \mathbf{d}'^b_B\right)$ can be allowed to fall into the same point in the superimposed constellation:

$$s_A\left(\mathbf{d}_A^b\right) + s_B\left(\mathbf{d}_B^b\right) = s'_A\left(\mathbf{d}'^b_A\right) + s'_B\left(\mathbf{d}'^b_B\right), \qquad (8)$$

if and only if they belong to the same WNC function output, i.e., iff $f\left(\mathbf{d}_A^b, \mathbf{d}_B^b\right) = f\left(\mathbf{d}'^b_A, \mathbf{d}'^b_B\right)$, where $f(.,.)$ is the hierarchical WNC function [13]. In this case the overlapping symbols produce the *same hierarchical output*, which can be successfully resolved by the destination, if matching HSI information is available (Table 1). The particular choice of the scaling coefficients $L_{A,n}^b$, $L_{B,n}^b$ generally depends on the particular choice of the hierarchical WNC function[2]. In the rest of this paper we discuss the constellation design for the *bit-wise exclusive-or* WNC function $\left(f\left(\mathbf{d}_A^b, \mathbf{d}_B^b\right) = \mathbf{d}_A^b \oplus \mathbf{d}_B^b\right.$, where $\oplus$ is the bit-wise XOR operation$\left.\right)$ [3].

Suitable overlapping of constellation symbols at the relay is provided by letting both sources transmit at the same power level, i.e. $L_{A,n}^b = L_{B,n}^b$ [11]. It can be easily shown that if these superposition modulation (7) coefficients are chosen as $L_{A,n}^b = L_{B,n}^b = 3^{\lfloor n/2 \rfloor}$ (for even n) and $L_{A,n}^b = L_{B,n}^b = \mathrm{j}\, 3^{\lfloor n/2 \rfloor}$ (for odd n), the received superimposed constellation at the relay will have overlaps only among those constellation symbols (8) which have the same hierarchical WNC output i.e. $f\left(\mathbf{d}_A^b, \mathbf{d}_B^b\right) = f\left(\mathbf{d}'^b_A, \mathbf{d}'^b_B\right) = \mathbf{d}_A^b \oplus \mathbf{d}_B^b = \mathbf{d}'^b_A \oplus \mathbf{d}'^b_B$.

3.2 Constellation design: superposed part

The superposed information given by $\mathbf{d}_A^s$, $\mathbf{d}_B^s$ has to be decoded separately by the relay (Table 1), and hence a source constellation avoiding all potential overlaps in the superimposed constellation (observed by the relay node) is desirable. In an AWGN channel ($h_{AR} = h_{BR} = 1$ in Fig. 1), this can be simply achieved by mapping the information bits to the real-valued Amplitude Shift Keying (ASK) constellation at S_A and purely imaginary-valued ASK constellation at S_B (or vice versa). Hence, in AWGN, the resulting superimposed constellation at the relay will form a regular QAM constellation without any overlapping of constellation symbols. For the assumed superposition modulation (7), this can be easily achieved by setting the coefficients as $L^s_{A,n} = 2^n$ and $L^s_{B,n} = \mathrm{j} \cdot 2^n$.

Algorithm 1 Constellation design for fixed N_b, N_s.

Input: (N_b, N_s)
for $n_s \in \{0, \ldots, N_s - 1\}$ ▷ "superposed levels"
  $L^s_{A,n_s} \leftarrow 2^{n_s}$, $L^s_{B,n_s} \leftarrow \mathrm{j}\, 2^{n_s}$
end
for $n_b \in \{0, \ldots, N_b - 1\}$ ▷ "basic levels"
  $L^b_{A,n_b} = L^b_{B,n_b} \leftarrow 2^{N_s} \cdot 3^{\lfloor n_b/2 \rfloor}$ for even n_b
  $L^b_{A,n_b} = L^b_{B,n_b} \leftarrow \mathrm{j}\, 2^{N_s} \cdot 3^{\lfloor n_b/2 \rfloor}$ for odd n_b
end
$\alpha^2 = \sum_{n_s=0}^{N_s-1} |L^s_{i,n_s}|^2 + \sum_{n_b=0}^{N_b-1} |L^b_{i,n_b}|^2$ ▷ "norm. coefficient"
for $i \in \{A, B\}$ and $\forall \mathbf{d}_i = [\mathbf{d}^b_i; \mathbf{d}^s_i]$, $d^b_{i,n_b}, d^s_{i,n_s} \in \{0, 1\}$:
  for $n_s \in \{0, N_s - 1\}$, $n_b \in \{0, N_b - 1\}$
    $s^s_{i,n_s} = L^s_{i,n_s} \left(2d^s_{i,n_s} - 1\right)$
    $s^b_{i,n_b} = L^b_{i,n_b} \left(2d^b_{i,n_b} - 1\right)$
  end
  $s_i = \mathscr{A}^i\left([\mathbf{d}^b_i; \mathbf{d}^s_i]\right) = \alpha^{-1}\left(\sum_{n_s=0}^{N_s-1} s^s_{i,n_s} + \sum_{n_b=0}^{N_b-1} s^b_{i,n_b}\right)$ ▷ "constellation symbols"
end
Output: Constellation alphabets $\mathscr{A}^A$, $\mathscr{A}^B$.

3.3 Systematic constellation design algorithm

To accommodate both the basic and superposed information streams into a joint source constellation, the scaling coefficients of basic information streams $\left(L^b_n\right)$ must be pre-scaled by a factor 2^{N_s} to ensure that they lie above the largest scaling coefficient of the superposed part. The basic stream (WNC) is thus preferred over the superposed one (routing) in order to maximize the overall throughput of the system. The systematic constellation design for arbitrary N_b, N_s is summarized in Algorithm 1.

The basic principle of source constellation design is illustrated in Fig. 3 ($N_b = 2$, $N_s = 1$), and some example sets of source constellations with constant $N_b + N_s = 2$ are depicted in Fig. 4.

4 Performance analysis I: uncoded system

For the sake of simplicity of the theoretical performance analysis, we assume that source → relay channels in WBN are AWGN[3] (i.e., $h_{iR} = 1$ for $i \in \{A, B\}$). We evaluate the throughput analytically as $(N_b + N_s) \cdot \left(1 - P^{\mathrm{FER}}_e\right)$, where P^{FER}_e denotes the frame error rate. Due to the system symmetry, we evaluate the throughput of $S_A \rightarrow D_A$ only. We set the frame length to $M = 768$ symbols (to match the HW evaluation) and the relay output mapper $\mathscr{A}^R$ will be defined as a conventional $(2^{N_b + 2N_s})$-QAM constellation. All performance metrics in the symmetric WBN are evaluated as a function of the actual SNRs $(\gamma_{\mathrm{MAC}}, \gamma_{\mathrm{HSI}}, \gamma_{\mathrm{BC}})$.

4.1 Analytical performance evaluation

Since all constellations points lie on a regular QAM grid, it is straightforward to evaluate the exact Symbol Error Rate (SER) for all channels analytically. The symbol error probability P^{SER}_e can be approximated by its upper-bound[4] $P^{\mathrm{SER}}_{e,\mathrm{UB}} = 1 - \left(\left(1 - P^{\mathrm{MAC}}_e\right)\left(1 - P^{\mathrm{HSI}}_e\right)\left(1 - P^{\mathrm{BC}}_e\right)\right)$, where P^{MAC}_e, P^{HSI}_e and P^{BC}_e are the probabilities of symbol error in the MAC, HSI, and BC channels (respectively). Since the system is memoryless, the overall FER is given by $P^{\mathrm{FER}}_e = 1 - \left(1 - P^{\mathrm{SER}}_e\right)^M$, which can be approximated by $P^{\mathrm{FER}}_{e,\mathrm{UB}} = 1 - \left(1 - P^{\mathrm{SER}}_{e,\mathrm{UB}}\right)^M$. Nearest neighbor pairwise error approximation can be used for efficient error rate evaluation. The lower-bound of the throughput can be then evaluated analytically as:

$$T_{\mathrm{LB}} = (N_b + N_s) \cdot \left(1 - P^{\mathrm{FER}}_{e,\mathrm{UB}}\right). \tag{9}$$

4.2 Hardware implementation

In addition to the analytical evaluation of throughput, we validate the results in a real-world setup. The hardware performance is evaluated using Ettus Research Universal Software Radio Peripherals (USRPs) which are computer-hosted Software Defined Radios (SDRs). The USRPs connect to a host computer via a Gigabit Ethernet link, which the host-based software uses to receive and transmit the baseband IQ stream. The host controls the USRP using USRP Hardware Driver (UHD) commands, including setting parameters such as amplifier gain. The USRP then performs the necessary baseband processing then up/down-conversion and transmission/reception. The Ettus Research N210 USRP is used, which consists of a motherboard and a radio frequency (RF) daughterboard; more details can be found in [26]. In these experiments the XCVR2450 daughterboards are used, these are dual-band transceivers with 100 mW output at 2.4–2.5 GHz and 50 mW output at 4.9–5.85 GHz. In this paper, GNU-Radio [27], which is an open-source toolkit, is used as the host-processing software.

4.2.1 *OFDM parameters*

An OFDM solution is employed with $N_{FFT} = 64$ subcarriers. There are 48 data subcarriers, with a DC carrier, 11 guard carriers and 4 pilot carriers at −21, −7, 7, and 21. A CP of $\frac{1}{4}$, i.e., 16 samples, is employed and a center frequency of $f_c = 2.4$ GHz is used. The USRP N210 is limited (by the Gigabit Ethernet link) to support an RF bandwidth up to 25 MHz with 14-bit resolution, though a 1 MHz bandwidth, i.e., 1 MS/s sampling rate, is used here. Each packet contains 16 OFDM symbols of payload with 48 data carriers per symbol and constellation parameters N_b, N_s are set to the desired values. The payload is random data, and no channel coding is currently implemented in the system.

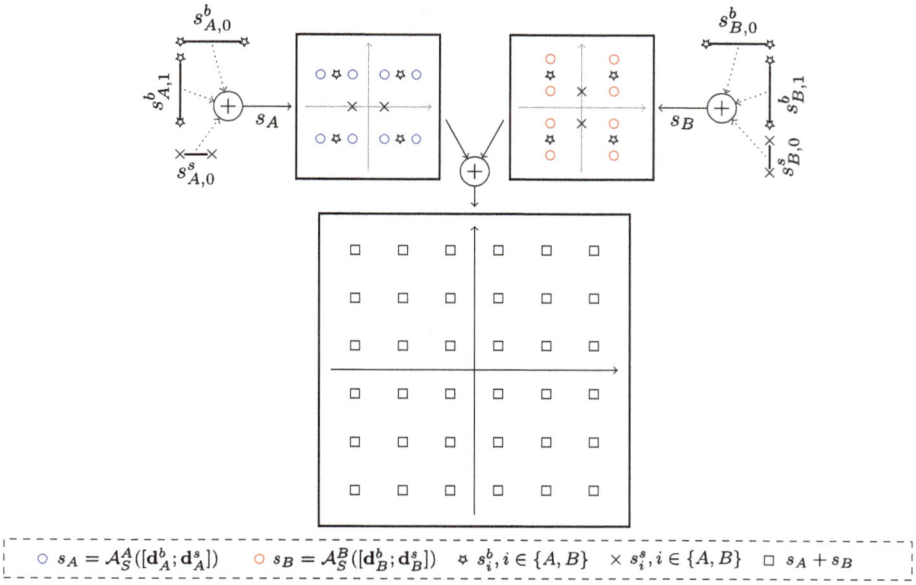

Fig. 3 Source constellation design example for $N_b = 2$, $N_s = 1$. Resulting constellations are depicted as blue circles (S_A output constellation), red circles (S_B output constellation), and squares (received superimposed constellation at R). Hierarchical function is $f\left(\mathbf{d}_A^b, \mathbf{d}_B^b\right) = \mathbf{d}_A^b \oplus \mathbf{d}_B^b$

4.2.2 *Node placement, synchronization, and channel estimation*

A 5-node, 2 × 1 × 2 topology with two sources, one relay and two destinations, is used as shown in Fig. 5. The USRPs each have a single 3dBi omni-directional antenna (at 2.4–2.5 GHz and 4.9–5.85 GHz). The first source transmitter (S_A) USRP is connected to a powerful Core i7 laptop, running a single instance of GNURadio, by a Gigabit Ethernet link. The second source transmitter (S_B) USRP is connected to the first by the MIMO expansion

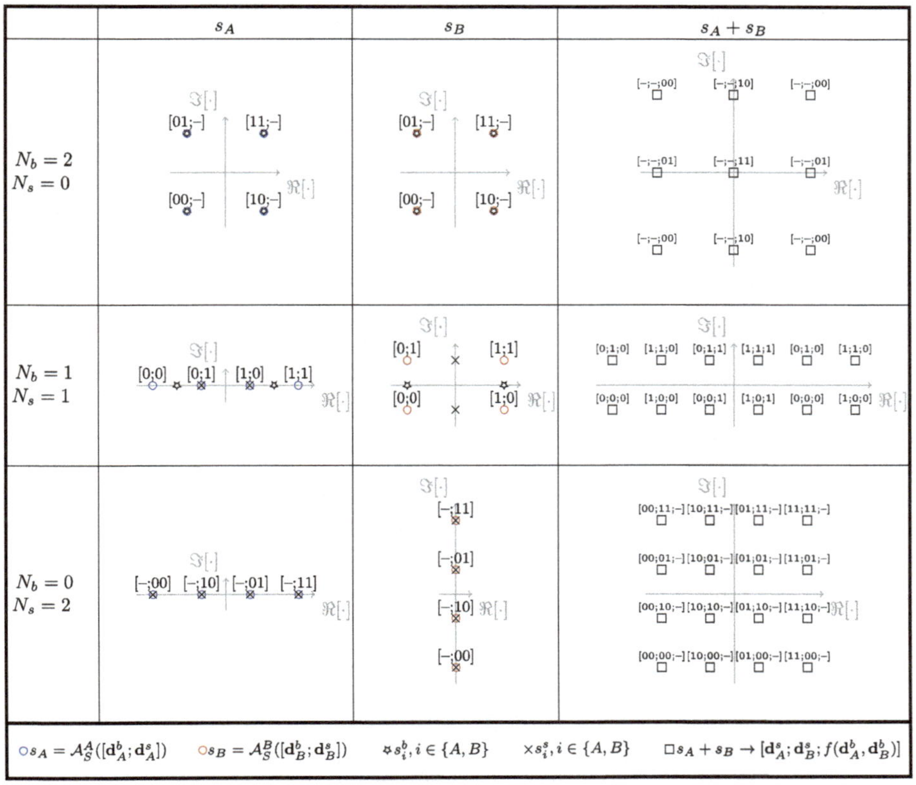

Fig. 4 Proposed constellation design for $(N_b, N_s) = \{(2, 0)\,; (1, 1)\,; (0, 2)\}$

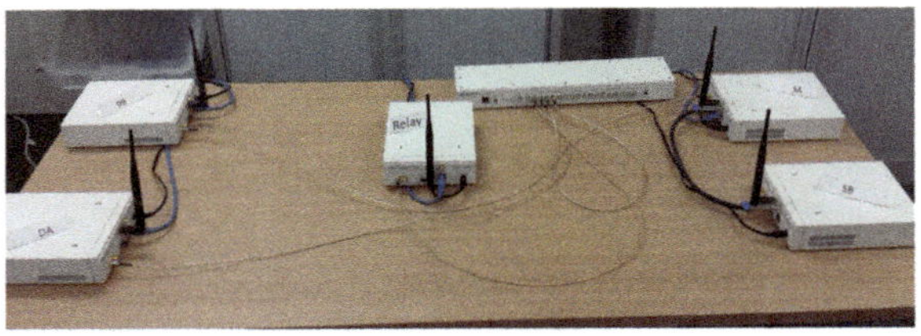

Fig. 5 Setup for HW evaluation (Ettus Research USRPs). Source transmissions are prerotated [23–25] to imitate the AWGN channel conditions (as used in the numerical evaluation). To allow a strict control of HSI channels SNR in the HW setup, the HSI channels are emulated by adding a Gaussian noise to the respective source signal and the resulting emulated HSI is passed to destinations via UDP. This also avoids the problem of node visibility where direct links $S_A \rightarrow D_A$ and $S_B \rightarrow D_B$ exist in the laboratory environment

cable. This provides the second transmitter with baseband data from GNURadio as well as synchronization signals from the first transmitter. The relay, D_A, and D_B USRPs are connected to further laptops running GNURadio by Gigabit Ethernet links. More details about synchronization, carrier frequency offset and timing offset estimation can be found in [28]. In this paper, the MIMO cable is used to synchronize the frequency of the two source USRPs, as well as to achieve sample-level timing synchronization. The latter avoids asynchronism in the superimposed constellations at the relays; more details about the impact of asynchronism in WNC can be found in [29]. Furthermore, an Ettus Octoclock clock distribution system [30] is used to provide external frequency synchronization signals to all nodes. This allows us to show system performance, and compare with the theoretical and simulation cases, without the impact of asynchronism affecting the results. Results with additional real-world impairments are a focus for future work, as discussed in the final section of this paper. Channel tracking using pilots is implemented as detailed in [28]; however, due to the static indoor environment (used to mimic the AWGN channel used in the theoretical work as closely as possible), the channel does not change significantly over the frame duration. The USRPs are closely located in line-of-sight to ensure the channel between them is fairly static, without fading, thus mimicking the channel used in the theoretical work. Moreover, as cabled synchronization is used, it is desirable to minimize the cable length. Low SNRs are achieved between the USRPs using low transmit powers.

4.3 Uncoded system throughput

To analyze the performance of the proposed constellations, we evaluate the throughput for fixed γ_{MAC} and γ_{BC} as a function of γ_{HSI}. We analyze all permissible constellations for a fixed $N_b + N_s = 2$ (see Fig. 4) and compare their performance with the throughput T_{ref} of the *reference* "conventional" multi-user detection with regular QAM constellation. Note again that the hierarchical function is set to the bit-wise XOR, i.e., $f\left(\mathbf{d}_A^b, \mathbf{d}_B^b\right) = \mathbf{d}_A^b \oplus \mathbf{d}_B^b$ in the evaluations.

In addition to the analytically evaluated throughput (lower-bound) T_{LB} (9) we also include the result of a numerical Monte-Carlo evaluation of throughput (T_{sim}) over 10^4 frames (Figs. 6, 7, 8, and 9) and throughput T_{USRP} evaluated by the HW real-world setup (Figs. 8 and 9). The particular related assumptions employed in the evaluations are summarized in Table 2 to improve the reader's perception. One can see that the reference scenario with conventional QPSK (4-QAM) provides zero throughput (T_{ref}) in the whole range of analyzed SNRs in all plots[5]. The performance of conventional WNC (in the specific 4-ary case the conventional WNC constellations are identical to $(N_b, N_s) = (2, 0)$ [4]) is, on the other hand, inevitably limited to high γ_{HSI} regions in WBN as it requires perfect HSI to enable the decoding of WNC functions at the destination nodes [15]. Remarkably, the novel proposed constellations provide a *non-zero throughput* even in the *low-to-medium SNR* region of HSI channels. The resulting SNR and throughput gains are highlighted in Figs. 6, 7, 8, and 9. Note also extremely close agreement of HW performance evaluation T_{USRP} in all cases (Figs. 8 and 9). Even in this simple case ($N_b + N_s$ fixed to 2) there are several interesting phenomena which can be observed in the throughput performance:

- The constellations with ($N_b = 0$, $N_s > 0$) assume full multi-user detection at the relay and hence they do not exploit the HSI channels (Table 1). Consequently, their throughput does not depend on γ_{HSI}. However, their performance worsens with decreasing γ_{MAC} (compare Figs. 6 and 8) and deteriorates quite dramatically with decreasing γ_{BC} (compare Figs. 8 and 9).
- The constellations with ($N_b > 0$, $N_s = 0$) require a reliable HSI channel and hence their throughput is zero for low γ_{HSI} and grows up to N_b for high γ_{HSI} and sufficiently high γ_{MAC}, γ_{BC} (all Figs. 6, 7, 8, and 9).
- The performance of all constellations with ($N_b > 0$, $N_s > 0$) is influenced by all system SNRs (γ_{MAC}, γ_{BC}, γ_{HSI}). Remarkably, for some particular values of γ_{MAC}, γ_{BC} these constellations significantly outperform those with $N_b = 0$ or $N_s = 0$ in the medium γ_{HSI} region (Figs. 7 and 9). Note that this is a direct consequence of superior ρ^2_{HSI} performance of these constellations[6] for constant $N_b + N_s$, i.e., for constant bits/constellation symbol transmitted by the sources.

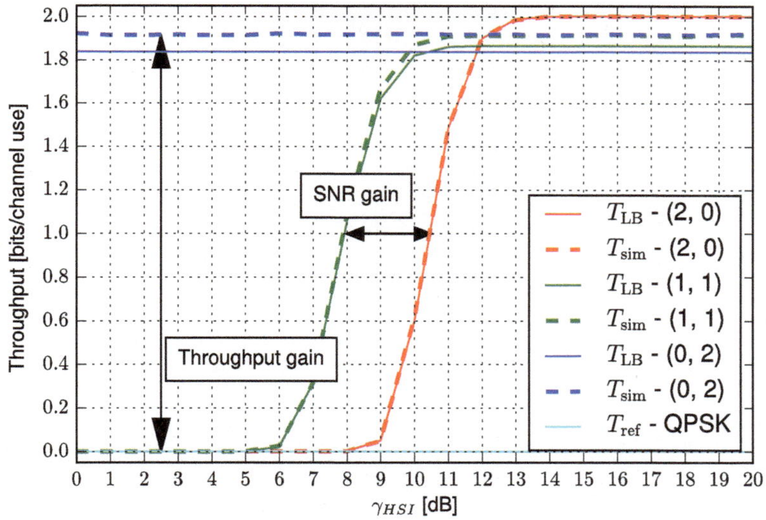

Fig. 6 Comparison of throughput as a function of γ_{HSI} for $\gamma_{MAC} = 16$ dB, $\gamma_{BC} = 20$ dB and given (N_b, N_s)

5 Adaptive constellation design

As discussed in the previous section, the throughput of the system depends heavily on the actual SNR conditions (given by $\gamma_{MAC}, \gamma_{BC}, \gamma_{HSI}$). By itself, this observation calls naturally for an *adaptive constellation design*. To generate a multi-source constellation able to *maximize the throughput* given the current SNR conditions in the WBN system, the source nodes S_A, S_B must be acquainted with the particular mapping operation $(\hat{\gamma}_{MAC}, \hat{\gamma}_{BC}, \hat{\gamma}_{HSI}) \mapsto (N_b^{\mathrm{I}}, N_s^{\mathrm{I}})$, providing the constellation parameters $N_b^{\mathrm{I}}, N_s^{\mathrm{I}}$ (input arguments to the Algorithm 1) for the given (available) SNR estimates[7] $(\hat{\gamma}_{MAC}, \hat{\gamma}_{BC}, \hat{\gamma}_{HSI})$. The desired *SNR mapping regions* (a three-dimensional look-up table in WBN) can be generated numerically in advance and stored at both source nodes to enable the adaptive constellation mapping.

Due to the numerical complexity of Monte-Carlo evaluation of throughput T_{sim} we use the analytically evaluated lower bound T_{LB} (9) to generate the SNR mapping regions for a wide range of $\gamma_{MAC}, \gamma_{BC}, \gamma_{HSI}$. The optimal constellation parameters (maximizing the throughput T_{LB} for a given SNR triplet $\gamma_{MAC}, \gamma_{BC}, \gamma_{HSI}$) are found by performing an exhaustive search over the constellation parameters as $\left(N_b^{\mathrm{I}}, N_s^{\mathrm{I}}\right) = \arg\max_{(N_b,N_s)} T_{LB}(\gamma_{MAC}, \gamma_{BC}, \gamma_{HSI})$.

The SNR mapping regions, including the resulting throughput for the optimal choice of $N_b^{\mathrm{I}}, N_s^{\mathrm{I}}$, are shown as a function of $\gamma_{MAC}, \gamma_{HSI}$ for $\gamma_{BC} \in$

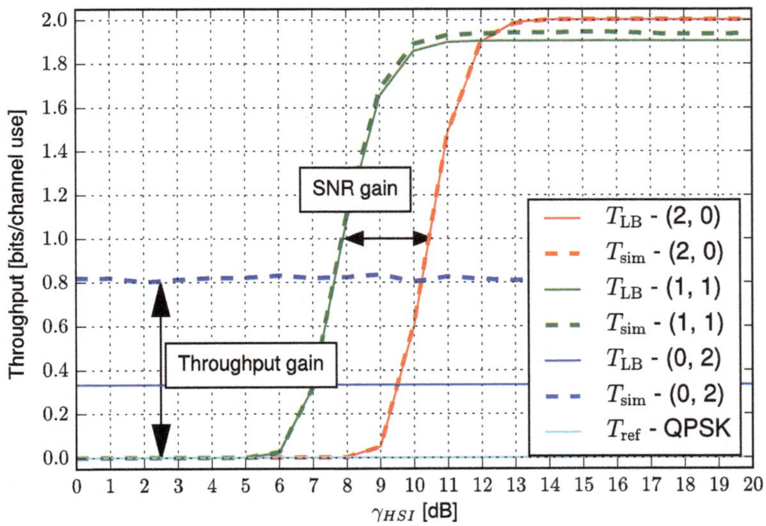

Fig. 7 Comparison of throughput as a function of γ_{HSI} for $\gamma_{MAC} = 17$ dB, $\gamma_{BC} = 17$ dB and given (N_b, N_s)

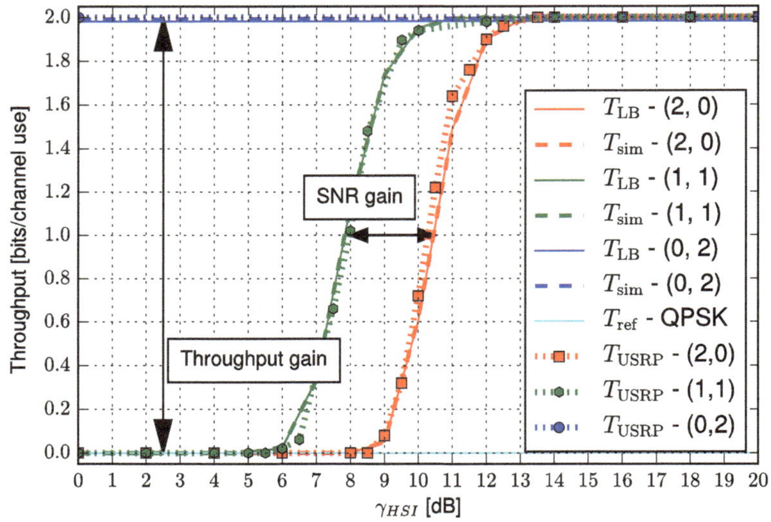

Fig. 8 Comparison of throughput as a function of γ_{HSI} for $\gamma_{\text{MAC}} = 20$ dB, $\gamma_{\text{BC}} = 20$ dB and given (N_b, N_s)

$\{15\,\text{dB}, 17\,\text{dB}, 20\,\text{dB}, 25\,\text{dB}\}$ in Fig. 10. Several interesting phenomena can be again observed in the figure:

- Since the transmission schemes with $(N_b = 0,\ N_s > 0)$ constellations can ignore the HSI channels (Table 1), these constellations are optimal in the low γ_{HSI} region ($\gamma_{\text{HSI}} < 8$ dB), regardless of the other channels' SNRs $\gamma_{\text{MAC}}, \gamma_{\text{BC}}$.
- The constellations with $(N_b > 0,\ N_s = 0)$ provide the best performance in a substantial part of the medium-to-high γ_{HSI} regions, depending on the actual $\gamma_{\text{MAC}}, \gamma_{\text{BC}}$.
- The "combined" constellations $(N_b > 0,\ N_s > 0)$ provide the maximal throughput (for sufficiently high γ_{BC}) in relatively large regions of $\gamma_{\text{MAC}}, \gamma_{\text{HSI}}$ (see the dominant throughput regions of $(N_b^{\text{I}}, N_s^{\text{I}}) = (2, 1)$ for $\gamma_{\text{BC}} = \{20\,\text{dB}, 25\,\text{dB}\}$). Note that these constellations have a great potential for application in systems with limited or partial HSI (see [15]).
- The BC channel SNR (γ_{BC}) can significantly limit the performance of the system, even in the case where both $\gamma_{\text{MAC}}, \gamma_{\text{HSI}}$ are relatively high (compare the results for $\gamma_{\text{BC}} = 15$ dB and $\gamma_{\text{BC}} = 20$ dB for $\gamma_{\text{MAC}} = \gamma_{\text{HSI}} \gtrsim 15$ dB). This behavior originates in the fact that ρ_{BC}^2 decreases with growing $N_b + 2N_s$ and hence low γ_{BC} essentially limits the maximum efficiently applicable cardinality of the relay output ($2^{N_b+2N_s}$-QAM) constellation. Equivalently, the

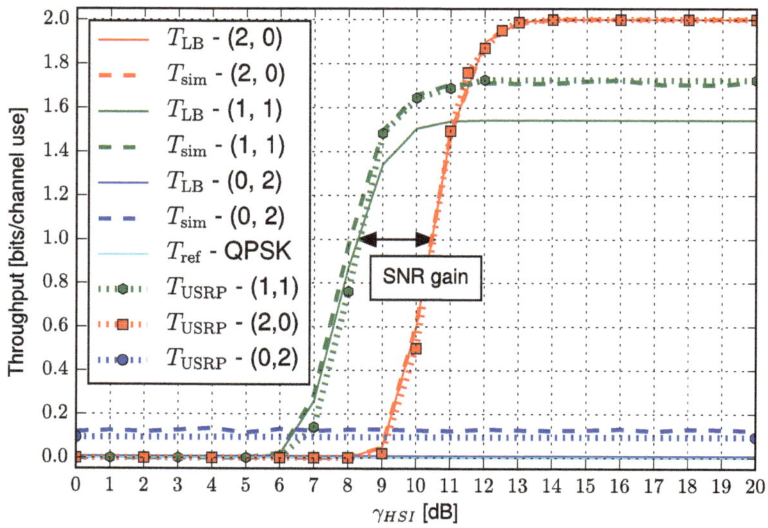

Fig. 9 Comparison of throughput as a function of γ_{HSI} for $\gamma_{\text{MAC}} = 20$ dB, $\gamma_{\text{BC}} = 16$ dB and given (N_b, N_s)

Table 2 Summary of assumptions in the uncoded system analysis

	T_{LB}	T_{sim}	T_{USRP}
Communication channel	AWGN	AWGN	Wireless channel
Time synchronization	Perfect	Perfect	External (Octoclock)
Phase pre-rotation	Perfect	Perfect	Feedback channel (UDP)
Direct link	Not considered	Not considered	Bridged by cable
Channel symmetry	Perfect	Perfect	Node placement
Throughput evaluation	Lower bound	Numerical	Numerical

optimal source constellation parameters $(N_b^{\mathrm{I}}, N_s^{\mathrm{I}})$ can become limited by the bottleneck induced by weak relay→destination (BC) channels.

- The constellations with odd N_b are rarely dominant over the other strategies with identical $N_b + N_s$. This is a direct consequence of their relatively poor ρ^2_{MAC} Euclidean distance properties.

To support a viability of the adaptive constellation mapping scheme, we again evaluate the system performance in the HW real-world setup. Since the HW evaluation is heavily time-consuming, we analyze the throughput T_{USRP} only for $\gamma_{\mathrm{BC}} = 20\,\mathrm{dB}$ and $0\,\mathrm{dB} \leq \gamma_{\mathrm{MAC}}, \gamma_{\mathrm{HSI}} \leq 30\,\mathrm{dB}$ with a step-size 3 dB. The measured T_{USRP} performance is available in Fig. 11. Close agreement between the T_{USRP} and T_{LB} can be observed again in the whole range of analyzed γ_{MAC}, γ_{HSI} (see Fig. 11).

6 Channel coding

We have shown that Algorithm 1 is a strong tool for the design of source constellations suitable for arbitrary channel SNRs in the WBN system. However, even

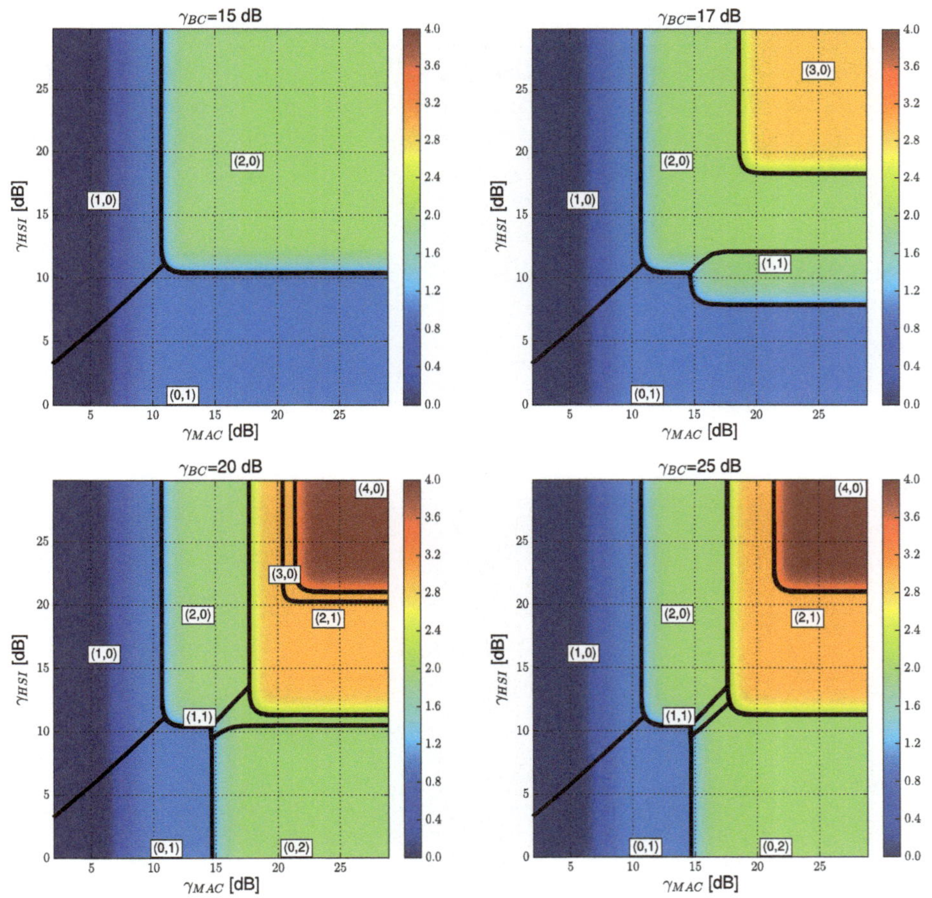

Fig. 10 Throughput performance T_{LB} and SNR mapping regions (including the optimal const. parameters $(N_b^{\mathrm{I}}, N_s^{\mathrm{I}})$) for $\gamma_{\mathrm{BC}} \in \{15\,\mathrm{dB}, 17\,\mathrm{dB}, 20\,\mathrm{dB}, 25\,\mathrm{dB}\}$

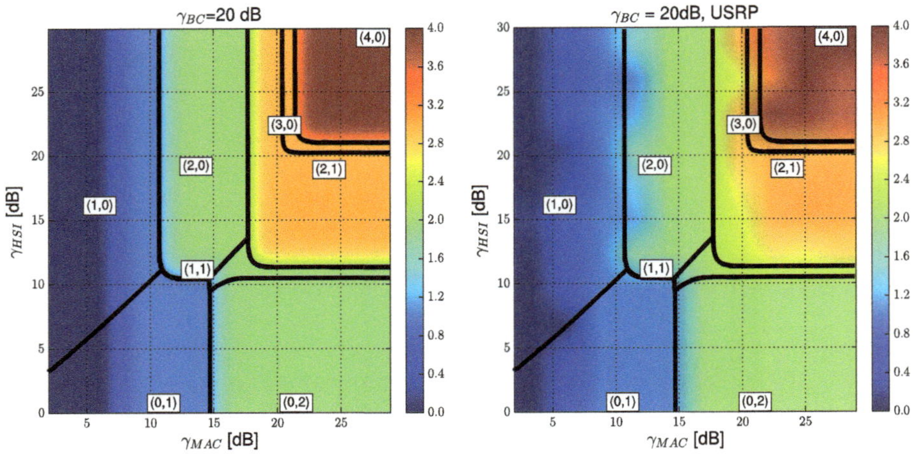

Fig. 11 Comparison of T_{LB} and the throughput performance evaluated in a real-world adaptive HW setup (T_{USRP}) for $\gamma_{\text{BC}} = 20$ dB. The SNR mapping regions (including the optimal const. parameters (N_b^l, N_s^l)) used in the HW evaluation are also emphasized in the figure

though the proposed constellations provide promising performance in the uncoded scenario, the extension to a coded system is desirable for practical wireless applications.

Considering the communication in uncoded WBNs (Fig. 2), there are many ways to improve its reliability through channel coding. In this paper, we propose one particular solution, based on the assumption that each source node splits its data into the basic and superposed part and then encodes these parts separately by two constituent channel encoders[8]. We believe that this approach provides the best insight into the channel-coding extension of WBN systems and hence we focus on its development in the rest of this section.

6.1 Per-link channel coding scheme for symmetric WBN with superposition modulations

In general, it is desirable to protect *all individual transmissions* in the system with error correction coding[9]. Accordingly, the relay node has to decode the desired data (including the particular WNC function) from its observation and then it should perform an additional re-encoding in order to protect the subsequent relay → destinations transmission. The resulting channel coding scheme (see Fig. 12) is introduced in the following text.

6.1.1 *Source processing*

Source S_A (likewise for S_B) wants to transmit a binary data word $\mathbf{D}_A$ of length $|\mathbf{D}_A| = |\mathbf{D}_B| = k_{\text{D}}$ to its respective

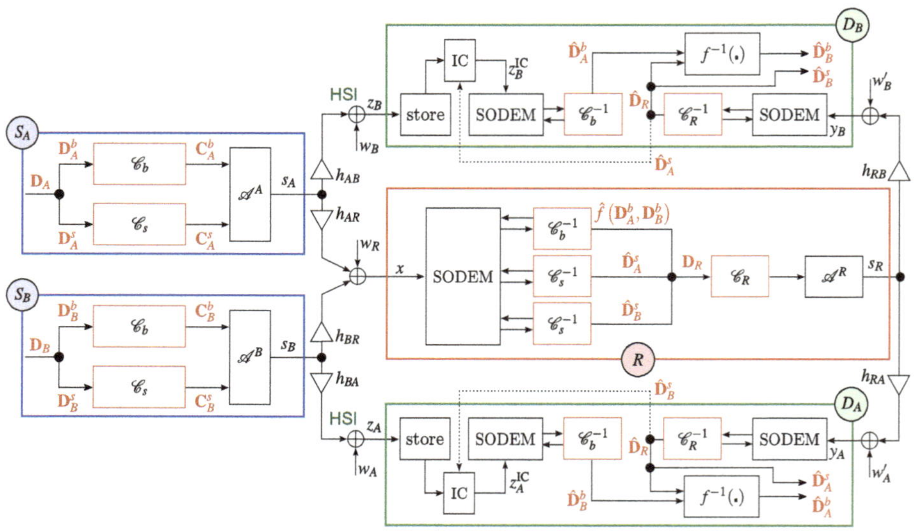

Fig. 12 Relaying scheme in the encoded WBN system. SODEM stands for a soft-output demodulator, IC is the interference canceler. The channel encoders/decoders which are appended to the uncoded system are emphasized

destination D_A. Each data word $\mathbf{D}_A$ is split into the basic and superposed part (similarly as in the uncoded case), i.e.,:

$$\mathbf{D}_A = \left[\mathbf{D}_A^b, \mathbf{D}_A^s\right] = \left[d_{A,0}^b, d_{A,1}^b, \ldots d_{A,k_\mathrm{D}^b-1}^b, d_{A,0}^s, d_{A,1}^s, \ldots d_{A,k_\mathrm{D}^s-1}^s\right],$$

where $d_{A,n}^b, d_{A,n}^s \in \{0,1\}$ and $k_\mathrm{D}^b, k_\mathrm{D}^s$ are the lengths of the basic and superposed data sub-words $\mathbf{D}_A^b, \mathbf{D}_A^s$ (respectively).

Subsequently, individual $\mathbf{D}_A^b, \mathbf{D}_A^s$ are encoded by two separate linear binary encoders $\mathscr{C}_b(\cdot), \mathscr{C}_s(\cdot)$ (assumed identical at both sources), producing the constituent codewords as:

$$\mathscr{C}_b\left(\mathbf{D}_A^b\right) = \mathbf{C}_A^b = \left[\mathbf{c}_{A,0}^b, \mathbf{c}_{A,1}^b, \ldots, \mathbf{c}_{A,\mathcal{N}-1}^b\right], \tag{10}$$

$$\mathscr{C}_s\left(\mathbf{D}_A^s\right) = \mathbf{C}_A^s = \left[\mathbf{c}_{A,0}^s, \mathbf{c}_{A,1}^s, \ldots, \mathbf{c}_{A,\mathcal{N}-1}^s\right], \tag{11}$$

where $\mathcal{N}$ is the number of codeword symbols (equivalently the number of channel uses required to transmit the codeword) and

$$\mathbf{c}_{A,j}^b = \left[c_{A,j_0}^b, \ldots, c_{A,j_{N_b-1}}^b\right],$$

$$\mathbf{c}_{A,j}^s = \left[c_{A,j_0}^s, \ldots, c_{A,j_{N_s-1}}^s\right]$$

are codeword symbols of length N_b (respectively N_s) bits. The dimensionality of binary codewords is thus $\left|\mathbf{C}_A^b\right| = \left|\mathbf{C}_B^b\right| = n_\mathrm{C}^b = \mathcal{N} \cdot N_b, \left|\mathbf{C}_A^s\right| = \left|\mathbf{C}_B^s\right| = n_\mathrm{C}^s = \mathcal{N} \cdot N_s$.

The constituent codewords $\mathbf{C}_A^b, \mathbf{C}_A^s$ are then forwarded to the joint constellation mapper $\mathscr{A}^A(\cdot)$ (provided by Algorithm 1), which produces a sequence of (N_s+N_b)-bit constellation symbols $s_{A,j} = \mathscr{A}^A\left(\left[\mathbf{c}_{A,j}^b; \mathbf{c}_{A,j}^s\right]\right)$. Note that in the encoded system it is the codeword symbols that are mapped to the desired constellation (instead of data symbols), but since $\left|\mathbf{c}_A^b\right| = \left|\mathbf{c}_B^b\right| = N_b$ and $\left|\mathbf{c}_A^s\right| = \left|\mathbf{c}_B^s\right| = N_s$, it is sufficient to formally substitute the codeword symbol $\mathbf{c}_i = \left[\mathbf{c}_i^b; \mathbf{c}_i^s\right]$ for the data symbol $\mathbf{d}_i = \left[\mathbf{d}_i^b; \mathbf{d}_i^s\right]$ in Algorithm 1 (lines 10, 15). The resulting constellation symbols are successively transmitted by S_A (simultaneously with $s_{B,j}$ from S_B) towards the relay node.

6.1.2 Relay processing

The crucial part of the encoded WBN system processing is based at the relay node. Like the uncoded case, the relay must decode jointly the *superposed* data subwords $(\mathbf{D}_A^s, \mathbf{D}_B^s)$ along with the *WNC function* of *basic* data sub-words $\left(f\left(\mathbf{D}_A^b, \mathbf{D}_B^b\right) = \mathbf{D}_A^b \oplus \mathbf{D}_B^b\right)$. While both $\mathbf{D}_A^s, \mathbf{D}_B^s$ can be decoded straightforwardly by the relay (using conventional single user decoders $\mathscr{C}_s^{-1}$), the fundamental question is how to decode the WNC function $f\left(\mathbf{D}_A^b, \mathbf{D}_B^b\right)$ since both $\mathbf{D}_A^b, \mathbf{D}_B^b$ are encoded separately at sources S_A, S_B[10]. Fortunately, as proven in [13], the WNC function $f\left(\mathbf{D}_A^b, \mathbf{D}_B^b\right)$ can be decoded directly from the relay observation (5), if the particular source data words are encoded by the same linear encoder $\mathscr{C}_b$, since

$$\mathscr{C}_b\left(f\left(\mathbf{D}_A^b, \mathbf{D}_B^b\right)\right) = \mathscr{C}_b\left(\mathbf{D}_A^b \oplus \mathbf{D}_B^b\right) = \mathscr{C}_b\left(\mathbf{D}_A^b\right) \oplus \mathcal{C}_b\left(\mathbf{D}_B^b\right) = \mathbf{C}_A^b \oplus \mathbf{C}_B^b, \tag{12}$$

where $\mathbf{C}_A^b \oplus \mathbf{C}_B^b$ itself is again a valid codeword and hence the WNC function $f\left(\mathbf{D}_A^b, \mathbf{D}_B^b\right)$ can be decoded directly from the relay observation (using a conventional single user decoder $\mathcal{C}_b^{-1}$).

After decoding all required data streams $\left(\hat{\mathbf{D}}_A^s, \hat{\mathbf{D}}_B^s, \hat{f}\left(\mathbf{D}_A^b, \mathbf{D}_B^b\right)\right)$, the relay creates a joint data word $\mathbf{D}_R = \left[\hat{\mathbf{D}}_A^s, \hat{\mathbf{D}}_B^s, \hat{f}\left(\mathbf{D}_A^b, \mathbf{D}_B^b\right)\right]$ and re-encodes it as $\mathscr{C}_R(\mathbf{D}_R) = \mathbf{C}_R = \left[\mathbf{c}_{R,0}, \mathbf{c}_{R,1}, \ldots, \mathbf{c}_{R,\mathcal{N}-1}\right]$. The encoded data are then mapped to the output 2^{N_R}-QAM (where $N_R = 2N_s + N_b$) constellation symbols $s_{R,j} = \mathscr{A}^R\left(\mathbf{c}_{r,j}\right), j \in \{0, 1, \ldots, \mathcal{N}-1\}$ and broadcast by the relay to both destination nodes.

6.1.3 Destination processing

As in the uncoded case, destinations D_A, D_B firstly store the signal received in the MA phase, and then, after decoding the relay data word $\mathbf{D}_R$ (from the signal received in the BC phase), both destinations can perform IC to remove the unintended superposed codewords from the MA phase signal to get the desired HSI and then finally recover the desired data (see Fig. 12 for the details).

6.2 Analysis of transmission rates

The last step in the design of a feasible modulation and coding scheme for WBN (see Fig. 12) is to identify the range of permissible transmission rates for the basic (r_b), superposed (r_s) and relay output (r_R) data streams. To achieve this goal, we analyze the constellation constrained (CC) capacities [31] of the proposed source constellations, including the CC capacity of the conventional relay output 2^{N_R}-QAM alphabets (used in the BC phase of communication).

The transmission rates are defined in bits per channel use, and hence $r_i = N_i \cdot k_\mathrm{D}^i / n_\mathrm{C}^i$, where N_i is the number of bits per channel symbol/use in the data stream i, $i \in \{b, s, R\}$. In accordance with the definition of source and relay codewords, we assume that each communication round requires $2\mathcal{N}$ channel uses, where the length of both MA and BC phase is set to $\mathcal{N}$ channel uses. Note that the assumption that BC and MA phases have identical length is generally only suboptimal (see e.g., [1, 15]), but nevertheless, it is a common assumption in practical systems and hence, we limit our attention to this particular case.

6.2.1 Relay observation

After splitting the source data streams, the whole system can be interpreted as a 4-user system where the relay has to decode the data $\mathbf{D}_A^s, \mathbf{D}_B^s, f\left(\mathbf{D}_A^b, \mathbf{D}_B^b\right)$ from the *virtual 3-user* $\left(S_A^s, S_B^s, S_{AB}^b\right)$ *multiple-access channel* observation (see Fig. 12). Unfortunately, since one of the users (S_{AB}^b) is only virtual, we cannot simply claim that the region of achievable rates in this virtual channel can be derived directly from the conventional cut-set bound analysis (see e.g., [18, 32]), but rather a careful information-theoretic analysis would be required to identify the exact rate region. However, such analysis is far beyond the scope of this paper, and hence, for simplicity reasons, we only conjecture that the eligible source rates (r_b, r_s) are limited by the conventional CC multiple-access capacity region [18, 32], which can be defined by the following set of mutual informations $I(.;.)$:

$$2r_s + r_b \leq I\left(x; \mathbf{c}_A^s, \mathbf{c}_B^s, f\left(\mathbf{c}_A^b, \mathbf{c}_B^b\right)\right), \tag{13}$$

$$2r_s \leq I\left(x; \mathbf{c}_A^s, \mathbf{c}_B^s | f\left(\mathbf{c}_A^b, \mathbf{c}_B^b\right)\right), \tag{14}$$

$$r_s + r_b \leq I\left(x; \mathbf{c}_A^s, f\left(\mathbf{c}_A^b, \mathbf{c}_B^b\right) | \mathbf{c}_B^s\right), \tag{15}$$

$$r_s + r_b \leq I\left(x; \mathbf{c}_B^s, f\left(\mathbf{c}_A^b, \mathbf{c}_B^b\right) | \mathbf{c}_A^s\right), \tag{16}$$

$$r_s \leq I\left(x; \mathbf{c}_A^s | \mathbf{c}_B^s, f\left(\mathbf{c}_A^b, \mathbf{c}_B^b\right)\right), \tag{17}$$

$$r_s \leq I\left(x; \mathbf{c}_B^s | \mathbf{c}_A^s, f\left(\mathbf{c}_A^b, \mathbf{c}_B^b\right)\right), \tag{18}$$

$$r_b \leq I\left(x; f\left(\mathbf{c}_A^b, \mathbf{c}_B^b\right) | \mathbf{c}_A^s, \mathbf{c}_B^s\right), \tag{19}$$

where (13), (14)–(16), and (17)–(19) are the corresponding third, second, and first order cut-set bounds (respectively). As we show later, the CC MAC region (as defined by (13)–(21)) provides a reasonable estimate of the maximal achievable source transmission rate pair (r_b, r_s) in WBN.

Since the superposed data sub-words $\mathbf{D}_A^s, \mathbf{D}_B^s$ are mapped to orthogonal ASK sets (Algorithm 1), the 2nd order cut-set bound $I\left(x; \mathbf{c}_A^s, \mathbf{c}_B^s | f\left(\mathbf{c}_A^b, \mathbf{c}_B^b\right)\right)$ (14) is equivalent to $I\left(x; \mathbf{c}_A^s | \mathbf{c}_B^s, f\left(\mathbf{c}_A^b, \mathbf{c}_B^b\right)\right) + I\left(x; \mathbf{c}_B^s | \mathbf{c}_A^s, f\left(\mathbf{c}_A^b, \mathbf{c}_B^b\right)\right)$ (defined in Eqs. (17) and (18)), which can be proven easily by the chain rule for mutual information of orthogonal signal sets (see [18]). Moreover, due to the assumed WBN system symmetry, the cut-set bounds defined in Eqs. (17) and (18) are identical, and the same is true for the pair of cut-set bounds defined in (15) and (16). Consequently, the CC multiple-access capacity region is completely defined by the four inequalities in (13), (15), (17), and (19).

6.2.2 Effective HSI channels

In addition to the successful relay decoding, the transmission rate of the basic information stream r_b must guarantee that the HSI (given by the unintended data sub-words $\mathbf{D}_A^b$ at D_B and $\mathbf{D}_B^b$ at D_A) can be decoded at both destinations (after perfect IC of superposed data codewords $\mathbf{C}_A^s$ at D_B and $\mathbf{C}_B^s$ at D_A from the stored MA phase signal). Note that since we assume that the relay has already successfully decoded $f\left(\mathbf{D}_A^b, \mathbf{D}_B^b\right)$ at this step, it can be efficiently exploited in the decoding process, resulting in the following upper-bounds on the source rate r_b:

$$r_b \leq I\left(z_B^{\mathrm{IC}}; \mathbf{c}_A^b \middle| f\left(\mathbf{c}_A^b, \mathbf{c}_B^b\right)\right), \tag{20}$$

$$r_b \leq I\left(z_A^{\mathrm{IC}}; \mathbf{c}_B^b \middle| f\left(\mathbf{c}_A^b, \mathbf{c}_B^b\right)\right), \tag{21}$$

where $z_A^{\mathrm{IC}}, z_B^{\mathrm{IC}}$ are the efficient HSI observations (after perfect IC) at D_A, D_B (respectively)—see Fig. 12. Both (20) and (21) are identical in the symmetric WBN system.

6.2.3 Relay broadcast channel

The relay rate r_R must guarantee that its data word $\mathbf{D}_R = \left[\hat{\mathbf{D}}_A^s, \hat{\mathbf{D}}_B^s, \hat{f}\left(\mathbf{D}_A^b, \mathbf{D}_B^b\right)\right]$ with length $|\mathbf{D}_R| = 2k_{\mathbf{D}}^s + k_{\mathbf{D}}^b = \mathcal{N}(2r_s + r_b)$ can be sent to both destinations in $\mathcal{N}$ channel uses. In addition to this, to enable successful decoding of the relay data at both destinations, the relay transmission rate r_R must be below the CC capacities of the corresponding relay→destination channels. Thus, the relay rate should be set to $r_R = r_b + 2r_s$, which gives us the last two inequalities for r_b, r_s:

$$r_R = 2r_s + r_b \leq I\left(y_A; \mathbf{c}_R\right), \tag{22}$$

$$r_R = 2r_s + r_b \leq I\left(y_B; \mathbf{c}_R\right), \tag{23}$$

where both cut-set bounds (22), (23) are identical due to the system symmetry.

6.3 Transmission rate region for fixed source constellation

To identify the region $(r_b, r_s) \in \mathscr{R}^{(N_b, N_s)}$ of eligible transmission rates for a fixed (N_b, N_s) source constellation, we can evaluate numerically[11] the set of relevant cut-set bounds (i.e., (13), (15), (17), (19), (20), and (22)) for given SNR conditions in the symmetric WBN system $(\gamma_{\mathrm{MAC}}, \gamma_{\mathrm{BC}}, \gamma_{\mathrm{HSI}})$. An example analysis of the source transmission rate region $\mathscr{R}^{(2,1)}$ for the $(N_b = 2, N_s = 1)$ source constellation is visualized in Fig. 13 for two different SNR setups. The cut-set bound rate region $\mathscr{R}^{(N_b, N_s)}$ is emphasized by the shaded area and the rate pair with maximal $r_b + r_s$ is identified in the figure. We have assumed that the source → relay channels are AWGN (or equivalently that perfect source phase prerotation [23–25] is implemented) in the numerical evaluation.

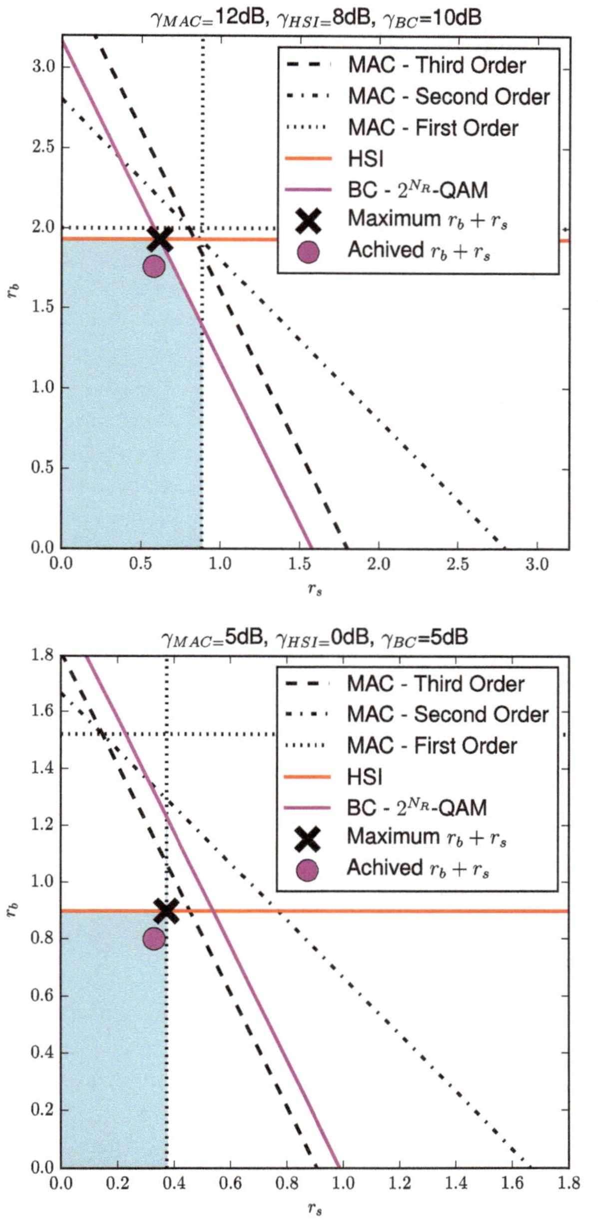

Fig. 13 Numerically evaluated cut-set bounds (mutual information for finite input constellations) for $(N_b = 2, N_s = 1)$ source constellations and 16-QAM at the relay. The particular SNR conditions are available in the titles of both sub-figures

7 Performance analysis II: adaptive encoded system

As shown in Fig. 13, the performance of the encoded WBN system is, like in the uncoded case, influenced by the immediate SNR conditions (given by $\gamma_{\mathrm{MAC}}, \gamma_{\mathrm{BC}}, \gamma_{\mathrm{HSI}}$) and hence an *adaptive modulation and coding scheme* is again of interest. However, before we introduce the*SNR mapping operation* for the encoded system, we show that the cut-set bound rate regions $\mathscr{R}^{(N_b,N_s)}$ (as introduced in the preceding section) provide a reasonable approximation of achievable source transmission rate pairs in the symmetric WBN system. In the whole section, we assume (for simplicity reasons) that perfect source phase pre-rotation [23–25] is implemented in the system, virtually reverting the source → relay channels to the AWGN case.

7.1 Numerical simulation of encoded WBN system throughput

We implement all constituent channel encoders in the WBN system (see Fig. 12) as parallel concatenated turbo codes with random interleavers of length $\mathcal{N} = 10^5$ symbols. By proper puncturing [33] of the channel encoders' outputs, we are able to set the source transmission rates (separately for the basic and superposed streams) to an almost arbitrary value in the range ($1/3 \rightarrow 1$). We set the initial source transmission rates to the maximum $r_b + r_s \in \mathscr{R}^{(N_b,N_s)}$ (as evaluated in Fig. 13) and by proper adjustment of the puncturing operation we successively decrease the source transmission rates r_b, r_s until the SER performance drops to zero. As one can see in Fig. 13, the achieved transmission rate pairs are relatively close to the maximal permissible values and hence we can consider the cut-set bound rate region $\mathscr{R}^{(N_b,N_s)}$ (as discussed in Section 6.3) to be a reasonable approximation of the achievable source transmission rate region.

7.2 Maximal throughput of the adaptive modulation-coding scheme

As in the uncoded case, the source nodes S_A, S_B in the adaptive modulation-coding WBN system must be aware of the particular mapping operation $(\hat{\gamma}_{\mathrm{MAC}}, \hat{\gamma}_{\mathrm{BC}}, \hat{\gamma}_{\mathrm{HSI}}) \mapsto (N_b^{\mathrm{II}}, N_s^{\mathrm{II}}; r_b^{\mathrm{II}}, r_s^{\mathrm{II}})$, providing the set of optimal (in the sense of maximal throughput) constellation/encoder parameters for the given SNR conditions $(\hat{\gamma}_{\mathrm{MAC}}, \hat{\gamma}_{\mathrm{BC}}, \hat{\gamma}_{\mathrm{HSI}})$. Note that, in general, all adaptive system parameters are functions of the actual SNR conditions in the system. However, we mostly omit the explicit notation in the following text to aid readability.

The desired adaptive system parameters can be obtained in the following two-step procedure. First of all, we analyze the cut-set bound rate regions $\mathscr{R}^{(N_b,N_s)}$ (as discussed in Section 6.3) to identify the maximal encoded system throughput $T_C^{(N_b,N_s)}$ for a fixed source constellation (N_b, N_s):

$$T_C^{(N_b,N_s)}(\gamma_{\mathrm{MAC}}, \gamma_{\mathrm{BC}}, \gamma_{\mathrm{HSI}}) = \max_{r_b,r_s \in \mathscr{R}^{(N_b,N_s)}} (r_b + r_s), \tag{24}$$

$$= r_b^{\max}(N_b, N_s) + r_s^{\max}(N_b, N_s). \tag{25}$$

Then, an exhaustive search over the constellation parameters N_b, N_s can be performed to find the maximum encoded system throughput $T_C^{\max}$ for the given SNR conditions $(\gamma_{\text{MAC}}, \gamma_{\text{BC}}, \gamma_{\text{HSI}})$:

$$T_C^{\max}(\gamma_{\text{MAC}}, \gamma_{\text{BC}}, \gamma_{\text{HSI}}) = T_C^{(N_b^{\text{II}}, N_s^{\text{II}})} = \max_{(N_b, N_s)} T_C^{(N_b, N_s)}, \quad (26)$$

where $(N_b^{\text{II}}, N_s^{\text{II}})$ are the *optimal source constellation parameters*. The corresponding *optimal source transmission rates* (25) are given simply by $r_b^{\text{II}} = r_b^{\max}(N_b^{\text{II}}, N_s^{\text{II}})$, $r_s^{\text{II}} = r_s^{\max}(N_b^{\text{II}}, N_s^{\text{II}})$.

The SNR mapping regions (including the throughput $T_C^{\max}$) for the adaptive modulation and coding scheme in WBN are evaluated as a function of $\gamma_{\text{MAC}}, \gamma_{\text{HSI}}$ in Fig. 14 for $\gamma_{\text{BC}} \in \{7.5\,\text{dB}, 15\,\text{dB}\}$. As one can see in Fig. 15, the increased reliability of transmissions on all individual channels in WBN (as compared to the uncoded system) translates directly into a significant enhancement of throughput performance in the whole range of observed SNRs. This observation justifies the viability of the proposed channel coding scheme.

7.3 Robustness analysis

In this section, we show that the proposed modulation and coding scheme is viable even if the real-world conditions induce some deviations from the system model assumed in this paper. From a practical point of view, we would be interested mainly in the situation, where a direct channel between a source and its desired destination is present (see Fig. 16). Even though this "direct" channel can be relatively weak, it inevitably affects the destinations' HSI observations. In addition to this, we analyze the system where source phase pre-rotation is switched off (or ineffective due to rapid channel dynamics), and thus the relay faces a multiple-access channel with a varying source channel phase offset. This assumption is valid for a practical system where the applicability of source phase pre-rotation is limited due to the unavailable ideal feedback channel or because of the undesirable delay induced by the feedback in the pre-rotation operation. To demonstrate the robustness of the proposed modulation and coding strategy, we compare the WBN performance results achieved by the Monte-Carlo numerical evaluation (see section 7.1) with USRP HW measurements. While in the numerical evaluation we still keep the idealized assumptions as defined by our system model (i.e. no direct source→destination channels, perfect channel phase pre-rotation), in the HW measurements we now allow all data communication to be performed through real wireless channels (compare with the HW setup in Section 4.2), introducing inevitably some direct channel observation at both destinations. In addition, we allow the source phase pre-rotation to be switched off in the HW measurements. The channel SNRs $\gamma_{\text{MAC}}, \gamma_{\text{HSI}}$ in the HW measurements can be controlled by the source transmit power adjustment and similarly, particular values of γ_{BC} can be set up by the adjustment of the relay transmit power.

To perform a comparison of the idealized numerical evaluation with the HW robustness measurements, we evaluate the range of SNRs where a zero error rate performance can be achieved for fixed constellations and channel encoder rates (N_b, N_s, r_b, r_s). To achieve this, we identify the triplet of SNRs $(\gamma_{\text{MAC}}, \gamma_{\text{HSI}}, \gamma_{\text{BC}})$ providing a zero-error rate performance at both destinations. This SNR triplet then defines the lower-most corner of the SNR region where a zero error rate performance can be achieved (see Fig. 17). We can conclude that:

- The presence of direct links requires only slightly increased values of $(\gamma_{\text{MAC}}, \gamma_{\text{HSI}}, \gamma_{\text{BC}})$ to ensure a zero error-rate performance for the fixed (N_B, N_S, r_b, r_s). However, it is evident that a modification of the proposed modulation and coding scheme could be desirable to fully harness the information transmitted over direct source destination channels.
- The source channel phase asynchrony in the system with direct links further degrades the performance on the sources → relay channel and thus an additional

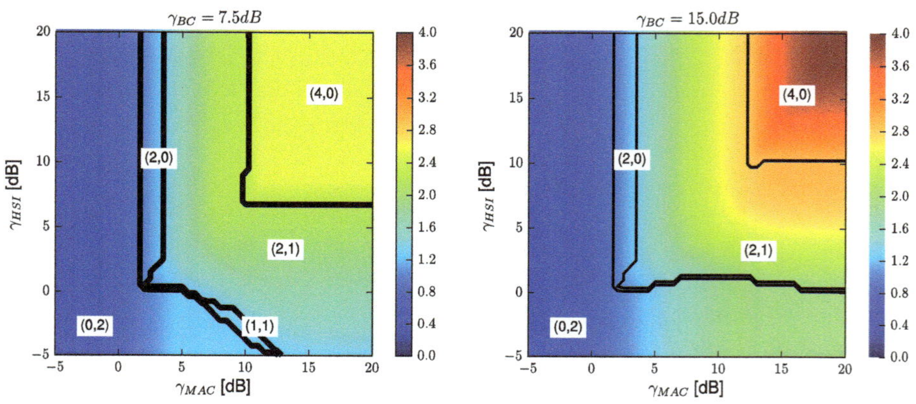

Fig. 14 Maximal encoded system throughput $T_C^{\max}$ for $\gamma_{\text{BC}} = 7.5$ dB and $\gamma_{\text{BC}} = 15$ dB

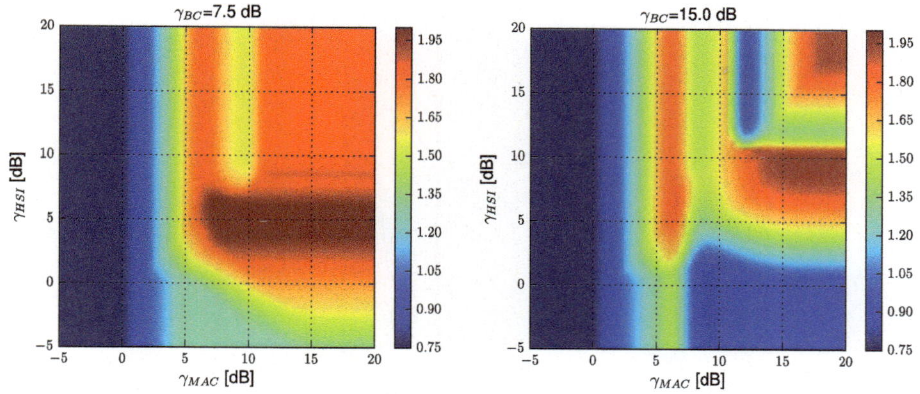

Fig. 15 Throughput enhancement $\Delta T_C = T_C^{\max} - T_{\text{LB}}$ [bits/channel use] of the coded over uncoded WBN system for $\gamma_{\text{BC}} = 7.5$ dB and $\gamma_{\text{BC}} = 15$ dB. Note that $0.75 \leq \Delta T_C \leq 2$ in the analyzed range of channel SNRs

increase of the source transmission power (and consequently γ_{MAC}) is required to provide a zero-error rate. However, γ_{BC} is not affected in this case, as the channel pre-rotation does not have any impact on the relay → destination performance.

7.4 Benchmarking with the state of the art three-step scenarios

For a more advanced comparison of the proposed method, we consider two conventional *three-step* communication scenarios.

7.4.1 Reference three-step scenarios

In the first scenario, both sources transmit their information one by one, but the relay is allowed to combine the received (fully decoded) data packets into the network coded packet $f(d_A, d_B) = d_A \oplus d_B$ and HSI channels are necessary to retrieve the desired information at both sources. This is the traditional network coding approach (see Fig. 18).

The second scenario is equivalent to conventional routing, where the sources subsequently transmit their data packet towards the relay. The relay then sends complete source information (no NC coding is applied) and thus the HSI channels are completely ignored (see Fig. 19).

7.4.2 Implementation details for reference scenarios

To provide a fair comparison, in both reference scenarios, we assume that the sources transmit with twice the power, as their transmit time is essentially halved when compared with the two-step scenarios. We assume QAM(N) constellation and adaptively look for N, r maximizing the throughput as in the proposed Section 7.2.

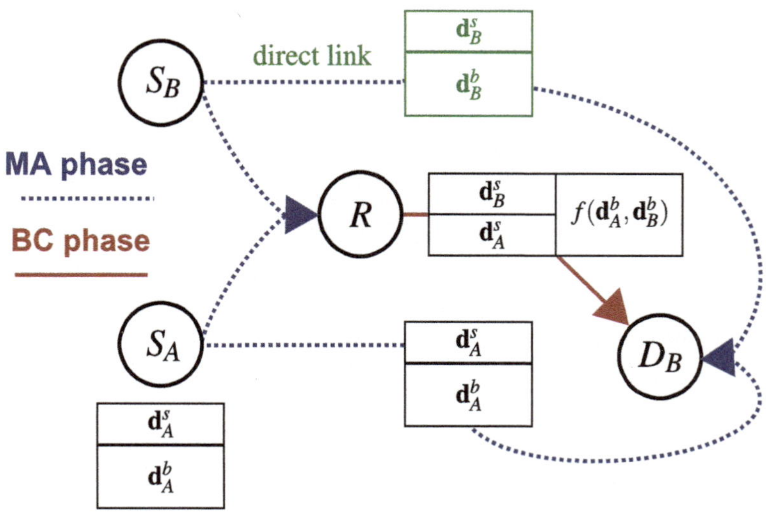

Fig. 16 Direct channel at destination D_B in the robustness analysis (likewise, a direct channel is assumed to be present also at D_A)

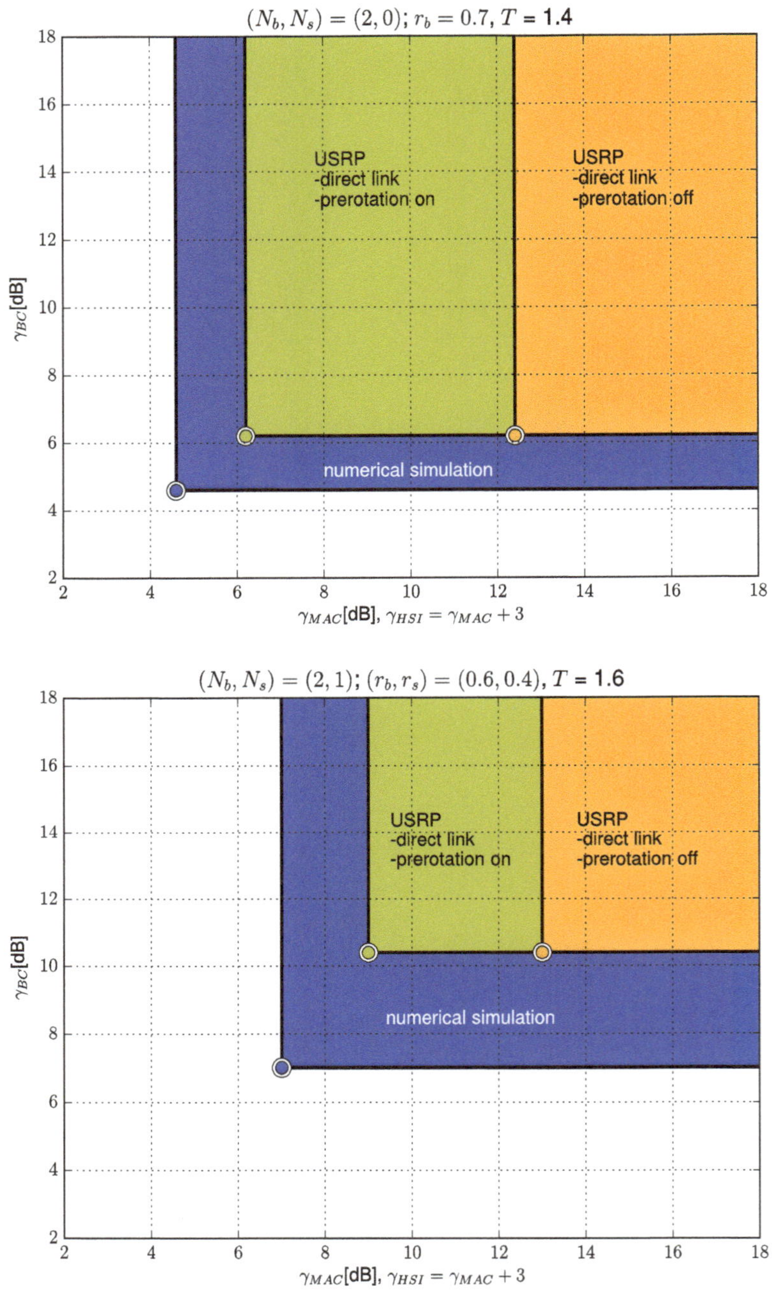

Fig. 17 Robustness analysis for two fixed (N_b, N_s, r_b, r_s) scenarios

7.4.3 Maximum transmission rates for the three-step reference scenarios

The theoretical limits for the considered three-step scenarios can easily be derived (see [15]) as

$$r_{\text{route}}^{3\text{step}} \leq \min\left\{\frac{2\text{Ca}(\gamma_{MAC})}{2}, \frac{2\text{Ca}(\gamma_{BC})}{2}\right\} \quad (27)$$

$$r_{\text{NC}}^{3\text{step}} \leq \min\left\{\frac{2\text{Ca}(2\gamma_{MAC})}{2}, \frac{2\text{Ca}(\gamma_{HSI})}{2}, \text{Ca}(\gamma_{BC})\right\}, \quad (28)$$

where $\text{CA}(\gamma) = \log_2(1+\gamma)$ denotes the conventional point-to-point channel capacity for Gaussian channels [18].

These bounds serve as an initial step for the tuple (rate, alphabet cardinality) optimization as in Section 6.2 but also as an ultimate theoretical bound of the individual reference strategies.

7.4.4 Numerical results

Within this experiment, we evaluated the following situations on our *USRP platform*:

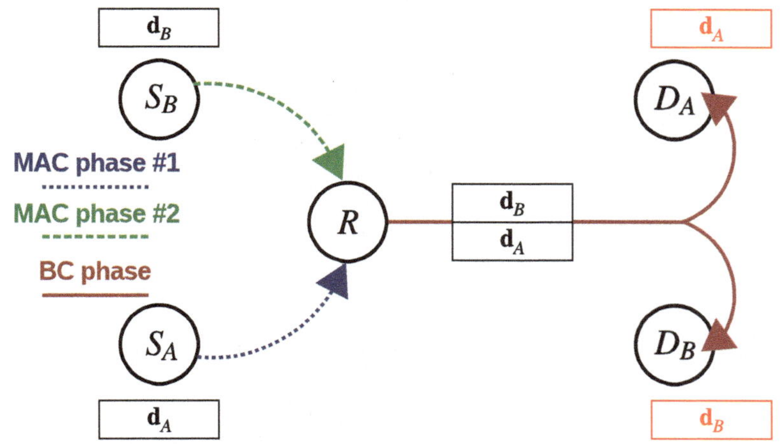

Fig. 18 Three-step reference scenario with conventional network coding

- The proposed adaptive constellation design from Section 7.2 based on the optimized tuple (r_b, r_s, N_b, N_s) for each measured point (HW-pre-rotation on).
- The proposed adaptive constellation design with assumptions violated (pre-rotation off).
- Both three-step reference scenarios introduced in Section 7.4.1.
- Fundamental theoretical limits for the conventional reference three-step scenarios.

In all measured cases, we manually selected the tuple maximizing the throughput. Our starting point was the theoretical analysis, but this analysis does not take into account the imperfect channel code and available discrete channel code rates (0.4, 0.5, …, 0.9). We also considered $N_b \leq 2, N_s \leq 1$ for simplicity[12].

We considered two setups:

- Strong HSI and equal MA and BC. This scenario is suited for standard wireless network coding, because it can efficiently utilize the strong HSI channel. See Fig. 20 for the corresponding result. As one can see, the proposed framework significantly outperforms both reference three-step scenarios as well the corresponding theoretical limits even if the pre-rotation is off! The bottleneck for the three-step routing is the BC channel because it needs to pass twice the amount of data from relay to destinations compared to all other schemes. The three-step NC approach does not suffer from this and therefore its performance is a bit better, nevertheless we pay for orthogonality in the MA stage. The performance of the proposed framework is superior to both reference models.
- Strong BC channel refers (among others) to a case with high relay power. The result is shown in Fig. 21.

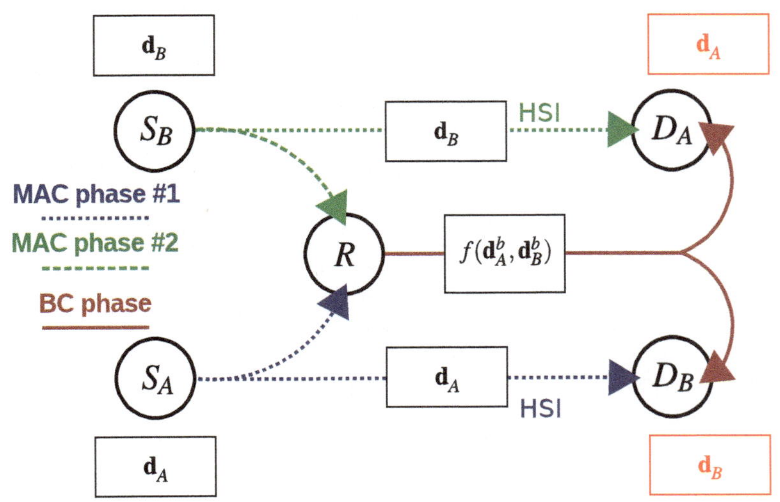

Fig. 19 Three-step reference scenario with conventional routing

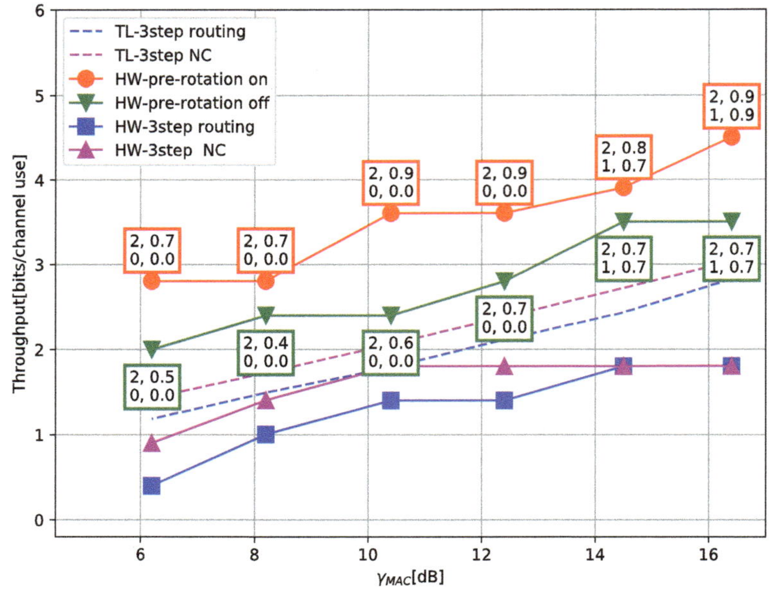

Fig. 20 Performance comparison for the setup $\gamma_{BC} \approx \gamma_{MAC}$ and $\gamma_{HSI} \approx \gamma_{MAC}$+3dB. The numbers in boxes correspond with particular values (first row:(N_b, r_b), second row:(N_s, r_s)) achieving a given maximal throughput

The three-step routing works much better when compared to the previous case, because of the strong BC channel. The conventional three-step network coding scheme suffers with a weak HSI channel. The proposed design again significantly outperforms both reference three-step scenarios. We should stress, that the performance gain is also achieved with *partial usage* of the weak HSI channel.

The saturation of the reference three-step strategies for high SNRs is a consequence of weaker performance of the underlaying channel code, where the rate of the code is adjusted by means of puncturing technique.

8 Conclusions

In this paper, we have proposed a systematic algorithm for the design of two-source constellations in the WBN

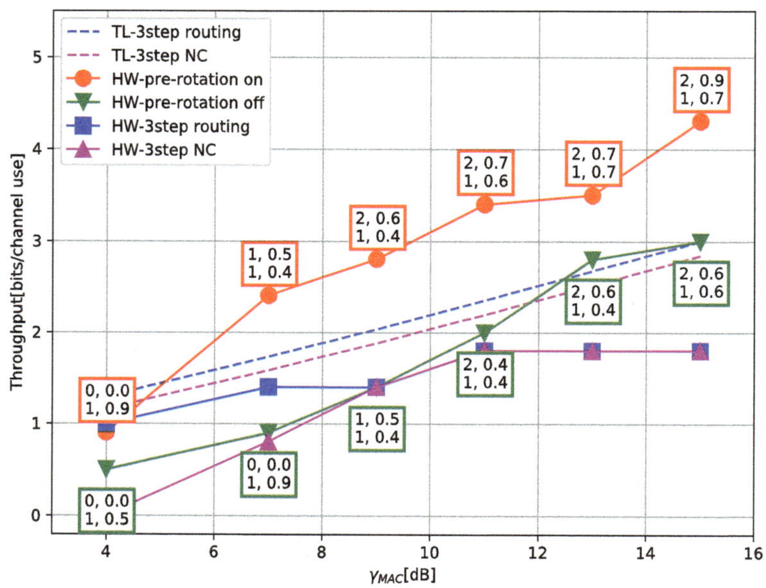

Fig. 21 Performance comparison for the setup $\gamma_{BC} \approx \gamma_{MAC}$+5dB and $\gamma_{HSI} \approx \gamma_{MAC}$-1dB. The numbers in boxes correspond with particular values (first row:(N_b, r_b), second row:(N_s, r_s)) achieving a given maximal throughput

system. The Algorithm 1 allows the matching of the constellation parameters to virtually any SNR conditions in the system, enabling a direct implementation of the adaptive constellation mapping scheme. The viability of the proposed constellation design was verified by means of numerical analysis, simulations and through real-world HW experiments (based on Ettus Research USRPs), including a HW-based robustness analysis.

Even in the uncoded scenario, the proposed constellations are able to significantly boost the WBN system throughput, and hence even their uncoded implementation can be interesting, especially in some basic wireless networks where channel coding is avoided to reduce complexity (e.g., in sensor networks). Furthermore, we have shown that the proposed source constellation design can be readily extended with appropriately designed channel encoding/decoding operations, significantly improving the reliability (including the throughput) of the resulting adaptive modulation-coding scheme.

In future work, we would like to focus on a full adaptive HW implementation of the proposed system, where the impact of several other parameters (e.g., channel fading and dynamics, asymmetry of channel gains, etc.) can significantly affect the system performance and possibly even require some modifications of the adaptive system design.

Nomenclature

WBN Wireless Butterfly Network
WNC Wireless Network Coding
HW Hardware
SNR Signal to Noise Ratio
SC Superposition Coding
HSI Hierarchical Side Information
DEM Demodulator
IC Interference Cancellation
MA Multiple Access
AWGN Additive White Gaussian Channel
ASK Amplitude Shift Keying
QAM Quadrature Amplitude Modulation
SER Symbol Error Rate
FER Frame Error Rate
SDR Software Defined Radio
USRP Universal Software Radio Peripherals
UHD USRP Hardware Driver
CP Cyclic Prefix
ODFD Orthogonal Frequency Division Multiplexing
DC Direct Current
MIMO Multiple Input Multiple Output
QPSK Quadrature Phase Shift Keying
UDP User Datagram Protocol
BC Broadcast Channel
MAC Multiple Access Channel
SODEM Soft Output Demodulator
CC Constellation Constrained

Endnotes

[1] Modification of the system model for the coded case will be provided in Section 6, together with a discussion on the design of individual channel encoders for particular information streams in the system.

[2] Modulo-sum [11, 34, 35] and bit-wise exclusive-or (XOR) [3, 23] hierarchical functions are commonly discussed in the WNC literature, however, unlike the bit-wise XOR, implementation of channel coding for the Modulo-sum mapping is a standalone research problem and hence it is out of the scope of this paper.

[3] Note that this also includes the system with perfect source phase prerotation [23–25], which can be obtained by a generalization of the scaling coefficients as $L'_n = L_n e^{-\mathrm{j}\measuredangle h_{SR}}$, where h_{SR} is the corresponding source→relay channel.

[4] The evaluated $P_{e,\mathrm{UB}}^{SER}$ is an upper-bound of P_e^{SER}, since an error in MA/BC phase does not necessarily induce an error in the overall communication chain of the uncoded system. This is especially true if the error occurs in the superposed bits $\mathbf{d}_B^s$ used for interference cancellation processing at D_A.

[5] In AWGN channel, the resulting superimposed constellation has too many overlaps for the reference source QPSK constellations, preventing a successful multi-user decoding of both of the separate user data streams in the uncoded system.

[6] It can be shown that for constant $N_b + N_s$ the minimum Euclidean distance ρ_{HSI}^2 grows with decreasing N_b, resulting in a more efficient utilization of "imperfect" HSI channels [15, 36] and consequently in a more reliable overheard HSI for the constellations with lower (non-zero) N_b.

[7] The estimated SNR parameters ($\hat{\gamma}_{\mathrm{MAC}}, \hat{\gamma}_{\mathrm{BC}}, \hat{\gamma}_{\mathrm{HSI}}$) must be delivered to both sources by a feedback channel from the relay and/or destinations. The particular implementation of this feedback transmission is beyond the scope of this paper and hence (for the sake of simplicity) we assume that these parameters are made available at both sources by a genie. Note that in the HW evaluation, the required feedback is provided via UDP connection.

[8] In a more general approach, each source data word is split into N_b+N_s parallel data sub-words that are encoded separately using the principles of multilevel coding (see e.g. [37]). However, the development of a general multilevel coding scheme for WBN is beyond the scope of this paper.

[9] Alternatively, a primitive end-to-end coding scheme can be designed, performing the channel coding solely at sources S_A, S_B and channel decoding (potentially computationally intensive) solely at destinations D_A, D_B, while leaving the relay R to operate on a symbol-by-symbol basis (as in the uncoded system). Unfortunately, this approach is only sub-optimal, as in this case the particular transmissions on source→relay and relay→destination channels are not individually protected by channel coding.

[10] Note that separate decoding of $\mathbf{D}_A^b$, $\mathbf{D}_B^b$ and subsequent evaluation of $f\left(\mathbf{D}_A^b, \mathbf{D}_B^b\right)$ (i.e. conventional joint/multi-user decoding concatenated with traditional network coding [38]) can significantly limit the performance of the system [13].

[11] The (conditional) probability density function $p\left(x|\mathbf{c}_A^s, \mathbf{c}_B^s, f\left(\mathbf{c}_A^b, \mathbf{c}_B^b\right)\right)$ which is required for the evaluation of particular mutual information in (13)-(21) can be obtained by a proper marginalization operation $p\left(x|\mathbf{c}_A^s, \mathbf{c}_B^s, f\left(\mathbf{c}_A^b, \mathbf{c}_B^b\right)\right) = \Pr\left\{f\left(\mathbf{c}_A^b, \mathbf{c}_B^b\right)\right\} \cdot \sum_{\mathbf{c}_A^b, \mathbf{c}_B^b : f\left(\mathbf{c}_A^b, \mathbf{c}_B^b\right)} p\left(x|\mathbf{c}_A^s, \mathbf{c}_B^s, \mathbf{c}_A^b, \mathbf{c}_B^b\right)$, where $p\left(x|\mathbf{c}_A^s, \mathbf{c}_B^s, \mathbf{c}_A^b, \mathbf{c}_B^b\right)$ is the likelihood function. For more details on the evaluation of CC capacity for finite input constellations see e.g. [13, 32].

[12] The higher order strategies like ($N_b = 4, N_s = 0$) performed similarly as considered ($N_b = 2, N_s = 1$) in our simulations. Note that much stronger solution based on the modulo-sum channel coding and wireless network coding function is available [35], nevertheless implementation of the non-binary channel code suitable for the USRP is challenging.

Funding
This work has been performed in the framework of the European research project DIWINE, which is partly funded by the European Union under its FP7 ICT Objective 1.1 - The Network of the Future. Work of Pavel Prochazka and Tomas Uricar was also partially supported by the Grant Agency of the Czech Technical University in Prague, SGS project SGS16/094/OHK3/1T/13.

Declarations
This work was partly reported in IEEE ICC [11].

Competing interests
The authors declare that they have no competing interests.

Author details
[1] Czech Technical University in Prague, Technicka 2, 16627 Prague, Czech Republic. [2] Toshiba—Telecommunications Research Laboratory, 32 Queen Square, Bristol, BS1 4ND, UK.

References

1. P Popovski, H Yomo, in *Proc. IEEE Internat. Conf. on Commun. (ICC)*. Physical network coding in two-way wireless relay channels (IEEE Conference Publications, Piscataway, 2007)
2. B Nazer, M Gastpar, Reliable physical layer network coding. Proc. IEEE. **99**(3), 438–60 (2011)
3. S Zhang, S-C Liew, L Lu, Physical-layer network coding: Tutorial, survey, and beyond. Phys. Commun. **2012**, 1–39 (2012). doi:10.1016/j.phycom.2012.05.002
4. T Koike-Akino, P Popovski, V Tarokh, in *Proc. IEEE Global Telecommunications Conf. (GlobeCom)*. Denoising maps and constellations for wireless network coding in two-way relaying systems (IEEE Conference Publications, Piscataway, 2008)
5. J Liu, M Tao, Y Xu, in *Proc. IEEE Internat. Conf. on Commun. (ICC)*. Pseudo exclusive-OR for LDPC coded two-way relay block fading channels (IEEE Conference Publications, Piscataway, 2011), pp. 1–5
6. K Yasami, A Razi, A Abedi, Analysis of channel estimation error in physical layer network coding. IEEE Commun. Lett. **15**(10), 1029–31 (2011). doi:10.1109/LCOMM.2011.082011.110301
7. T Uricar, J Sykora, Non-uniform 2-slot constellations for bidirectional relaying in fading channels. IEEE Commun. Lett. **15**(8), 795–7 (2011)
8. S Zhang, S-C Liew, PP Lam, in *Proc. IEEE Inf. Theory Workshop (ITW)*. On the synchronization of physical-layer network coding (IEEE Conference Publications, Piscataway, 2006), pp. 404–408
9. F Rossetto, M Zorzi, in *Proc. IEEE Workshop on Signal Processing Advances in Wireless Communications*. On the design of practical asynchronous physical layer network coding (IEEE Conference Publications, Piscataway, 2009), pp. 469–473
10. Y Huang, S Wang, Q Song, L Guo, A Jamalipour, Synchronous physical-layer network coding: a feasibility study. IEEE Trans. Wireless Commun. **12**(8), 4048–4057 (2013)
11. P Prochazka, T Uricar, D Halls, J Sykora, in *Proc. IEEE Int. Conf. on Commun. (ICC)*. Relaying in butterfly networks: superposition constellation design for wireless network coding (IEEE Conference Publications, London, 2015), pp. 1–7
12. B Nazer, M Gastpar, in *Proc. IEEE Internat. Symp. on Inf. Theory (ISIT)*. Compute-and-forward: Harnessing interference with structured codes (IEEE Conference Publications, Piscataway, 2008), pp. 772–776
13. J Sykora, A Burr, Layered design of hierarchical exclusive codebook and its capacity regions for HDF strategy in parametric wireless 2-WRC. IEEE Trans. Veh. Technol. **60**(7), 3241–52 (2011)
14. T Uricar, B Qian, J Sykora, WH Mow, in *Proc. IEEE Wireless Commun. Network. Conf. (WCNC)*. Superposition coding for wireless butterfly network with partial network side-information (IEEE Conference Publications, Shanghai, 2013), pp. 1–6
15. T Uricar, B Qian, J Sykora, WH Mow, Wireless (physical layer) network coding with limited hierarchical side-information: Maximal sum-rates in 5-node butterfly network. IEEE Trans. Wireless Commun. **13**(10), 5582–95 (2014)
16. IW-H Ho, SC Liew, L Lu, in *Proc. IEEE Internat. Symp. on Inf. Theory (ISIT)*. Feasibility study of physical-layer network coding in 802.11p VANETs, (2014), pp. 646–650. doi:10.1109/ISIT.2014.6874912
17. P Prochazka, WNC tutorial—constellation design for butterfly network. http://pavel.prochazka.info/tutorial/const_design_WNC. Accessed 28 Dec 2016
18. TM Cover, JA Thomas, *Elements of information theory*. (John Wiley & Sons, The Atrium, Southern Gate, Chichester PO19 8SQ, UK, 1991)
19. TS Han, K Kobayashi, A new achievable rate region for the interference channel. IEEE Trans. Inf. Theory. **27**, 49–60 (1981)
20. N Shende, BS Rajan, in *Proc. National Conference on Communications (NCC)*. On the achievable rate of AWGN relay channel with finite input constellations (IEEE Conference Publications, Piscataway, 2012), pp. 1–5
21. P Popovski, E de Carvalho, Improving the rates in wireless relay systems through superposition coding. IEEE Trans. Wireless Commun. **7**(12), 4831–6 (2008)
22. PA Hoeher, T Wo, Superposition modulation: myth and facts. IEEE Communications Mag. **49**(12), 110–6 (2011)
23. T Koike-Akino, P Popovski, V Tarokh, Optimized constellations for two-way wireless relaying with physical network coding. IEEE J. Sel. Areas Commun. **27**(5), 773–87 (2009)
24. S Zhang, S-C Liew, Applying physical-layer network coding in wireless networks. EURASIP J. on Wireless Comm. and Netw. **2010**, 1–12 (2010). doi:10.1155/2010/870268

25. A Li, Y Yan, H Kayama, An enhanced denoise-and-forward relaying scheme for fading channel with low computational complexity. IEEE Signal Process. Lett. **15**, 857–860 (2008)
26. E Research, N200/N210 Product overview. https://www.ettus.com/content/files/07495_Ettus_N200-210_DS_Flyer_HR_1.pdf. Accessed 12 Feb 2016
27. GNURadio, GNURadio overview. http://gnuradio.org/redmine/projects/gnuradio. Accessed 12 Feb 2016
28. T Hynek, D Halls, J Sykora, in *Proc. European Wireless Conf. (EW)*. Practical implementation of cloud initialization procedure for wireless physical layer network coding clouds (VDE Conference Publications, Berlin, 2014), pp. 1–6
29. D Halls, W Thompson, S Galimberti, S Savazzi, T Hynek, J Sykora, J Coon, Evaluation of practical system constraints, DIWINE deliverable D5.12 (2014). Technical report, FP7 ICT Objective 1.1 - The Network of the Future
30. E Research, OctoClock-G Product overview. https://www.ettus.com/product/details/OctoClock-G. Accessed 12 Feb 2016
31. E Biglieri, *Coding for wireless channels*. (Springer, 233 Spring Street, New York, NY 10013-1578,USA, 2010)
32. J Harshan, BS Rajan, On two-user gaussian multiple access channels with finite input constellations. IEEE Trans. Inf. Theory. **57**(3), 1299–327 (2011)
33. CB Schlegel, LC Perez, *Trellis and turbo coding: iterative and graph-based error control coding*. (John Wiley & Sons, The Atrium, Southern Gate, Chichester PO19 8SQ, UK, 2015)
34. M Hekrdla, J Sykora, in *Vehicular Technology Conference (VTC Spring), 2013 IEEE 77th*. On indexing of lattice-constellations for wireless network coding with modulo-sum decoding (IEEE Conference Publications, Piscataway, 2013), pp. 1–6
35. P Prochazka, Nonbinary channel coded physical layer network coding over modulo-sum algebraic ring structures. IEEE Commun. Lett. **20**(3), 538–41 (2016)
36. T Uricar, J Sykora, Non-uniform 2-slot constellations for relaying in butterfly network with imperfect side information. IEEE Commun. Lett. **16**(9), 1369–72 (2012)
37. U Wachsmann, RFH Fischer, JB Huber, Multilevel codes: theoretical concepts and practical design rules. IEEE Trans. Inf. Theory. **45**(5), 1361–91 (1999)
38. RW Yeung, S-YR Li, N Cai, Z Zhang, *Network Coding Theory*. (NOW Publishers, 2600 AD Delft, The Netherlands, 2006)

Compressive cyclostationary spectrum sensing with a constant false alarm rate

Andreas Bollig[1*], Anastasia Lavrenko[2], Martijn Arts[1] and Rudolf Mathar[1]

Abstract

Spectrum sensing is a crucial component of opportunistic spectrum access schemes, which aim at improving spectrum utilization by allowing for the reuse of idle licensed spectrum. Sensing a spectral band before using it makes sure the legitimate users are not disturbed. To that end, a number of different spectrum sensing method have been developed in the literature. *Cyclostationary* detection is a particular sensing approach that takes use of the built-in periodicities characteristic to most man-made signals. It offers a compromise between achievable performance and the amount of prior information needed. However, it often requires a significant amount of data in order to provide a reliable estimate of the cyclic autocorrelation (CA) function. In this work, we take advantage of the inherent sparsity of the cyclic spectrum in order to estimate CA from a low number of linear measurements and enable blind cyclostationary spectrum sensing. Particularly, we propose two compressive spectrum sensing algorithms that exploit further prior information on the CA structure. In the first one, we make use of the joint sparsity of the CA vectors with regard to the time delay, while in the second one, we introduce structure dictionary to enhance the reconstruction performance. Furthermore, we extend a statistical test for cyclostationarity to accommodate *sparse* cyclic spectra. Our numerical results demonstrate that the new methods achieve a near constant false alarm rate behavior in contrast to earlier approaches from the literature.

Keywords: Cyclostationarity, Spectrum sensing, Sparse recovery

1 Introduction

The scarcity of radio spectrum constitutes a major roadblock to current and future innovation in wireless communications. To alleviate this problem, it has been proposed to make spectral resources, which are currently underutilized, available for reuse under a paradigm that goes by the name of opportunistic spectrum access (OSA) [1]. Spectrum sensing is one of its core technologies. It allows an unlicensed transceiver, a so-called secondary user (SU), to access a licensed spectral band without interfering with the owner of the band's license, the so-called primary user (PU). The fundamental task in spectrum sensing is to decide between two hypotheses, the first of which states that the spectral band under investigation is free ($\mathcal{H}_0$), while the second asserts that it is occupied ($\mathcal{H}_1$). Considering the baseband signal $x(t)$ observed at a secondary system receiver, the two hypotheses can be written as

$$\begin{aligned} \mathcal{H}_0 : x(t) &= \eta(t), \\ \mathcal{H}_1 : x(t) &= s'(t) + \eta(t), \end{aligned} \qquad (1)$$

where $\eta(t)$ denotes receiver noise and $s'(t)$ stands for a PU signal after propagation effects.

A number of spectrum sensing algorithms have been proposed in the literature [2–4]. Broadly speaking, they can be divided into three major types, namely, energy detection, stochastic feature detection, and matched filter detection, where different types require different amounts of prior knowledge about the PU signal. While matched filter ([5], Ch. 4.3) detectors require the knowledge of the exact waveform of at least a part of the PU signal, energy detection [6] does not require any prior knowledge. Feature detectors are an in-between as they only make assumptions about structural or statistical properties of the signal. One of the stochastic features which lets an SU receiver discriminate between pure stationary noise ($\mathcal{H}_0$) and a communication signal contaminated

*Correspondence: bollig@ti.rwth-aachen.de
[1]Institute for Theoretical Information Technology, RWTH Aachen University, Kopernikusstraße 16, 52074 Aachen, Germany
Full list of author information is available at the end of the article

with noise ($\mathcal{H}_1$) is cyclostationarity. In contrast to pure stationary noise, most man-made signals vary periodically with time [7] and can thus be characterized as cyclostationary. Although the data contained in a modulated signal may be a purely stationary random process, the coupling with sine wave carriers, pulse trains, repeating, spreading, hopping sequences, and cyclic prefixes going along with its modulation causes a built-in periodicity [8].

The use of cyclostationarity for the purpose of spectrum sensing has been investigated from a variety of perspectives ranging from single node signal detection [9–16] to collaborative approaches that take use of the spatial diversity [17, 18]. One of the particularly well-known algorithms for cyclostationary spectrum sensing is the so-called *time-domain test (TDT)* as introduced in [9]. The test can decide between the presence and absence of cyclostationarity for a pre-specified potential cycle frequency α. It operates on the cyclic autocorrelation (CA), which, given an observed signal $x(t)$, is defined as [7]

$$R_x^\alpha(\tau) = \lim_{T\to\infty} \frac{1}{T} \int_{-T/2}^{T/2} x(t+\tau/2)x^*(t-\tau/2)e^{-j2\pi\alpha t}\mathrm{d}t \tag{2}$$

for a potential cycle frequency α and a delay τ. For purely stationary signals, $R_x^\alpha(\tau) = 0$ for all $\alpha \neq 0$, while for cyclostationary signals, $R_x^\alpha(\tau) \neq 0$ for some $\alpha \neq 0$. The α with non-zero CA coefficients are called cycle frequencies. The set of cycle frequencies caused by one of potentially multiple incommensurate second-order periodicities in a cyclostationary signal comprises the periodicity's fundamental cycle frequency (the reciprocal of the fundamental period) as well as its harmonics. Given the above information, we can rewrite the hypothesis test (1) as

$$\begin{aligned} &\mathcal{H}_0 : \forall\{\alpha \in \mathbb{R}|\alpha \neq 0\} : R_x^\alpha(\tau) = 0, \\ &\mathcal{H}_1 : \exists\{\alpha \in \mathbb{R}|\alpha \neq 0\} : R_x^\alpha(\tau) \neq 0. \end{aligned} \tag{3}$$

It is important to note that practically, instead of the statistical CA (2), one normally operates on the sample CA obtained from a limited number of signal samples. The coefficients of the sample CA are not constant but rather follow different probability distributions, depending on whether $\mathcal{H}_0$ or $\mathcal{H}_1$ is true. To account for this, the hypothesis test (3) is modified by considering different test statistic under $\mathcal{H}_0$ and $\mathcal{H}_1$. As a result, the TDT provides a typical constant false alarm rate (CFAR) performance.

Returning to (3), we note that the CA is zero on its whole support except the set of cycle frequencies and $\alpha = 0$. Therefore, it can be called sparse. The exploitation of sparsity in signal processing has a long history [19]. The recent years, however, have seen a vastly accelerated development of the field resulting in a new sampling paradigm called compressive sampling (CS) [20, 21]. It postulates that sparse or compressible signals, i.e., the signals that can be represented or well approximated by only a few non-zero coefficients in some domain, can be sampled and recovered from fewer number of measurements than traditionally required. A crucial observation here is that one can design an effective measurement strategy that is governed by the amount of the signals' *information content*, rather than its *ambient dimension*. To date, there is a large amount of powerful algorithms available that solve sparse recovery problems in the CS context ranging from optimization approaches to classical pursuits such as the orthogonal matching pursuit (OMP) [22] and more specialized algorithms such as the compressive sampling matching pursuit (CoSaMP) [23] for instance.

Multiple contributions have been made in the field of compressive cyclostationary spectrum sensing. The authors of [24] for instance formulate the estimation of the cyclic autocorrelation as a sparse recovery problem, which they solve using the OMP [22]. Based on the sparse estimate of the CA, they propose two detection methods that exploit different CA properties. The first one, called slot comparison method (SCM), compares the biggest CA components OMP finds in two consecutive blocks of samples. If for both blocks, the same discrete cycle frequencies are chosen, $\mathcal{H}_1$ is selected; otherwise, $\mathcal{H}_0$ is selected. The second detection algorithm is called symmetry method (SM). It exploits the fact, that for certain types of signals, the CA is symmetric around the direct current (DC) component. Although both of them present blind detectors meaning that they operate without prior knowledge of the cycle frequencies present in the signal, they do not allow for a CFAR performance, which is considered a desired detector feature [4]. Instead of the CA, the authors of [25] use the spectral correlation (SC), which is the Fourier transform of the CA over τ, for detecting multiple transmitters in a wideband signal using compressive sampling. In order to estimate the SC from compressed samples via CS, they established a direct linear relation between the compressed samples and the SC. Based on [25], the authors of [26] derive a method for recovering the SC from sub-Nyquist samples using a reduced complexity approach, for which they provide a closed-form solution. In [27], the modulated wideband converter (MWC) [28] is used to obtain the SC from sub-Nyquist samples to then apply cyclostationarity detection. Furthermore, it has been recently shown that under certain conditions, the CA can be efficiently recovered from a low number of samples even without enforcing the sparsity property [29–31]. This can be done by exploiting cross-correlations between different outputs of the compressive sampler. The main drawback of the aforementioned works is that providing the estimate of the entire cyclic spectrum, they still require the knowledge of the cycle frequencies for the detection step.

In this work, we employ a composite approach that combines the sparse recovery of CA from its compressive measurements for blind cycle frequency estimation with a CFAR TDT detection. This said the contribution of this paper is manifold. We propose two novel sparsity-aided CA estimation algorithms, both of which exploit *further* prior information about the CA in addition to its sparsity: the simultaneous OMP-based (SOber) and the dictionary assisted (Dice) compressive CA estimator. The first one exploits the joint sparsity of the CA vectors with regard to the time delay in order to recover the CA matrix for all delays simultaneously, while the second one takes advantage of the signal-induced structure of the CA by introducing structure dictionaries into the recovery process. In order to evaluate the performance of the proposed CA estimators, we derive a closed-form expression of the CA of *sampled* linearly modulated signals with rectangular pulse shape. Furthermore, we show how this expression can be used as prior information in the dictionary assisted approach. Note that the use of sparse recovery in the novel CA estimation approaches results in the automatic detection of signal's cycle frequencies. This in turn allows *blind* spectrum sensing by eliminating the integral need of the classical TDT for the perfect knowledge of the said cycle frequencies. However, the resulting sparse structure of the compressive CA estimates does not allow for the application of the traditional TDT since the noise statistics are missing. To compensate for this phenomenon, we develop a modified TDT and thus enable blind compressive cyclostationary spectrum sensing. Numerical tests show that the proposed method achieves a near CFAR behavior.

The remainder of this paper is structured as follows. Section 2.1 introduces the signal model and presents the classical method for CA estimation, while Section 2.2 presents the time-domain test based on the classical CA estimation. A CA estimator based on joint sparsity of multiple vectors is introduced in Section 3 and the CA estimator exploiting additional prior knowledge is described in Section 4. An extension of the TDT to accommodate sparse CA estimates is developed in Section 5. The numerical evaluation of the proposed estimation and detection approaches as well as the interpretation of the results is given in Section 6. Section 7 concludes the paper.

2 Cyclostationary spectrum sensing

2.1 System model and CA estimation

Consider a secondary system receiver that needs to decide whether a certain spectral band is occupied or free. It samples the baseband signal $x(t)$ uniformly with a sampling period T_e. This results in the vector of discrete samples $\mathbf{x}_{t_0} \in \mathbb{C}^N$, where

$$\mathbf{x}_{t_0} = [x(t_0), x(t_0 + T_e), \ldots, x(t_0 + (N-1)T_e)]^{\mathrm{T}}. \tag{4}$$

We assume that the vector $\mathbf{x}_{t_0}$ is discrete and zero mean, and due to the nature of man-made signals, it represents an (almost [32], Ch. 1.3) cyclostationary process [9]. The presence of stochastic periodicity in the samples and thus the presence of a man-made signal can be revealed by applying a detection algorithm such as the TDT to the CA of the samples. There are different ways of obtaining the CA from the baseband samples, one of which is the following (classical) estimator

$$\hat{R}^{a}_{x,t_0}(\nu) = \frac{1}{N} \sum_{n=0}^{N-1-\nu} x(t_0 + nT_e)x^*(t_0 + (n+\nu)T_e) \times e^{-j2\pi \frac{a}{N} n} e^{-j\pi \frac{a}{N} \nu}. \tag{5}$$

Evaluating this function results in the CA coefficient for the cycle frequency $\alpha = \frac{a}{NT_e}$ and the time delay $\tau = \nu T_e$, where a stands for the discrete cycle frequency and ν denotes the discrete time delay. Note that the factor $e^{-j\pi \frac{a}{N} \nu}$ remains constant throughout the sum. It is a phase shift necessary to maintain compatibility with the symmetric CA (2). The estimator (5) is biased but exhibits a smaller estimation variance than an unbiased one [9].

We define $\hat{\mathbf{r}}^{\nu}_{x}$ as an N length CA vector whose nth element is $\hat{R}^{n}_{x,t_0}(\nu)$, i.e.,

$$\hat{\mathbf{r}}^{\nu}_{x} = \left[\hat{R}^{0}_{x,t_0}(\nu), \ldots, \hat{R}^{N-1}_{x,t_0}(\nu)\right]^{\mathrm{T}}. \tag{6}$$

Subsequently, we re-write the estimation of the CA vector as a matrix-vector product. To do so, we need the (N element) *delay product* with time delay $\tau = \nu T_e$, which is given by

$$\mathbf{y}^{\nu}_{N} = \mathbf{x}_{t_0} \circ \mathbf{x}^{*}_{t_0+\nu T_e}, \tag{7}$$

where $\circ$ denotes component-wise multiplication. Note that since the receiver only takes N samples, $\mathbf{x}^{*}_{t_0+\nu T_e}$ is zero-padded at the end while $\mathbf{y}^{\nu}_{N}$ is a vector of length N. The CA vector is now given by

$$\hat{\mathbf{r}}^{\nu}_{x} = \frac{1}{N} F \mathbf{y}^{\nu}_{N}, \tag{8}$$

where $\mathbf{F}$ denotes the $(N \times N)$ discrete Fourier transform (DFT) matrix. The $N \times n_\nu$ CA matrix for time delays $\nu_1 T_e, \ldots, \nu_{n_\nu} T_e$ is given by

$$\hat{\mathbf{R}}_x = \left[\hat{\mathbf{r}}^{\nu_1}_{x}, \ldots, \hat{\mathbf{r}}^{\nu_{n_\nu}}_{x}\right] = \frac{1}{N} \mathbf{F} \mathbf{Y}_N, \tag{9}$$

with $\mathbf{Y}_N = \left[\mathbf{y}^{\nu_1}_{N}, \ldots, \mathbf{y}^{\nu_{n_\nu}}_{N}\right] \in \mathbb{C}^{N \times n_\nu}$.

2.2 The time-domain test (TDT) for cyclostationarity

Given the statistical CA, one could decide between $\mathcal{H}_0$ and $\mathcal{H}_1$ by testing it for being non-zero at the signal's inherent cycle frequencies according to (3). However, as mentioned in Section 1, instead of the statistical CA, we only have access to its estimation, the sample CA (which asymptotically converges to the statistical CA). This hinders the direct applicability of (3) for signal detection as coefficients of the sample CA are not constant anymore. In the seminal work [9], the probability distributions that the sample CA coefficients follow under $\mathcal{H}_0$ or $\mathcal{H}_1$ have been identified and a test for cyclostationarity based on this knowledge has been designed. The test is briefly described in the following.

Consider the $2n_\nu \times 1$ vector

$$\hat{\mathbf{r}}_{xx^*}(a_0) = \Big[\Re\Big\{\hat{\mathbf{R}}_x[a_0, \nu_1]\Big\}, \ldots, \Re\Big\{\hat{\mathbf{R}}_x[a_0, \nu_{n_\nu}]\Big\}, \Im\Big\{\hat{\mathbf{R}}_x[a_0, \nu_1]\Big\}, \ldots, \Im\Big\{\hat{\mathbf{R}}_x[a_0, \nu_{n_\nu}]\Big\}\Big]^{\mathrm{T}}, \tag{10}$$

which represents the concatenation of the real and the imaginary part of the row of $\hat{\mathbf{R}}_x$ corresponding to the discrete cycle frequency a_0. The frequency a_0 is the cycle frequency of interest, i. e., the one for the presence of which we want to test the signal. Given this vector, we can formulate the following non-asymptotic hypotheses

$$\begin{aligned} \mathcal{H}_0 : \hat{\mathbf{r}}_{xx^*}(a_0) &= \boldsymbol{\epsilon}_{xx^*}(a_0), \\ \mathcal{H}_1 : \hat{\mathbf{r}}_{xx^*}(a_0) &= \mathbf{r}_{xx^*}(a_0) + \boldsymbol{\epsilon}_{xx^*}(a_0), \end{aligned} \tag{11}$$

where $\mathbf{r}_{xx^*}(a_0)$ is the deterministic but unknown asymptotic counterpart of $\hat{\mathbf{r}}_{xx^*}(a_0)$ and $\boldsymbol{\epsilon}_{xx^*}(a_0)$ is the estimation error. Note that in contrast to the hypotheses from Eq. (3), this formulation considers the presence of cyclostationarity in the received signal for *one fixed* cycle frequency a_0.

Since $\mathbf{r}_{xx^*}(a_0)$ is non-random, the distribution of $\hat{\mathbf{r}}_{xx^*}(a_0)$ under $\mathcal{H}_0$ and $\mathcal{H}_1$ only differs in mean. As shown in [9], the estimation error $\boldsymbol{\epsilon}_{xx^*}(a_0)$ asymptotically follows a Gaussian distribution, i. e.,

$$\lim_{N\to\infty} \sqrt{N}\boldsymbol{\epsilon}_{xx^*}(a_0) \overset{\mathrm{D}}{=} \mathcal{N}(0, \Sigma_{xx^*}(a_0)), \tag{12}$$

where $\Sigma_{xx^*}(a_0)$ is the statistical covariance matrix of $\hat{\mathbf{r}}_{xx^*}(a_0)$ and $\overset{\mathrm{D}}{=}$ denotes convergence in distribution. The $2n_\nu \times 2n_\nu$ covariance matrix $\Sigma_{xx^*}(a_0)$ can be computed as [9]

$$\Sigma_{xx^*}(a_0) = \begin{bmatrix} \Re\left\{\frac{\mathbf{Q}+\mathbf{Q}^*}{2}\right\} & \Im\left\{\frac{\mathbf{Q}-\mathbf{Q}^*}{2}\right\} \\ \Im\left\{\frac{\mathbf{Q}+\mathbf{Q}^*}{2}\right\} & \Re\left\{\frac{\mathbf{Q}^*-\mathbf{Q}}{2}\right\} \end{bmatrix}, \tag{13}$$

where the (m, n)th entries of the $n_\nu \times n_\nu$ matrices $\mathbf{Q}$ and $\mathbf{Q}^*$ are given by

$$\mathbf{Q}(m, n) = S_{\mathbf{y}_N^{\nu_m}\mathbf{y}_N^{\nu_n}}(2a_0, a_0), \text{ and} \tag{14}$$

$$\mathbf{Q}^*(m, n) = S^*_{\mathbf{y}_N^{\nu_m}\mathbf{y}_N^{\nu_n}}(0, -a_0) \tag{15}$$

respectively. The term $S_{\mathbf{y}_N^{\nu_m}\mathbf{y}_N^{\nu_n}}(\cdot, \cdot)$ denotes the unconjugated, while the term $S^*_{\mathbf{y}_N^{\nu_m}\mathbf{y}_N^{\nu_n}}(\cdot, \cdot)$ denotes the conjugated cyclic spectrum of a signal. One way to estimate these is to determine the following frequency-smoothed periodograms:

$$\hat{S}_{\mathbf{y}_N^{\nu_m}\mathbf{y}_N^{\nu_n}}(2a_0, a_0) = \frac{1}{NL} \sum_{s=-\frac{L-1}{2}}^{\frac{L-1}{2}} W(s) \times \hat{\mathbf{R}}_x[a_0 - s, \nu_n]\, \hat{\mathbf{R}}_x[a_0 + s, \nu_m] \tag{16}$$

$$\hat{S}^*_{\mathbf{y}_N^{\nu_m}\mathbf{y}_N^{\nu_n}}(0, -a_0) = \frac{1}{NL} \sum_{s=-\frac{L-1}{2}}^{\frac{L-1}{2}} W(s) \times \hat{\mathbf{R}}_x^*[a_0 + s, \nu_n]\, \hat{\mathbf{R}}_x[a_0 + s, \nu_m], \tag{17}$$

where W is a normalized spectral window of odd length L. Looking at the Eqs. (16) and (17), it becomes clear why the cyclic spectrum is often referred to as the spectral correlation.

Given the estimated quantities described above, the following generalized likelihood ratio (GLR) test statistic can be derived [17]

$$\mathcal{T}_{xx^*} = N\hat{\mathbf{r}}_{xx^*}(a_0)\hat{\Sigma}^{-1}_{xx^*}(a_0)\hat{\mathbf{r}}^{\mathrm{T}}_{xx^*}(a_0). \tag{18}$$

The test statistic can be interpreted as a normalized energy. The inverse of the covariance matrix scales $\hat{\mathbf{r}}_{xx^*}(a_0)$ such that under $\mathcal{H}_0$, its entries follow a standard normal distribution. Thus, under $\mathcal{H}_0$, the test statistic asymptotically follows a central chi-squared distribution with $2n_\tau$ degrees of freedom i. e., $\lim_{N\to\infty} \mathcal{T}_{xx^*} \overset{\mathrm{D}}{=} \chi^2_{2n_\nu}$, while under $\mathcal{H}_1$, the test statistic asymptotically follows a non-central chi-squared distribution with unknown non-centrality parameter λ, i. e., $\lim_{N\to\infty} \mathcal{T}_{xx^*} \overset{\mathrm{D}}{=} \chi^2_{2n_\nu}(\lambda)$. Based on the above test statistic, we can design a CFAR detector with some false alarm rate P_{fa} by finding the corresponding decision threshold in the $\chi^2_{2n_\nu}$ tables. We cannot design a test based on a desired detection rate P_{d} since although $\mathbf{r}_{xx^*}(a_0)$ is deterministic, it depends on the type of signal emitted by the transmitter as well as the signal to noise ratio (SNR) at the receiver, both of which are assumed to be unknown.

The classical approach for cyclostationary spectrum sensing is to apply the TDT to the CA estimate from (5). However, to do so, one needs to know which cycle frequency to test beforehand, which eliminates the possibility of true *blind* spectrum sensing. One could sequentially test the received signal for all possible cycle frequencies. However, with high probability, the estimation noise at some cycle frequency would have a value above the decision threshold, leading to a false alarm.

3 Sparsity-aided CA estimation: simultaneous OMP-based estimator

As discussed in Section 1, for most man-made signals the CA is (asymptotically) sparsely occupied, containing spikes only at the DC component as well as the cycle frequencies of inherent signal periodicities and their harmonics. In this section, we take advantage of this inherent sparsity and cast the CA estimation as a joint sparse recovery problem. Since this method is able to detect the CA's support, it removes the traditional approach's requirement of knowing the cycle frequencies beforehand, enabling thus blind cyclostationarity-based spectrum sensing possible.

We begin by rewriting Eq. (9) as

$$\mathbf{Y}_N = N\mathbf{F}^{-1}\hat{\mathbf{R}}_x, \tag{19}$$

where $\mathbf{F}^{-1}$ is the inverse discrete Fourier transform (IDFT) matrix. Now, consider an $m \times N$ matrix $\mathbf{M}$, which consists of a selection of m rows of the $N \times N$ identity matrix $\mathbf{I}_N$. It represents the undersampling operation. Applying $\mathbf{M}$ to $\mathbf{x}_{t_0}$, we obtain an $m \times 1$ vector of compressive samples[1] $\bar{\mathbf{x}}_{t_0} = \mathbf{M}\mathbf{x}_{t_0}$. Now, we can calculate an m element delay product with time delays $\tau = \nu_i T_e$, i.e., $\mathbf{y}_m^\nu = \bar{\mathbf{x}}_{t_0} \circ \bar{\mathbf{x}}^*_{t_0+\nu T_e}$. Stacking all $\mathbf{y}_m^\nu$ together into one matrix $\mathbf{Y}_m$, we finally obtain

$$\mathbf{Y}_m = \mathbf{M}\mathbf{Y}_N = N\mathbf{M}\mathbf{F}^{-1}\hat{\mathbf{R}}_x, \tag{20}$$

where $\mathbf{Y}_m$ contains a selection of m coefficients of the delay products for different delays $\nu_i T_e$. Note that $\nu_i \in [0, N-1]$ are chosen such that $\bar{x}t_0 + \nu_i T_e$ is non-empty. We now want to recover $\hat{\mathbf{R}}_x$ from $\mathbf{Y}_m$ by solving the underdetermined inverse problem (20). To do so, we exploit our knowledge about the CA's sparsity.

The straightforward solution would be to solve the following optimization problem

$$\begin{aligned} &\min \left\| \mathrm{vec}\left\{\hat{\mathbf{R}}_x\right\} \right\|_{\ell_0} \\ &\text{s.t. } \mathbf{Y}_m = N\mathbf{M}\mathbf{F}^{-1}\hat{\mathbf{R}}_x, \end{aligned} \tag{21}$$

where $\|\cdot\|_{\ell_0}$ denotes the ℓ_0 "norm" [20], which is the number of non-zero entries in a vector, and vec$\{\cdot\}$ stands for the vectorization of a matrix, i. e., the concatenation of its columns to a single vector. Eq. 21 is known to be a non-convex combinatorial problem [20]. One way to solve it within a practically feasible amount of time is to substitute the ℓ_0-"norm" by its tightest convex relaxation, the ℓ_1 norm. With high probability, this produces the same result since for most large underdetermined systems of linear equations, the minimal ℓ_1-norm solution is also the sparsest solution [33]. Another way of solving (21) efficiently is by applying one of the many greedy sparse recovery algorithms that have been developed in the field of CS, such as, e.g., *orthogonal matching pursuit (OMP)* [22].

OMP is a greedy algorithm that iteratively determines a vector's support from an underdetermined system of linear equations and subsequently recovers the vector by solving a least-square problem. Using it, we could solve (21) for each column of $\hat{\mathbf{R}}_x$ individually (as in [24]), i. e., we could solve

$$\begin{aligned} &\min \left\| \hat{\mathbf{r}}_x^\nu \right\|_{\ell_0} \\ &\text{s.t. } \mathbf{y}_m^\nu = N\mathbf{M}\mathbf{F}^{-1}\hat{\mathbf{r}}_x^\nu, \end{aligned} \tag{22}$$

for each ν. Instead, we notice that the vectors $\hat{\mathbf{r}}_x^\nu\big|_{\nu=\nu_1}^{\nu_{n_\nu}}$ are jointly sparse with regard to the time delay. Therefore, stacking $\hat{\mathbf{r}}_x^\nu$ in $\hat{\mathbf{R}}_x$ results in a row-sparse matrix whose rows are (asymptotically) non-zero only at the indices corresponding to the cycle frequencies. In order to exploit this additional structure, we propose to use an extension of OMP called simultaneous orthogonal matching pursuit (SOMP) [34] to recover the CA matrix $\hat{\mathbf{R}}_x$ at once. The CA estimation based on SOMP is summarized in Algorithm 1, which we further refer to as SOber.

Algorithm 1: SOMP[34]-based CA estimator (SOber)

Input: $\mathbf{Y}_m, n_{\text{iter}}, \mathbf{A} = N\mathbf{M}\mathbf{F}^{-1}$
Output: $\hat{\mathbf{R}}_x$

1 $\hat{\mathbf{R}}_x = \mathbf{0}, S_0 = \emptyset$;
2 **for** $i = 1, \ldots, n_{iter}$ **do**
3 $\quad \mathbf{C} = (\mathbf{Y}_m - \mathbf{A}\hat{\mathbf{R}}_x)^{\mathsf{H}}\mathbf{A}$;
4 $\quad S_i = S_{i-1} \cup \underset{j\in\{1,\ldots,N\}}{\operatorname{argmax}} \left\{\|\mathbf{c}_j\|_{\ell_1}\right\}$;
5 $\quad$ **for** $k = 1, \ldots, n_\nu$ **do**
6 $\quad\quad \hat{\mathbf{r}}_x^{\nu_k} = \underset{\mathbf{z}\in\mathbb{C}^N}{\operatorname{argmin}} \left\{\left\|\mathbf{y}_m^{\nu_k} - \mathbf{A}\mathbf{z}\right\|_{\ell_2}, \operatorname{supp}(\mathbf{z}) \subset S_i\right\}$;

The goal of SOber is to find the indices of the atoms contained in $\mathbf{Y}_m$, i. e., the support of the columns of $\hat{\mathbf{R}}_x$, and subsequently recover the identified non-zero rows of $\hat{\mathbf{R}}_x$ by solving least-square problems. We start with an empty support S_0. Each iteration, one atom (a column of the matrix $\mathbf{A} = N\mathbf{M}\mathbf{F}^{-1}$) index is added to the support. The index is selected according to the sum of the absolute correlation values between the corresponding atoms and the delay products of different time delays (lines 3–4). Using the new support set S_i, a least-square problem is solved for each column in $\hat{\mathbf{R}}_x$ (lines 5–6). In each iteration, the atom index to be added to the index set is chosen according to the correlation between the residuum of $\mathbf{Y}_m$ and the atom set. Since every iteration adds one index to the support set, one usually chooses n_{iter} greater than or equal to the sparsity of the signal to be recovered. The difference between OMP (used in, e. g., [24]) and SOMP can be found in line 4, where SOMP jointly considers the amount of correlation between atoms and the delay products of *multiple* delays, while OMP would select the support of $\hat{\mathbf{r}}_x^\nu\big|_{l=1}^{n_\nu}$ for each l individually.

4 Sparsity-aided CA estimation: dictionary-assisted estimator

In Section 3, we have described a SOMP-based algorithm that estimates the cycle frequencies and the CA from fewer samples than required using the classic approach by taking into consideration the inherent sparsity of the CA. In this section, we develop an algorithm that makes use of additional prior knowledge about the signal's structure in the form of structure dictionaries to further enhance the cycle frequency and CA estimation. Like SOber, the new algorithm does not require the prior knowledge about the cycle frequencies contained in the signal.

One fact about the CA that could be exploited is that using a rectangular pulse shape, a linearly modulated signal's CA exhibits spikes not only at the signal's fundamental cycle frequency but also at the harmonics thereof. Another one is the symmetry of the CA around the DC component. First steps in this direction showing promising results have been taken in [35]. The drawback of the solution proposed in [35] is that the convex optimization problem used to recover the CA becomes huge for practical parameter choices, which results in a prohibitively large computational complexity. To circumvent this, we propose an OMP-based greedy algorithm that takes advantage of the additional prior knowledge while featuring a much smaller complexity than the optimization problem.

The proposed Dice algorithm (Algorithm 2) follows the same idea as the SOber algorithm (Algorithm 1) in that, it iteratively determines the support of the sparse CA and subsequently recovers it by solving an overdetermined least-square problem. However, in contrast to SOber, Dice facilitates the use of further prior knowledge in addition to the CA's sparsity in the recovery process. Thus, in addition to the inputs received by SOber, Dice needs a set of structure dictionaries $\mathring{\mathbf{D}}_l^{(\frac{N}{2})}|_{l=1}^{n_\nu}$, one dictionary for each delay value ν_l that is to be considered in the recovery process. Since the structure dictionaries do not necessarily model the DC component of the CA, it is added to the support set in the initialization phase in Dice (line 1). Instead of working with the amount of correlation between the residuum and the atoms directly as in SOber, the Dice algorithm computes combinations of these as dictated by the structure dictionaries in use (lines 3, 4). This way, the decision about the non-zero cycle frequencies (line 5) takes into account the structure of the CA. Additionally, instead of adding a single element to the support set per iteration, Algorithm 2 adds all indices to the support set that have a non-zero value in the selected dictionary word. The recovery step (cf. lines 6, 7) remains unchanged. Note that in Algorithm 2, the abs(·) operator stands for the element-wise absolute value of a matrix, while the selection operator $[\cdot]_{l:}$ denotes the lth row of a matrix.

Algorithm 2: Dictionary assisted CA estimator (Dice)

Input: $\mathbf{Y}_m, n_{\text{iter}}, \mathbf{A} = N\mathbf{MF}^{-1}, \mathring{\mathbf{D}}_l^{(\frac{N}{2})}|_{l=1}^{n_\nu}$
Output: $\hat{\mathbf{R}}_x$

1 $\hat{\mathbf{R}}_x = \mathbf{0}, S_0 = \{0\}$;
2 **for** $i = 1, \ldots, n_{iter}$ **do**
3   **for** $l = 1, \ldots, n_\nu$ **do**
4         $\left[\mathring{\mathbf{C}}\right]_{l:} = \text{abs}\left(\left(\mathbf{y}_m^{\nu_l} - \mathbf{A}\hat{\mathbf{r}}_x^{\nu_l}\right)^{\mathsf{H}}\mathbf{A}\right)\mathring{\mathbf{D}}_l^{(\frac{N}{2})}$;
5     $S_i = S_{i-1} \cup \left\{h \,\middle|\, [\mathring{\mathbf{D}}^{(\frac{N}{2})}]_{hj} \neq 0, j = \underset{j\in\{1,\ldots,\frac{N}{2}\}}{\text{argmax}} \left\{\|\mathring{\mathbf{c}}_j\|_{\ell_1}\right\}\right\}$;
6   **for** $k = 1, \ldots, n_\nu$ **do**
7         $\hat{\mathbf{r}}_x^{\nu_k} = \underset{\mathbf{z}\in\mathbb{C}^N}{\text{argmin}}\left\{\left\|\mathbf{y}_m^{\nu_k} - \mathbf{Az}\right\|_{\ell_2}, \text{supp}(\mathbf{z}) \subset S_i\right\}$;

In the following, we introduce two particular structure dictionaries that can be used with the proposed algorithm: (i) the dictionary that accounts for the symmetry of the CA and (ii) the dictionary that describes the harmonic structure of the CA as well as its shape.

4.1 Symmetry dictionary

Let $\mathbf{D}_{\text{sym}}^{(\frac{N}{2})} \in \{0,1\}^{\frac{N}{2}\times\frac{N}{2}}$ denote the symmetry dictionary. Its columns represent possible cycle frequencies contained in the set $a \in \{1, \ldots, \frac{N}{2}\}$. For simplicity, this set is chosen such that the frequencies contained in it lie at the center frequencies of the CA's DFT bins. An entry of the dictionary covers elements 1 to $\frac{N}{2}$ of $\hat{\mathbf{r}}_x^\nu$ which is indexed from 0 to $N-1$. The symmetry dictionary is simply given by the identity matrix, i. e., $\mathbf{D}_{\text{sym}}^{(\frac{N}{2})} = \mathbf{I}_{\frac{N}{2}}$. To model the whole vector $\hat{\mathbf{r}}_x^\nu$, the dictionary is extended to include the DC component, which is set to zero, as well as the negative cycle frequencies. Note that the DC component is set to zero because its value is independent of the presence of cyclostationarity. The resulting *full* dictionary is exemplarily given by

$$\mathring{\mathbf{D}}_{\text{sym}}^{(3)} = \begin{pmatrix} 0 & 0 & 0 \\ 1 & 0 & 0 \\ 0 & 1 & 0 \\ 0 & 0 & 1 \\ 0 & 1 & 0 \\ 1 & 0 & 0 \end{pmatrix}. \tag{23}$$

The circle above the symbol indicates that it is the full version of the dictionary, i. e., the one spanning the whole Fourier range. The *ones* in the matrix specify the locations of the non-zero coefficients in the CA fitting the format of (9). Note that in the case of the symmetry dictionary, all of $\mathring{\mathbf{D}}_l^{(\frac{N}{2})}|_{l=1}^{n_\nu}$ in Algorithm 2 are identical i. e., $\mathring{\mathbf{D}}_l^{(\frac{N}{2})}|_{l=1}^{n_\nu} = \mathring{\mathbf{D}}_{\text{sym}}^{(\frac{N}{2})}$.

4.2 Asymptotic CA and asymptotic dictionary

The symmetry structure dictionary exploits one of the facts we know about the CA. In order to explore an extreme in terms of prior knowledge, we create a dictionary that contains the maximum possible amount of prior information about the CA, i. e., the one containing the asymptotic CA itself. This requires knowledge of the analytic expression for the discrete asymptotic CA vector, which we derive in the following.

To assess the performance of different CA estimation algorithms, we employ common linearly modulated signals with symbol length T_s as described by the following equation ([7], Eq. 73)

$$s(t) = \sum_{n=-\infty}^{\infty} c_n p\,(t - nT_s + \phi). \tag{24}$$

Here, $p(t)$ is a deterministic finite-energy pulse, ϕ represents a fixed pulse-timing phase parameter and c_n stands for the nth symbol to be transmitted. We are now interested in an expression for the discrete asymptotic CA vector of the above signal type.

The fundamental cycle frequency of the built-in periodicity of the signal from (24) is $\frac{1}{T_s}$. Its continuous CA is given by ([7], Eq. 81)

$$R^{\alpha}_{s,T_s}(\tau) = \begin{cases} 0 & \text{for } \alpha \neq \frac{k}{T_s} \\ \frac{1}{T_s}\sum\limits_{n=-\infty}^{\infty} R_c(nT_s) r^{\alpha}_p(\tau - nT_s) e^{j2\pi\alpha\phi} & \text{otherwise,} \end{cases} \tag{25}$$

where $k \in \mathbb{Z}$ and $r^{\alpha}_{\alpha}(\tau)$ is defined as ([7], Eq. 82)

$$r^{\alpha}_{\alpha}(\tau) \triangleq \int_{-\infty}^{\infty} p\,(t + \tau/2) p^*\,(t - \tau/2)\, e^{-j2\pi\alpha t}\, dt. \tag{26}$$

The symbol $\mathbb{Z}$ denotes the set of integers, i. e., $k \in \{\ldots, -2, -1, 0, 1, 2, \ldots\}$.

We consider the case where c_n is a purely stationary random sequence. Thus, its autocorrelation $R_c(nT_s) = R^0_c(nT_s)$ is non-zero only at $n = 0$ (cf. (2)), reducing (25) to

$$R^{\alpha}_{s,T_s}(\tau) = \begin{cases} 0 & \text{for } \alpha T_s \notin \mathbb{Z} \\ \frac{\sigma_c^2}{T_s} r^{\alpha}_p(\tau) e^{j2\pi\alpha\phi} & \text{otherwise,} \end{cases} \tag{27}$$

where σ_c^2 is the average power of c_n. In the following, we assume a rectangular pulse shape of length T_s, i. e., $p(t) = \text{rect}(\frac{t}{T_s})$, which leads to $p(t + \frac{\tau}{2})p^*(t - \frac{\tau}{2}) = \text{rect}\left(\frac{t}{T_s - |\tau|}\right)$. Thus, applying the Fourier transform to (26) yields

$$R^{\alpha}_{s,T_s}(\tau) = \begin{cases} 0 & \text{for } \alpha T_s \notin \mathbb{Z} \\ \sigma_c^2 \frac{T_s - |\tau|}{T_s} \text{sinc}(\alpha(T_s - |\tau|)) e^{j2\pi\alpha\phi} & \text{otherwise,} \end{cases} \tag{28}$$

for $|\tau| \leq T_s$ where $\text{sinc}(x) = \frac{\sin(\pi x)}{\pi x}$. Note that the use of the absolute value of the delay stems from the fact that for a real symmetric pulse shape $p(t)$, the expression $p(t + \frac{\tau}{2})p^*(t - \frac{\tau}{2})$ is symmetric with respect to τ.

Equation 28 represents the CA of the continuous-time signal described by (24). The CA of the sampled version of (24) at its fundamental cycle frequency and the harmonics thereof is given by

$$R'^{a}_{s,n_s}(\nu)\Big|_{a=k\frac{N}{n_s}} = \frac{\sigma_c^2}{n_s} \frac{\sin(\pi \frac{a}{N}(n_s - |\nu|))}{\sin\left(\pi \frac{a}{N}\right)} e^{j2\pi \frac{a}{N} d_\phi}. \tag{29}$$

The derivation of this expression can be found in the appendix.

The coefficients of the closed-form expression (29) together with the alternative case $R'^{a}_{s,n_s}(\nu)\big|_{a \neq k\frac{N}{n_s}} = 0$ at different discrete cycle frequencies a are arranged in a vector $\mathbf{r}^{\nu}_{s,n_s}[a]$ matching the format of the DFT matrix, such that

$$\mathbf{r}^{\nu}_{s,n_s}[a] = \begin{cases} R'^{\alpha}_{s,n_s}(\nu) & \text{for } a \in \{0, \ldots, \frac{N}{2}\}, \\ R'^{(a-N)}_{s,n_s}(\nu) & \text{for } a \in \{\frac{N}{2} + 1, \ldots, N - 1\}. \end{cases} \tag{30}$$

Note that adding purely stationary noise to the signal $s(t)$ does not change its asymptotic CA (with the exception of $(a, \nu) = (0, 0)$, at which point the CA's value is the average power of signal and noise, cf. (2)) since the noise exhibits no inherent periodic behavior. Due to this fact, (30) can also be used as a reference for the CA of signals contaminated with additive white Gaussian noise (AWGN) with the exception mentioned.

Given (30), we can now construct the asymptotic dictionary:

$$\overset{\circ}{\mathbf{D}}{}^{(\frac{N}{2})}_{\text{asy},l} = \left[\frac{\text{abs}\left(\mathbf{r}^{\nu_l}_{s,n_s=\frac{N}{1}}\right)}{\left\|\mathbf{r}^{\nu_l}_{s,n_s=\frac{N}{1}}\right\|_{\ell_1}}, \ldots, \frac{\text{abs}\left(\mathbf{r}^{\nu_l}_{s,n_s=\frac{N}{N/2}}\right)}{\left\|\mathbf{r}^{\nu_l}_{s,n_s=\frac{N}{N/2}}\right\|_{\ell_1}} \right]. \tag{31}$$

Note that in contrast to the single symmetry dictionary, there is a whole set of asymptotic dictionaries, one for each delay value of interest. The columns of the dictionaries correspond to actual symbol lengths i. e., actual cycle frequencies. Thus, each column contains the absolute value of the normalized asymptotic CA of a cycle frequency candidate where the discrete symbol lengths $n_s \in \left\{\frac{N}{1}, \ldots, \frac{N}{N/2}\right\}$ correspond to the discrete cycle frequencies $a \in \{1, \ldots, N/2\}$. It is worth noting that in addition to its role as the basis of the second structure dictionary for Algorithm 2, the expression (30) serves as a reference for the direct comparison of different CA estimation methods in Section 6.

5 Cyclostationarity detection from sparse cyclic spectra

Both the SOMP-based (Algorithm 1) and the dictionary-assisted CA estimation (Algorithm 2) are able to recover the CA without knowing which cycle frequencies are contained in the signal beforehand. Furthermore, since the the (row) support of $\hat{\mathbf{R}}_x$ corresponds to the candidate cycle frequencies, its identification in Algorithm 1 (line 4) or in Algorithm 2 (line 5) can be interpreted as blind cyclostationary spectrum sensing by itself. However, under practical limitations on the number of samples available for CA estimation and attainable SNR levels, the support estimate is likely to contain errors, e.g., missed and/or falsely identified support entries. This calls for a further testing of the candidate cyclofrequencies that can be performed by applying the TDT method described in Section 2.2. However, the obtained sparse CA is not directly compatible with the traditional TDT because, since only few of the coefficients of $\hat{\mathbf{R}}_x$ are recovered and all other coefficients are set to zero, it is not possible to reliably estimate the covariance matrix $\hat{\Sigma}_{xx^*}$ under $\mathcal{H}_0$. To tackle this problem, we present a modification of the traditional TDT.

The traditional TDT is a CFAR detector, i. e., the probability density function (PDF) of its test statistic under $\mathcal{H}_0$ is asymptotically independent of any signal parameters like, e. g., the noise power. To achieve this, the TDT first estimates the CA noise covariance and then rescales the original CA by this estimate so that the scaled CA follows a *standard* Gaussian distribution. This is where the problem occurs. Although, we are ultimately only interested in the CA coefficients that are located at the signal's cycle frequencies, for the estimation of the noise covariance, we need the coefficients lying between the cycle frequencies, which only carry estimation noise. SOber and Dice do not recover these. Thus, we propose an extension to the TDT, the sparse TDT, to bridge this gap in the following.

To obtain optimal CA recovery performance, one would choose the sensing matrix $\mathbf{A}$ with minimum structure, i. e., the selection of the m entries of the delay product would be completely random. However, to tackle the aforementioned problem, we choose a combination of consecutive and random delay product elements. The consecutive part comprises the first $\lceil \beta m \rceil$ rows of $\mathbf{Y}_m$, where $\beta \in [0.01, 0.5]$ and $\lceil \cdot \rceil$ denotes the ceiling operation. The remainder of the rows of $\mathbf{Y}_m$ is a random selection of the remaining rows of $\mathbf{Y}_N$. The first step of the sparse TDT is to determine the classical CA estimation of the consecutive block of delay product elements. In the next step, the cycle frequency of interest a_0 is determined using either Algorithm 1 or Algorithm 2. Next, the covariance matrix for the cycle frequency a_0 corresponding to the N size CA $\left(\hat{\Sigma}^{(N)}_{xx^*}(a_0)\right)$ needs to be determined, where the superscript (N) indicates the corresponding CA size. It is obtained as

$$\hat{\Sigma}^{(N)}_{xx^*}(a_0) = \frac{\hat{\Sigma}^{(\lceil \beta m \rceil)}_{xx^*}(\lceil \beta \frac{m}{N} a_0 \rceil)}{\sqrt{\beta \frac{m}{N}}}, \tag{32}$$

where $\hat{\Sigma}^{(\lceil \beta m \rceil)}_{xx^*}$ is the covariance matrix corresponding to the $\lceil \beta m \rceil$ size CA estimated from the consecutive samples in the first step. The test statistic is subsequently evaluated as (cf. (18))

$$\mathcal{T}^{\text{sparse}}_{xx^*} = N \hat{\mathbf{r}}_{xx^*}(a_0) \left(\frac{\hat{\Sigma}^{(\lceil \beta m \rceil)}(\lceil \beta \frac{m}{N} a_0 \rceil)}{\sqrt{\beta \frac{m}{N}}} \right)^{-1} \hat{\mathbf{r}}^T_{xx^*}(a_0). \tag{33}$$

The consecutive sample ratio β is a trade-off parameter. The optimal sparse recovery performance is to be expected for the case that $\mathbf{A} = N\mathbf{MF}^{-1}$ has the smallest possible amount of structure, which here corresponds to the case where the set of known delay product elements is chosen completely at random i. e., for $\beta = 0$. Contrarily, the best estimation quality for the CA covariance matrix $\hat{\Sigma}_{xx^*}$ is achieved when all known delay product elements are consecutive i. e., for $\beta = 1$.

6 Numerical evaluation

In this section, we compare the performance of the methods presented in the preceding sections. The parameters used throughout this section are given in Table 1.

We begin by investigating the influence of the consecutive sample ratio β on the spectrum sensing performance. Figure 1 shows how the detection rate changes with β for an SNR of 0 dB and different false alarm rates. For all methods but the OMP, $\beta = 0.15$ seems to be a good choice. For the OMP, the detection rate increases monotonically with β. However, as can be seen below, even

Table 1 System parameters

Parameter	Symbol	Value(s)
Size of the CA vector - CS methods	N	4000
No. of known delay product elements	m	1000
Size of the CA vector - classic method	m	1000
Discrete time delays	ν	$\{1, 2, 3, 4\}$
Modulation type		BPSK
Discrete symbol length	n_s	8
Signal to noise ratio	SNR	$\{-4, \ldots, 4\}$dB
No. of Monte Carlo instances		10000
Consecutive sample ratio	β	$\{0.01, \ldots, 0.5\}$
Covariance estimation window type	W	Kaiser
Kaiser window parameter	α_{Kaiser}	10
Kaiser window length	L	201

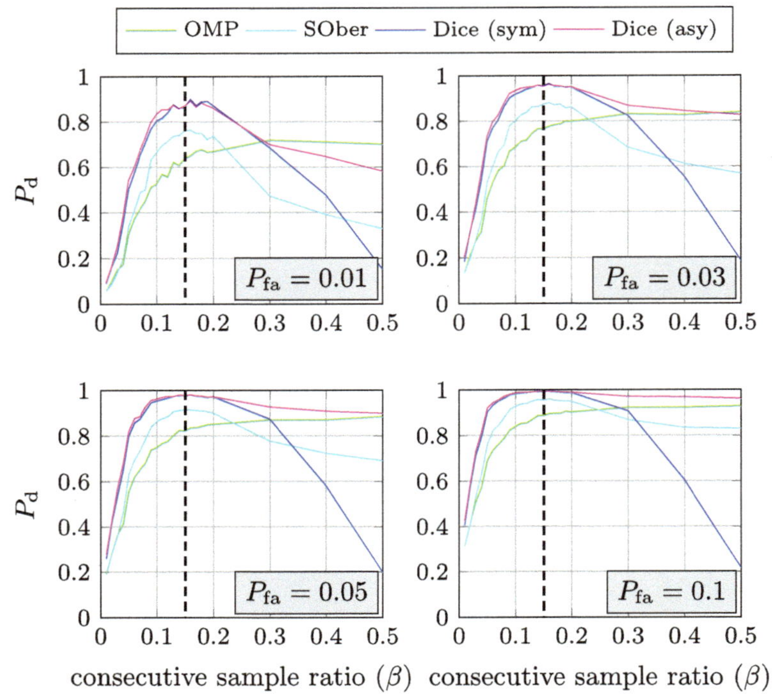

Fig. 1 Detection rate over consecutive sample ratio for different false alarm rates at 0 dB SNR

for the OMP, a high β is no good choice regarding other performance categories.

In Fig. 2, the best achievable detection rate, i. e., the detection rate for the individual best choice of β, of the different detectors is plotted over the receiver SNR for different false alarm rates. The term *oracle* expresses that a method has prior knowledge about the exact cycle frequencies contained in the signal. The classic method depends on this knowledge while for the sparse recovery, it reduces the CA recovery to solving the overdetermined

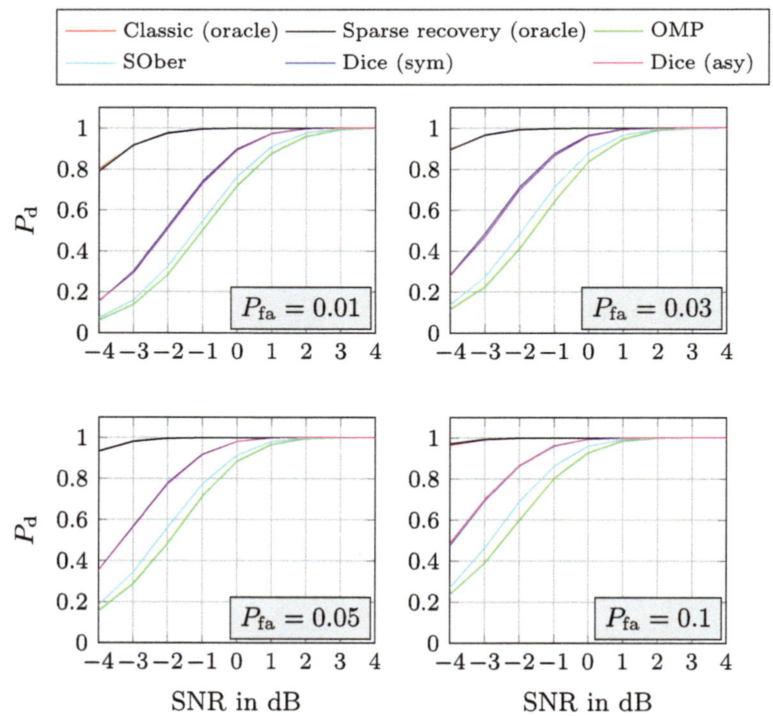

Fig. 2 Maximum detection rate (optimal individual consecutive sample ratio selection) over SNR for different false alarm rates

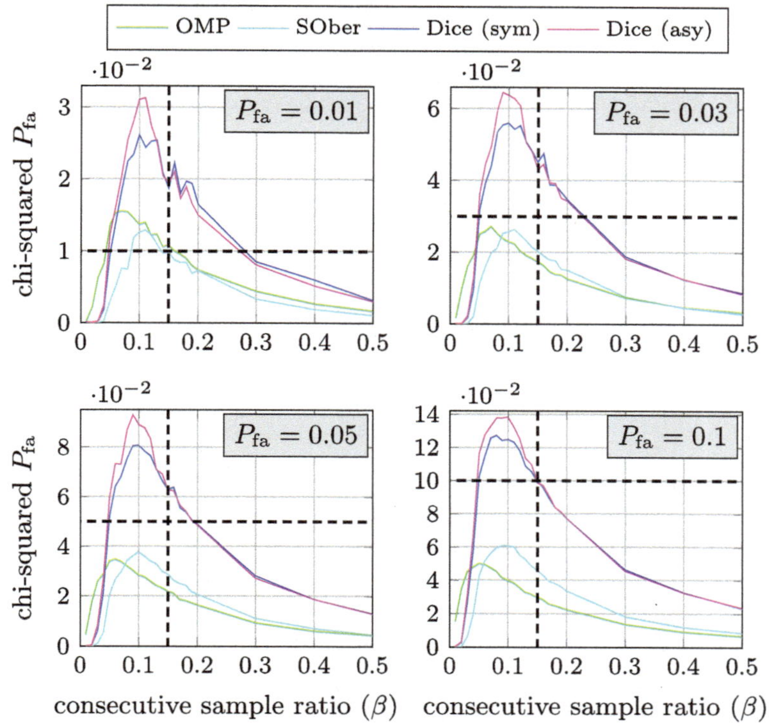

Fig. 3 False alarm rate that has to be selected according to the chi-squared distribution to obtain different actual false alarm rates over the consecutive sample ratio

least-square problem for the given support (cf. lines 5 and 6 in Algorithm 1 or lines 6 and 7 in Algorithm 2). As expected, the oracle methods outperform the methods which have to determine the CA support themselves by a large margin. Regarding the case of missing support knowledge, the Dice algorithm clearly outperforms the SOber algorithm as well as OMP. It is to be noted that both, Fig. 1 as well as Fig. 2 do not show a significant performance advantage of exploiting the full knowledge of the asymptotic CA (Dice (asy)) over just exploiting its symmetry property (Dice (sym)) for a sensible choice of β.

The lines in Fig. 3 show which false alarm rate according to the ideal chi-squared distribution has to be set in order to achieve 1, 3, 5, and 10% false alarm rate in the actual system. The dashed lines cross at the desired false alarm rate with $\beta = 0.15$. While the two Dice methods roughly keep within a 1% offset, OMP, and SOber show a decreasing degree of equivalence for an increasing false alarm rate. This indicates that using the chi-squared distribution for setting the decision threshold of the Dice algorithm is viable, which is an important observation. It means that in contrast to many other spectrum sensing

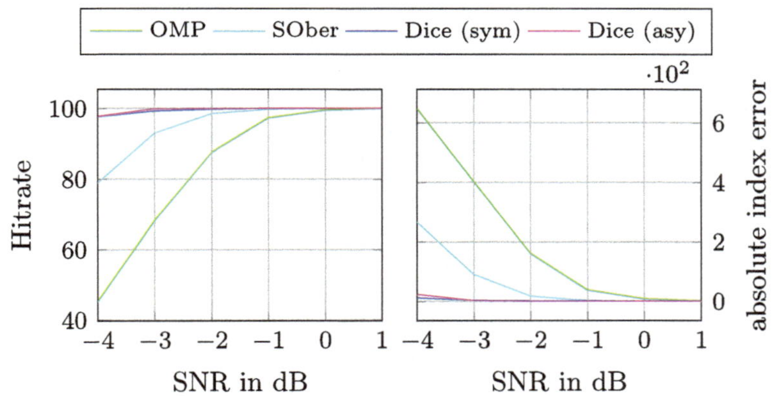

Fig. 4 *Left* hit rate over SNR, *right* absolute index error over SNR. Both at consecutive sample ratio 0.15

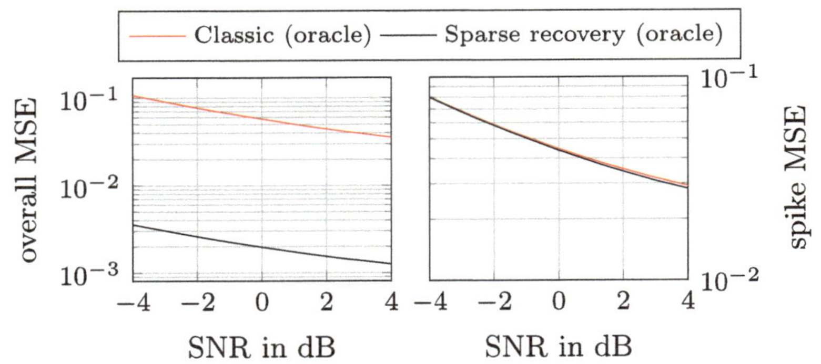

Fig. 5 MSE between the CAF estimation and the actual (analytic) value. *Left* over the whole support, *right* at the cycle frequencies. Both at consecutive sample ratio 0.15

algorithms, Dice approximately possesses a desirable feature called constant false alarm rate i. e., its test statistic is independent of system parameters like the receiver noise power.

Figure 4 shows how well the support of the CA is recovered by the different methods. Since different types of communication signals feature different cycle frequencies, this information can be used for system identification. The hit rate is the chance of exactly recovering the correct support while the absolute index error is the mean recovery error in terms of CA bins. Obviously, the Dice methods have superior support recovery capabilities.

The final performance category we evaluate is the CA estimation quality achievable by sparse recovery methods measured by the mean squared error (MSE). In the left graph of Fig. 5, the MSE over the whole CA is plotted while the right graph shows the MSE at the spikes of the CA, i. e., the MSE at the actual cycle frequencies. To determine the error, we use the analytic expression for the asymptotic CA vector as derived in Section 4.2, i. e., the MSE is defined as

$$\frac{\|\hat{\mathbf{R}}_x - \left[\mathbf{r}^{\nu_1}_{s,n_s}, \ldots, \mathbf{r}^{\nu_{n_\nu}}_{s,n_s}\right]\|_F^2}{N n_\nu}, \tag{34}$$

where $\|\cdot\|_F$ is the Frobenius norm. The sparse recovery method has a much lower *overall* MSE. This is caused by the fact that it sets all CA coefficients but the detected support to zero while the classical method results in a CA that features estimation noise between the spikes. Regarding the *spike* MSE, both methods seem to perform roughly equivalently.

7 Conclusions

Blind operation and constant false alarm rate (CFAR) are desirable characteristics of spectrum sensing algorithms. Unfortunately, cyclostationarity-based approaches typically only feature either one or the other. We showed that this can be changed by using sparse recovery methods in the CA estimation. Subsequently, we developed a way to use further prior knowledge in addition to sparsity for superior CA estimation. We derived a closed-form expression of the CA of *sampled* linearly modulated signals with rectangular pulse shape to be used both as prior information for the CA estimation and as a reference for comparison. Finally, we extended a well-known statistical test for cyclostationarity to accommodate sparse input. The results allow us to conclude that the proposed Dice algorithm in combination with the symmetric structure dictionary constitutes a viable alternative to the classical TDT for the case of missing prior information about the cycle frequencies contained in the signal.

Endnote

[1]Note that although we use the vector of Nyquist rate samples x_{t_0} to calculate $\bar{\mathrm{x}}_{t_0}$, we do so for notational convenience only. In practice, one can directly obtain the sub-Nyquist samples $\bar{\mathrm{x}}_{t_0}$ by means of non-uniform sampling for instance.

8 Appendix

8.1 Discrete asymptotic CA

The relation between the CA of the continuous time-domain signal $s(t)$, and its sampled counterpart $\{s(nT_e)\}$ is given by ([36], Ch. 11, Sec. C, Eq. (111))

$$\mathcal{R}'^{\alpha}_{s,T_s}(\nu T_e) = \sum_{l=-\infty}^{\infty} R^{\alpha+\frac{l}{T_e}}_{s,T_s}(\nu T_e) e^{j\pi l\nu}. \tag{35}$$

The sum over l reflects the infinite aliasing caused by the sampling. In the next step, we insert (28) into (35). Also, we express quantities in terms of the sampling period T_e, i. e., $T_s \rightarrow n_s T_e, \alpha \rightarrow \frac{a}{NT_e}, \phi \rightarrow d_\phi T_e$, with $n_s, a, N \in \mathbb{Z}$. This leads to

$$R'^{a}_{s,n_s}(\nu) = \begin{cases} 0 & \text{for } a \neq \frac{kN}{n_s} \\ \sigma_c^2 \frac{n_s-|\nu|}{n_s} e^{j2\pi \frac{a}{N} d_\phi} \sum_{l=-\infty}^{\infty} e^{j\pi l\nu} & \\ \cdot e^{j2\pi l d_\phi} \mathrm{sinc}((\frac{a}{N}+l)(n_s-|\nu|)) & \text{otherwise,} \end{cases} \tag{36}$$

for $|\nu| \leq n_s$, where n_s is the oversampling factor. In this step, we used the fact that for our assumptions all aliases of the fundamental cycle frequency and its harmonics lie on top of the actual fundamental cycle frequency and its harmonics, i. e., $(\alpha + \frac{l}{T_e})T_s \in \mathbb{Z}$ iff $\alpha T_s \in \mathbb{Z}$. Inserting the discrete quantities given above, we get $(\frac{a}{N} + l)n_s \in \mathbb{Z}$ iff $\frac{a}{N}n_s \in \mathbb{Z}$. Since $n_s \in \mathbb{Z}$ and $l \in \mathbb{Z}$, this always holds. To rule out any spectral leakage, we choose N as an integer multiple of n_s, since then, $a = k\frac{N}{n_s}$ is also an integer and thus the fundamental discrete cycle frequency and its harmonics hit center frequencies of frequency bins.

For $a = k\frac{N}{n_s}$ expression (36) can be shown to be

$$R'^{a}_{s,n_s}(\nu)\Big|_{a=k\frac{N}{n_s}} = \sigma_c^2 \frac{\sin(\pi\frac{a}{N}(n_s-|\nu|))}{\pi n_s} e^{j2\pi\frac{a}{N}d_\phi} \sum_{l=-\infty}^{\infty} \frac{(-1)^l}{\frac{a}{N}+l}. \tag{37}$$

To obtain (37) we used the definition of the sinc and exploited the facts that $e^{j\pi k} = (-1)^k$ for $k \in \mathbb{Z}$ and that $\sin(x + k\pi) = (-1)^k \sin(x)$ for $k \in \mathbb{Z}$. The pulse timing phase parameter d_ϕ was set to $\frac{n_s+1}{2}$. This has the following reason. In order to simplify the numerical evaluation, we want to choose ϕ such that the beginning of the observed receiver signal is aligned with the rectangular pulse shapes, i. e., we would set $\phi = \frac{T_s}{2}$. However, doing so would lead to the need to sample at the discontinuities caused by the instant change in amplitudes at the transition between symbols. To avoid this, we choose $\phi = \frac{T_s}{2} + \epsilon$, where $\epsilon \in (0, T_e)$. Note that (37) is the same for any $\epsilon \in (0, T_e)$. In order to ease the derivation, we can thus choose $\epsilon = \frac{T_e}{2}$, i. e., $d_\phi = \frac{n_s+1}{2}$.

The infinite series in (37) can be expressed as

$$\begin{aligned}\sum_{l=-\infty}^{\infty} \frac{(-1)^l}{\frac{a}{N}+l} &= \frac{N}{a} + \sum_{l=1}^{\infty} \frac{(-1)^l}{\frac{a}{N}+l} + \frac{(-1)^l}{\frac{a}{N}-l} \\ &= \frac{N}{a} + \frac{1}{2}\sum_{l=1}^{\infty} -\frac{1}{l+\frac{a}{2N}-\frac{1}{2}} + \frac{1}{l-\frac{a}{2N}-\frac{1}{2}} - \frac{1}{l-\frac{a}{2N}} + \frac{1}{l+\frac{a}{2N}} \\ &= \frac{N}{a} + \frac{1}{2}\sum_{l=1}^{\infty} -\frac{l+\frac{a}{2N}-\frac{1}{2}-\frac{a}{2N}+\frac{1}{2}}{l(l+\frac{a}{2N}-\frac{1}{2})} + \frac{l-\frac{a}{2N}-\frac{1}{2}+\frac{a}{2N}+\frac{1}{2}}{l(l-\frac{a}{2N}-\frac{1}{2})} \\ &\quad -\frac{l-\frac{a}{2N}+\frac{a}{2N}}{l(l-\frac{a}{2N})} + \frac{l+\frac{a}{2N}-\frac{a}{2N}}{l(l+\frac{a}{2N})} \\ &= \frac{N}{a} + \frac{1}{2}\sum_{l=1}^{\infty} \frac{\frac{a}{2N}-\frac{1}{2}}{l(l+\frac{a}{2N}-\frac{1}{2})} - \frac{-\frac{a}{2N}-\frac{1}{2}}{l(l-\frac{a}{2N}-\frac{1}{2})} + \frac{-\frac{a}{2N}}{l(l-\frac{a}{2N})} - \frac{\frac{a}{2N}}{l(l+\frac{a}{2N})}.\end{aligned} \tag{38}$$

The digamma function, denoted by $\psi(z)$, possesses a series expansion given by ([37], Eq. (6.3.16))

$$\psi(1+z) = -\gamma + \sum_{n=1}^{\infty} \frac{z}{n(n+z)} \text{ for } z \notin \{-1, -2, -3, \ldots\}, \tag{39}$$

where γ denotes the Euler-Mascheroni constant. We can thus simplify (38) by expressing it in terms of the digamma function as

$$\begin{aligned}\sum_{l=-\infty}^{\infty} \frac{(-1)^l}{\frac{a}{N}+l} &= \frac{N}{a} + \frac{1}{2}\left(\psi\left(\frac{1}{2}+\frac{a}{2N}\right) - \psi\left(\frac{1}{2}-\frac{a}{2N}\right)\right. \\ &\quad \left. +\psi\left(1-\frac{a}{2N}\right) - \psi\left(1+\frac{a}{2N}\right)\right).\end{aligned} \tag{40}$$

Since the reflection and the recurrence formulas of the digamma function are known to be ([37], Eq. (6.3.7))

$$\psi(1-z) - \psi(z) = \pi\cot(\pi z) \tag{41}$$

and ([37], Eq. (6.3.5))

$$\psi(1+z) = \psi(z) + \frac{1}{z}, \tag{42}$$

respectively, we obtain

$$\begin{aligned}&\psi\left(\frac{1}{2}+\frac{a}{2N}\right) - \psi\left(\frac{1}{2}-\frac{a}{2N}\right) \\ &= \psi\left(1-\left(\frac{1}{2}-\frac{a}{2N}\right)\right) - \psi\left(\frac{1}{2}-\frac{a}{2N}\right) \\ &= \pi\cot\left(\pi\left(\frac{1}{2}-\frac{a}{2N}\right)\right)\end{aligned} \tag{43}$$

and

$$\begin{aligned}&\psi\left(1-\frac{a}{2N}\right) - \psi\left(1+\frac{a}{2N}\right) \\ &= \psi\left(1-\frac{a}{2N}\right) - \psi\left(\frac{a}{2N}\right) - 2\frac{N}{a} \\ &= \pi\cot(\pi\frac{a}{2N}) - 2\frac{N}{a}.\end{aligned} \tag{44}$$

Inserting (43) and (44) into (40) results in

$$\begin{aligned}&\sum_{l=-\infty}^{\infty} \frac{(-1)^l}{\frac{a}{N}+l} = \frac{1}{2}\pi\left(\cot\left(\pi\left(\frac{1}{2}-\frac{a}{2N}\right)\right) + \cot\left(\pi\frac{a}{2N}\right)\right) \\ &= \frac{1}{2}\pi\left(\tan\left(\pi\frac{a}{2N}\right) + \cot\left(\pi\frac{a}{2N}\right)\right) = \frac{\pi}{\sin(\pi\frac{a}{N})}.\end{aligned} \tag{45}$$

Finally, substituting (45) into (37) gives us the expression (29).

Funding

This work was partly supported by the Deutsche Forschungsgemeinschaft (DFG) projects CoCoSa (grant MA 1184/26-1) and CLASS (grant MA 1184/23-1).

Competing interests

The authors declare that they have no competing interests.

Author details

[1]Institute for Theoretical Information Technology, RWTH Aachen University, Kopernikusstraße 16, 52074 Aachen, Germany. [2]Institute for Information Technology, TU Ilmenau, 98684 Ilmenau, Germany.

References

1. Q Zhao, BM Sadler, A survey of dynamic spectrum access. IEEE Signal Process. Mag. **24**(3), 79–89 (2007)
2. T Yücek, H Arslan, A survey of spectrum sensing algorithms for cognitive radio applications. IEEE Commun. Surv. Tutorials. **11**(1), 116–130 (2009)
3. Y Zeng, Y-C Liang, AT Hoang, R Zhang, A review on spectrum sensing for cognitive radio: challenges and solutions. EURASIP J. Adv. Signal Process. **2010** (2010)
4. E Axell, G Leus, EG Larsson, HV Poor, Spectrum sensing for cognitive radio: state-of-the-art and recent advances. IEEE Signal Process. Mag. **29**(3), 101–116 (2012)

5. SM Kay, *Fundamentals of Statistical Signal Processing, Vol, II: Detection Theory*. (Prentice Hall, Upper Saddle River, 1998)
6. H Urkowitz, Energy detection of unknown deterministic signals. Proc. IEEE. **55**(4), 523–531 (1967)
7. WA Gardner, Exploitation of spectral redundancy in cyclostationary signals. IEEE Signal Process. Mag. **8**(2), 14–36 (1991)
8. WA Gardner, Signal interception: a unifying theoretical framework for feature detection. IEEE Trans. Commun. **36**(8), 897–906 (1988)
9. AV Dandawate, GB Giannakis, Statistical tests for presence of cyclostationarity. IEEE Trans. Signal Process. **42**(9), 2355–2369 (1994)
10. WA Gardner, A Napolitano, L Paura, Cyclostationarity: half a century of research. Signal Process. **86**(4), 639–697 (2006)
11. K Kim, IA Akbar, KK Bae, J-S Um, CM Spooner, JH Reed, in *New Frontiers in Dynamic Spectrum Access Networks, 2007. DySPAN 2007. 2nd IEEE International Symposium On*. Cyclostationary approaches to signal detection and classification in cognitive radio (IEEE, Dublin, 2007), pp. 212–215
12. J Chen, A Gibson, J Zafar, in *Cognitive Radio and Software Defined Radios: Technologies and Techniques, 2008 IET Seminar On*. Cyclostationary spectrum detection in cognitive radios (IET, London, 2008), pp. 1–5
13. PD Sutton, KE Nolan, LE Doyle, Cyclostationary signatures in practical cognitive radio applications. IEEE J. Selected Areas Commun. **26**(1), 13–24 (2008)
14. CM Spooner, RB Nicholls, Spectrum sensing based on spectral correlation. Cogn. Radio Technol. **2**, 593–634 (2009)
15. Z Khalaf, A Nafkha, J Palicot, in *Circuits and Systems (MWSCAS), 2011 IEEE 54th International Midwest Symposium On*. Blind cyclostationary feature detector based on sparsity hypotheses for cognitive radio equipment (IEEE, Seoul, 2011), pp. 1–4
16. A Napolitano, Cyclostationarity: new trends and applications. Signal Process. **120**, 385–408 (2016)
17. J Lundén, V Koivunen, A Huttunen, HV Poor, Collaborative cyclostationary spectrum sensing for cognitive radio systems. IEEE Transa. Signal Process. **57**(11), 4182–4195 (2009)
18. M Derakhshani, T Le-Ngoc, M Nasiri-Kenari, Efficient cooperative cyclostationary spectrum sensing in cognitive radios at low SNR regimes. IEEE Trans. Wireless Commun. **10**(11), 3754–3764 (2011)
19. DL Donoho, Scanning the technology. Proc. IEEE. **98**(6), 910–912 (2010)
20. DL Donoho, Compressed sensing. IEEE Trans. Inf. Theory. **52**(4), 1289–1306 (2006)
21. EJ Candès, MB Wakin, An introduction to compressive sampling. IEEE Signal Proc. Mag. **25**(2), 21–30 (2008)
22. JA Tropp, AC Gilbert, Signal recovery from random measurements via orthogonal matching pursuit. IEEE Trans. Inf. Theory. **53**(12), 4655–4666 (2007)
23. D Needell, JA Tropp, CoSaMP: iterative signal recovery from incomplete and inaccurate samples. Commun. ACM. **53**(12), 93–100 (2010)
24. Z Khalaf, J Palicot, in *Cognitive Communication and Cooperative HetNet Coexistence*. New blind free-band detectors exploiting cyclic autocorrelation function sparsity (Springer, Berlin, 2014), pp. 91–117
25. Z Tian, Y Tafesse, BM Sadler, Cyclic feature detection with sub-Nyquist sampling for wideband spectrum sensing. IEEE J. Selected Topics Signal Process. **6**(1), 58–69 (2012)
26. E Rebeiz, V Jain, D Cabric, in *IEEE International Conference on Communications (ICC)*. Cyclostationary-based low complexity wideband spectrum sensing using compressive sampling, (Ottawa, 2012), pp. 1619–1623
27. D Cohen, E Rebeiz, V Jain, YC Eldar, D Cabric, in *IEEE International Workshop on Computational Advances in Multi-Sensor Adaptive Processing (CAMSAP)*. Cyclostationary feature detection from sub-Nyquist samples, (San Juan, 2011), pp. 333–336
28. M Mishali, YC Eldar, From theory to practice: sub-Nyquist sampling of sparse wideband analog signals. IEEE J. Selected Topics Signal Process. **4**(2), 375–391 (2010)
29. G Leus, Z Tian, in *Computational Advances in Multi-Sensor Adaptive Processing (CAMSAP), 2011 4th IEEE International Workshop On*. Recovering second-order statistics from compressive measurements (IEEE, San Juan, 2011), pp. 337–340
30. DD Ariananda, G Leus, in *Acoustics, Speech and Signal Processing (ICASSP), 2014 IEEE International Conference On*. Non-uniform sampling for compressive cyclic spectrum reconstruction (IEEE, Florence, 2014), pp. 41–45
31. D Cohen, YC Eldar, Sub-Nyquist cyclostationary detection for cognitive radio. IEEE Trans. Signal Process. **65**(11), 3004–3019 (2017)
32. A Napolitano, *Generalizations of Cyclostationary Signal Processing: Spectral Analysis and Applications*. (John Wiley & Sons, Hoboken, 2012)
33. DL Donoho, For most large underdetermined systems of linear equations the minimal ℓ_1-norm solution is also the sparsest solution. Commun. Pure Appl. Math. **59**, 797–829 (2004)
34. JA Tropp, AC Gilbert, MJ Strauss, Algorithms for simultaneous sparse approximation. Part I: Greedy pursuit. Signal Process. **86**(3), 572–588 (2006)
35. A Bollig, R Mathar, in *IEEE International Conference on Acoustics, Speech and Signal Processing (ICASSP)*. Dictionary-based reconstruction of the cyclic autocorrelation via ℓ_1-minimization for cyclostationary spectrum sensing, (Vancouver, 2013)
36. WA Gardner, *Statistical Spectral Analysis: A Nonprobabilistic Theory*. (Prentice-Hall, New Jersey, 1986)
37. M Abramowitz, IA Stegun, *Handbook of Mathematical Functions with Formulas, Graphs, and Mathematical Tables*. (Courier Corporation, New York, 1964)

14

Adaptive robust time-of-arrival source localization algorithm based on variable step size weighted block Newton method

Chee-Hyun Park[1] and Joon-Hyuk Chang[2*]

Abstract

We propose a line-of-sight (LOS)/non-line-of-sight (NLOS) mixture source localization algorithms that utilize the weighted block Newton (WBN) and variable step size WBN (VSSWBN) method, in which the weighting matrix is determined in the form of the inverse of the squared error or as an exponential function with a negative exponent. The proposed WBN and VSSWBN algorithms converge in two iterations; thus, the required number of extra samples in the transient period is negligible. Also, we perform an analysis of the mean square error (MSE) of the weighted block Newton method. To verify the superiority of the proposed methods, the MSE performances are compared via extensive simulation.

Keywords: Block Newton method, Adaptive filter, Weighting matrix, Variable step size, Non-line-of-sight (NLOS)

1 Introduction

The aim of the source localization system is to find a geometrical point of intersection using the measurements from each receiver, such as the time difference of arrival (TDOA), time of arrival (TOA), or received signal strength (RSS). Localizing a point source in which passive and stationary sensors are used has been a popular research issue in the areas of radar, sonar, global positioning system, video conferencing, and telecommunication. Two key issues need to be resolved in source localization. The first is that the wireless localization systems must be able to cope with rapidly varying channel conditions. To do this, adaptive filters have been important in adaptive source localization methods to stabilize localization performance under varying channel conditions [1–3]. However, these adaptive source localization algorithms were designed for the Gaussian noise situation and their localization performance is severely degraded in impulsive noise environments, such as Gaussian mixture and Student's t distribution. Also, the convergence rate of the least mean square (LMS) algorithm adopted in [1–3] may be much slower in some adverse environments. The second key issue is the challenge of localization in environments of line-of-sight (LOS)/non-line-of-sight (NLOS) mixture. LOS/NLOS mixture scenarios occur when there are obstructions between transmitters and receivers located in indoor environments and outdoor situations such as urban areas. The localization performance of traditional approaches, which assume only LOS conditions, is severely degraded under NLOS conditions; thus, mitigation of NLOS errors has become an urgent task and has been extensively investigated in the last decade. In general, research of the LOS/NLOS mixture problem for localization takes one of two approaches: (1) the constrained least squares (LS) method using optimization, such as semidefinite relaxation and second-order cone relaxation [4–9] and (2) localization based on the "NLOS identify and discard" [10–14]. Although localization using the optimization method has relatively high accuracy, the computational load is intensive. Localization using the "NLOS identify and discard" method also has relatively high accuracy when the LOS/NLOS mixture sensors are perfectly separated from the LOS sensors. However, the complete classification of LOS sensors and LOS/NLOS mixture sensors is nearly impossible, making it evident that classification error exists and the resulting

*Correspondence: jchang@hanyang.ac.kr
[2]Department of Electronic Engineering, Hanyang University, Seoul 133-791, Republic of Korea
Full list of author information is available at the end of the article

false classification incurs drastically increased localization error. Furthermore, when the number of sensors is large, the number of cases to be calculated is increased; thus, the computational burden is much high. Although our research results deal with the case in which the positions of sensors are accurately known, some recent works assume the unknown coordinates of sensors [15–18].

The motivation for this paper is as follows. Adaptive localization methods have been proposed for positioning in changing environments [1–3]. However, conventional adaptive localization algorithms, which minimize the squared error sum, are not robust to non-Gaussian noise (impulsive noise) situations, so they must be adapted to impulsive noise conditions. Our proposed algorithm minimizes the weighted squared error sum instead of the squared error sum, and the weighting matrix in the proposed algorithm counteracts the adverse effects of the LOS/NLOS mixture sensor. Namely, the weight is small when the sample is an outlier, which means an outlier-corrupted measurement is prevented from entering into the minimization of the cost function. Also, the convergence rate of the LMS method is slower than that of the Newton method and this slow convergence rate increases the number of samples in the transient period of the adaptive filter and this is clearly a waste of resources. Thus, we propose a robust weighted block Newton (WBN) and variable step size WBN (VSSWBN) algorithms that use the Newton method instead of the LMS method. As can be seen from simulation results, the proposed WBN and VSSWBN methods converge in two iterations. Therefore, it is possible to neglect the number of additional samples in the transient period of the adaptive localization algorithm.

The organization of this paper is as follows. Section 2 explains the LOS/NLOS mixture source localization problem to be solved in this paper. In Section 3, the details of the existing localization methods are addressed. The proposed adaptive localization methods using the weighting matrix and variable step size are addressed in Section 4. The MSE analysis of the proposed WBN algorithm is performed in Section 5. The estimation performances of the proposed methods are evaluated via simulation results in Section 6, comparing them with those of the existing algorithms. Finally, the conclusion is presented in Section 7.

2 Problem formulation

The main goal of the TOA-based source localization method is to accurately determine the position of a source using multiple circles whose centers are the locations of sensors. In the LOS/NLOS mixture source localization context, the measurement equation is represented as

$$r_i = d_i + n_i = \sqrt{(x - x_i)^2 + (y - y_i)^2} + n_i, \qquad (1)$$

where $n_i \sim (1-\epsilon)N(0,\sigma_1^2) + \epsilon N(\mu_2,\sigma_2^2), \sigma_1^2 \ll \sigma_2^2,\ i = 1,2,\ldots,M,$ with M denoting the number of sensors [7, 19–21]. Also, r_i is the measured distance between the source and the ith sensor and d_i is the range (distance) model between the source and the ith sensor. The measurement noise n_i is modeled as a Gaussian mixture distribution, where the LOS noise is distributed according to $N\left(0,\sigma_1^2\right)$ with a probability $(1-\epsilon)$ and the NLOS noise distributed by $N\left(\mu_2,\sigma_2^2\right)$ with a probability of ϵ. It is assumed that while the statistics of the inlier can be obtained, the mean and variance of the outlier distribution are unknown. In practice, σ_1^2 can be estimated by observing the energy bins in an absence of the transmitted signal and indeed the sample variance is usually adopted. Here, ϵ $(0 \le \epsilon \le 1)$ is the contamination ratio (i.e., fraction of contamination) which is a small number (typically smaller than 0.1) [7, 19–21]. Also, $[x \;\; y]^T$ is the true source position and $[x_i \;\; y_i]^T$ is the position of the ith sensor. Note that, throughout this paper, a lowercase boldface letter denotes a vector, an uppercase boldface letter indicates a matrix and the superscript T signifies the vector/matrix transpose. The purpose of this paper is to determine the source position that minimizes the MSE of the position estimate.

3 Review of the existing TOA localization methods

In this section, we briefly discuss the block LMS algorithm, M, and LMedS estimators in terms of the formulation of the source localization.

3.1 Block LMS source localization [1]

Squaring (1) and rearranging yield the following equation:

$$\begin{aligned} &x_i x + y_i y - 0.5R + m_i = 0.5\left(x_i^2 + y_i^2 - r_i^2\right), \\ &i = 1,2,\ldots,M, \end{aligned} \qquad (2)$$

where $R = x^2 + y^2, m_i = -d_i n_i - \frac{1}{2}n_i^2$. For convenience, (2) can be simply represented in a matrix form as

$$\mathbf{A}\mathbf{x} + \mathbf{q} = \mathbf{b}, \qquad (3)$$

where $\mathbf{q} = [\, m_1, \cdots, m_M]^T$, $\mathbf{x} = [\, x \; y \; R\,]^T$,

$$\mathbf{A} = \begin{pmatrix} x_1 & y_1 & -0.5 \\ \vdots & \vdots & \vdots \\ x_M & y_M & -0.5 \end{pmatrix} \text{ and } \mathbf{b} = \frac{1}{2}\begin{pmatrix} x_1^2 + y_1^2 - r_1^2 \\ \vdots \\ x_M^2 + y_M^2 - r_M^2 \end{pmatrix}.$$

The block LMS location estimate is obtained iteratively as follows:

$$\mathbf{x}^{(k+1)} = \mathbf{x}^{(k)} + \mu \mathbf{A}^T \mathbf{e}^{(k)} \qquad (4)$$

where $\mathbf{e}^{(k)} = \mathbf{b}_k - \mathbf{A}\mathbf{x}^{(k)}$, superscript is the iteration number, and μ is a positive step size (see [1] for details). Note that the input signal in the conventional LMS algorithm was substituted with $\mathbf{A}^T$ in the block LMS method.

3.2 LMedS estimator

Classical LS regression consists of minimizing the sum of the squared residuals. In the LMedS algorithm, the sum is replaced by the median [22] of the squared residuals. The LMedS estimator can resist the effect of nearly 50% of the contamination in the data. The algorithm used to obtain a solution with this method can be summarized as follows [23]:

(1) Calculate the m subsets of three measurements.
(2) For each subset S, compute a location by trilateration $\mathbf{P}_S$.
(3) For each solution $\mathbf{P}_S$, the residues $\mathbf{R}_S$ are obtained as

$$\mathbf{R}_S = \left[(r_1 - \hat{d}_1)^2, (r_2 - \hat{d}_2)^2, \cdots, (r_M - \hat{d}_M)^2\right] \quad (5)$$

where r_i is the same with (1), $\hat{d}_i = \sqrt{\left(x_{ps} - x_i\right)^2 + \left(y_{ps} - y_i\right)^2}$ and the median of the residuals $\mathbf{R}_S$ is computed.
(4) The solution $\mathbf{P}_S$, which gives the minimum median, is determined as the source location.

3.3 *M*-estimator

The M-estimator is a class of robust estimator that has been considered for NLOS mitigation purposes. The location estimate using the M-estimator is obtained in the following:

$$\hat{\mathbf{x}}^{\mathrm{M}} = \min_{\mathbf{x}} \sum_{i=1}^{M} \rho \left\{ \frac{b_i - \mathbf{a}_i^T \mathbf{x}}{s} \right\} \quad (6)$$

where $\mathbf{a}_i^T$ is the ith row vector of $\mathbf{A}$, $\mathbf{x}$ is the position parameter to be estimated, and b_i is the sample in the ith sensor. The standard deviation in (6) is an unknown value, so it should be estimated. The median absolute deviation (MAD), represented as s, is used as the estimate of the standard deviation and it is defined as

$$s = 1.483 \ \mathrm{med}_i \left[\left| \left\{ b_i - \mathbf{a}_i^T \mathbf{x} \right\} - \mathrm{med}_k \left\{ b_k - \mathbf{a}_k^T \mathbf{x} \right\} \right| \right] \quad (7)$$

where med is the abbreviation of the median. Also, $\rho(\cdot)$ is defined as follows:

$$\rho(\nu) = \begin{cases} \nu^2/2 & |\nu| \le \eta \\ \eta|\nu| - \eta^2/2 & |\nu| > \eta. \end{cases} \quad (8)$$

Then, the position estimate using the M-estimator is obtained by using the Newton method as follows:

$$\hat{\mathbf{x}}^{(k+1)} = \hat{\mathbf{x}}^{(k)} + s(\mathbf{A}^T\mathbf{A})^{-1}\mathbf{A}^T \psi \left\{ \frac{\mathbf{b} - \mathbf{A}\hat{\mathbf{x}}^{(k)}}{s} \right\} \quad (9)$$

where

$$\psi(\nu) = \begin{cases} \nu & |\nu| \le \eta \\ \eta \cdot \mathrm{sign}(\nu) & |\nu| > \eta, \end{cases} \quad (10)$$

and sign$(\cdot)$ is the sign function defined as $\mathrm{sign}(\nu) = \begin{cases} 1, & \text{if } \nu > 0 \\ 0, & \text{if } \nu = 0 \\ -1, & \text{if } \nu < 0. \end{cases}$

4 Proposed adaptive robust localization methods

In this paper, the LOS/NLOS mixture state is divided into the LOS and LOS/NLOS states. The LOS state denotes the case where the contamination ratio is zero (ε=0) and the LOS/NLOS state is the condition in which $0 < \varepsilon \le 1$. The adaptive robust localization algorithms in this paper can be represented as follows.

4.1 WBN method

This proposed algorithm modifies the block LMS source localization algorithm [1] to robustify the block LMS algorithm against outliers by using a weighting matrix which is given as the inverse of the square error or an exponential function with a negative exponent. Also, we adopt the block Newton method instead of the block LMS algorithm because the convergence rate of the Newton method is known to be much faster than that of the LMS algorithm [24]. The simulation results show that the MSE performance of the WBN algorithm converged in two iterations. The cost function to be minimized is defined as follows:

$$(\mathbf{b} - \mathbf{A}\mathbf{x})^T \mathbf{Q}^{-1} (\mathbf{b} - \mathbf{A}\mathbf{x}). \quad (11)$$

The solution of the WBN method is represented as follows:

$$\begin{aligned} x^{(k+1)} &= \mathbf{x}^{(k)} - \mu {\mathbf{H}^{(k)}}^{-1} \mathbf{g}^{(k)} \\ &= \mathbf{x}^{(k)} + \mu (\mathbf{A}^T \mathbf{Q}^{(k)} \mathbf{A})^{-1} \mathbf{A}^T \mathbf{Q}^{(k)} \mathbf{e}^{(k)} \end{aligned} \quad (12)$$

where $\mathbf{H}^{(k)}$ is the Hessian matrix, $\mathbf{g}^{(k)}$ is the gradient vector of the cost function in the kth iteration, $\mathbf{Q}^{(k)} = \mathrm{diag}\left[\frac{1}{(b_{1,k} - \mathbf{a}_1^T \mathbf{x}^{(k)})^2} \cdots \frac{1}{(b_{M,k} - \mathbf{a}_M^T \mathbf{x}^{(k)})^2}\right]$ or $\mathrm{diag}\left[e^{-\frac{(b_{1,k} - \mathbf{a}_1^T \mathbf{x}^{(k)})^2}{2\zeta}} \cdots e^{-\frac{(b_{M,k} - \mathbf{a}_M^T \mathbf{x}^{(k)})^2}{2\zeta}}\right]$, ζ is a tuning parameter to be determined through offline work, $\mathbf{e}^{(k)} = \mathbf{b}_k - \mathbf{A}\mathbf{x}^{(k)}$, $\mathbf{b}_k = \left[b_{1,k} \cdots b_{M,k}\right]^T$, and $b_{i,k}$ denotes the sample of the ith sensor at the kth iteration. We use the weighting matrix $\mathbf{Q}^{(k)}$ to alleviate the adverse effect of the LOS/NLOS mixture components. Note that the weights are small for samples contaminated by outliers because the squared residuals are large and they are large for samples in the LOS state because the squared residuals are small. The advantage of using the weighting matrix ($\mathbf{Q}^{(k)}$) is that it alleviates the adverse effects of outliers. Meanwhile, the disadvantages are that the estimation performance may be inferior to the LOS-based algorithm in the LOS situation and the weight can diverge to infinity when the squared residual is close to zero. In this case, the small positive value can be added to the squared residual for the stability of the algorithm.

4.2 WCBN method

The clipped LMS method has been widely used to reduce the complexity of the LMS algorithm [25]. The computational complexity is reduced by clipping the input data to their polarity bits because the clipped LMS algorithm is a multiplication-free method. Following this algorithm, the matrix $\mathbf{A}$ is quantized by a sign function because $\mathbf{A}^T$ in the block Newton method corresponds to the input signal in the conventional LMS algorithm. Then, the proposed WCBN algorithm is obtained as follows:

$$\begin{aligned}\mathbf{x}^{(k+1)} &= \mathbf{x}^{(k)} + \mu \left(\text{sign}\left(\mathbf{A}^T\right)\mathbf{Q}^{(k)}\text{sign}(\mathbf{A})\right)^{-1} \\ &\quad \text{sign}\left(\mathbf{A}^T\right)\mathbf{Q}^{(k)}\mathbf{e}^{(k)}. \end{aligned} \tag{13}$$

4.3 VSSWBN algorithm

The variable step size adaptive algorithm has been used to improve the MSE performance of the fixed step size adaptive method [26–28]. This algorithm updates the step size by minimizing the MSE cost function and adopting this technique to the WBN method yields the following iterative equation:

$$\begin{aligned}\mu^{(k)} &= \mu^{(k-1)} - \frac{\rho}{2}\frac{\partial J^{(k)}}{\partial \mathbf{x}^{(k)T}} \cdot \frac{\partial \mathbf{x}^{(k)}}{\partial \mu^{(k-1)}} \\ &= \mu^{(k-1)} + \rho\left(\mathbf{e}^{(k)T}\mathbf{Q}^{(k)}\mathbf{A}\right)\left(\mathbf{A}^T\mathbf{Q}^{(k-1)}\mathbf{A}\right)^{-1} \\ &\quad \times \left(\mathbf{A}^T\mathbf{Q}^{(k-1)}\mathbf{e}^{(k-1)}\right) \end{aligned} \tag{14}$$

$$\mu^{(k)} = \begin{cases} \mu_{\max}, & \mu^{(k)} \geq \mu_{\max} \\ \mu_{\min}, & \mu^{(k)} < \mu_{\min} \\ \mu^{(k)}, & \text{othewise} \end{cases} \tag{15}$$

$$\mathbf{x}^{(k+1)} = \mathbf{x}^{(k)} + \mu^{(k)}\left(\mathbf{A}^T\mathbf{Q}^{(k)}\mathbf{A}\right)^{-1}\mathbf{A}^T\mathbf{Q}^{(k)}\mathbf{e}^{(k)} \tag{16}$$

where $J^{(k)} = \mathbf{e}^{(k)T}\mathbf{Q}^{(k)}\mathbf{e}^{(k)}$, $\mathbf{e}^{(k)} = \mathbf{b}_k - \mathbf{A}\mathbf{x}^{(k)}$, and $\mu_{\max}, \rho$ are generally selected through experiments to provide the maximum convergence rate preserving steady state misadjustment error small. The value of $\mu_{\min}$ is determined as the level of steady-state misadjustment and the required tracking capabilities. The length of the transient period of the adaptive source localization method is desired to be short because the extra samples are consumed until the algorithm converges compared to the existing robust localization algorithm. Therefore, the step size is determined as large as possible in the proposed method to aid fast convergence. In our simulation results, the VSSWBN algorithm showed a superior MSE performance and a similar convergence rate compared to the WBN method for large step sizes (both algorithms converged in two iterations). However, the MSE performances of the VSSWBN and WBN algorithms were similar for small or medium step sizes.

5 Performance analysis

5.1 MSE performance analysis

This section presents the MSE analysis of the proposed algorithm. We derive the MSE of the WBN algorithm instead of the VSSWBN method for convenience of analysis. The state error vector is represented as

$$\begin{aligned} x^{(k+1)} - \mathbf{x}^o &= \mathbf{x}^{(k)} - \mathbf{x}^o + \mu\left(\mathbf{A}^T\mathbf{Q}^{(k)}\mathbf{A}\right)^{-1}\mathbf{A}^T\mathbf{Q}^{(k)} \\ &\quad \times \left(\mathbf{b}_k - \mathbf{A}\mathbf{x}^{(k)}\right) \\ &= (1-\mu)\left(\mathbf{x}^{(k)} - \mathbf{x}^o\right) + \mu\left(\mathbf{A}^T\mathbf{Q}^{(k)}\mathbf{A}\right)^{-1} \\ &\quad \mathbf{A}^T\mathbf{Q}^{(k)}\left(\mathbf{b}_k - \mathbf{A}\mathbf{x}^o\right) \end{aligned} \tag{17}$$

where $\mathbf{x}^o$ is the true source position. The steady-state error vector ($\mathbf{v}^{(\infty)}$) can be attained as

$$\begin{aligned} &\mathbf{v}^{(k+1)} \\ &= (1-\mu)\mathbf{v}^{(k)} + \mu(\mathbf{A}^T\mathbf{Q}^{(k)}\mathbf{A})^{-1}\mathbf{A}^T\mathbf{Q}^{(k)}(\mathbf{b}_k - \mathbf{A}\mathbf{x}^o) \\ &= (1-\mu)^k\mathbf{v}^{(0)} + \mu\sum_{j=0}^{k-1}\left\{(1-\mu)^{(k-j-1)}(\mathbf{A}^T\mathbf{Q}^{(j)}\mathbf{A})^{-1}\mathbf{A}^T\right. \\ &\qquad \left.\mathbf{Q}^{(j)}(\mathbf{b}_j - \mathbf{A}\mathbf{x}^o)\right\} \end{aligned} \tag{18}$$

$$\mathbf{v}^{(\infty)} \simeq \lim_{k\to\infty}\mu\sum_{j=0}^{k-1}\left\{(1-\mu)^{(k-j-1)}(\mathbf{A}^T\mathbf{Q}^{(j)}\mathbf{A})^{-1}\mathbf{A}^T\mathbf{Q}^{(j)}\right. \\ \left.(\mathbf{b}_j - \mathbf{A}\mathbf{x}^o)\right\}.$$

The MSE is the same with the trace of the steady-state error covariance matrix. Therefore, the MSE is represented as follows:

$$\begin{aligned} &\text{tr}\left\{E\left[\mathbf{v}^{(\infty)}\mathbf{v}^{(\infty)T}\right]\right\} \\ &\simeq \lim_{k\to\infty}\text{tr}\left\{\mu^2\sum_{j=0}^{k-1}\sum_{q=0}^{k-1}\left[(1-\mu)^{(k-q-1)}(\mathbf{A}^T\mathbf{Q}^{(q)}\mathbf{A})^{-1}\mathbf{A}^T\mathbf{Q}^{(q)}\right.\right. \\ &\quad \times E\left[\left(\mathbf{b}_q - \mathbf{A}\mathbf{x}^o\right)\left(\mathbf{b}_j - \mathbf{A}\mathbf{x}^o\right)^T\right]\mathbf{Q}^{(j)}\mathbf{A}\left(\mathbf{A}^T\mathbf{Q}^{(j)}\mathbf{A}\right) \\ &\quad \left.\left.\times (1-\mu)^{(k-j-1)}\right]\right\} \\ &= \lim_{k\to\infty}\text{tr}\left[\mu^2\sum_{l=0}^{k-1}\left\{(1-\mu)^{2(k-l-1)}\left(\mathbf{A}^T\mathbf{Q}^{(l)}\mathbf{A}\right)^{-1}\right.\right. \\ &\quad \left.\left.\times\mathbf{A}^T\mathbf{Q}^{(l)}\mathbf{R}_l\mathbf{Q}^{(l)}\mathbf{A}(\mathbf{A}^T\mathbf{Q}^{(l)}\mathbf{A})^{-1}\right\}\right] \\ &\simeq \text{tr}\left[\mu^2\left\{\left(\mathbf{A}^T\mathbf{Q}^{(k-1)}\mathbf{A}\right)^{-1}\mathbf{A}^T\mathbf{Q}^{(k-1)}\mathbf{R}_{k-1}\mathbf{Q}^{(k-1)}\right.\right. \\ &\quad \left.\left.\times\mathbf{A}\left(\mathbf{A}^T\mathbf{Q}^{(k-1)}\mathbf{A}\right)^{-1}\right\}\right] \\ &\simeq \text{tr}\left[\mu^2(\mathbf{A}^T\mathbf{Q}^{(k-1)}\mathbf{A})^{-1}\right] \end{aligned} \tag{19}$$

where

$$\begin{aligned}\mathbf{R}_l &= E\left[\left(\mathbf{b}_l - \mathbf{A}\mathbf{x}^o\right)\left(\mathbf{b}_l - \mathbf{A}\mathbf{x}^o\right)^T\right] \\ &= \text{diag}\left[R^l_{1,1}\cdots R^l_{i,i}\cdots R^l_{M,M}\right] \\ R^l_{i,i} &= \begin{cases} d_i^2\sigma_1^2, & \text{if i} \in \text{LOS sensor;} \\ (1-\epsilon)d_i^2\sigma_1^2 + \epsilon(d_i^2\sigma_2^2 + \frac{1}{2}\sigma_2^4 + \mu_2^2\sigma_2^2), & \text{if i} \in \text{LOS/NLOS mixture sensor;} \end{cases}\end{aligned} \tag{20}$$

and $\mathbf{Q}^{(l)}$ was treated as the constant matrix for the ease of derivation. The WBN algorithm converges when $0 < \mu < 1$ from (18). In the derivation of the last term of (19), it is assumed that $\mathbf{Q}^{(k-1)} \simeq \mathbf{R}_{k-1}^{-1}$ ($\mathbf{Q}$ is the instantaneous form of $\mathbf{R}^{-1}$) and the summation is approximately equal to $\mu^2\left\{(\mathbf{A}^T\mathbf{Q}^{(k-1)}\mathbf{A})^{-1}\mathbf{A}^T \mathbf{Q}^{(k-1)}\mathbf{R}_{k-1}\mathbf{Q}^{(k-1)}\mathbf{A}(\mathbf{A}^T\mathbf{Q}^{(k-1)}\mathbf{A})^{-1}\right\}$ because $0 < 1 - \mu < 1$. Note that the inverse of the squared error is relatively large in the LOS sensor and is small when the outliers exist. Then, although σ_2 is much larger than σ_1, the MSE is nearly constant with respect to the standard deviation of NLOS noise and bias because the effect of large error standard deviation of LOS/NLOS mixture sensor in $\mathbf{R}_l$ is attenuated by the weighting matrix $\mathbf{Q}^{(l)}$ (see (19) and (20)).

5.2 Computational complexity analysis

We compared the computational complexity for the localization algorithms. The computational complexities of localization methods are represented in the Table 1, and M is the number of sensors, and N is the number of unknown variables to be estimated. The inverse and multiplication operations for the matrix were mainly considered because they are computationally intensive. The computational complexity of the VSSWBN method was higher than that of the WBN algorithm because it requires the additional computation of the step size. Also, the computational complexity of the WCBN method was lower than those of WBN and VSSWBN algorithms.

6 Simulation results

In this section, we compare the MSE performances of the proposed LOS/NLOS mixture source localization methods with those of the M-estimator [22, 29] and LMedS estimator [23]. In these simulation settings, the source was assumed to be located within a 400-m^2 region to determine the performance over the entire area. Note that the number of sensors used in this experiment was seven. Next, 30 different source locations were generated with a uniform distribution and sensors fixed. Five hundred Monte-Carlo simulations were performed for each given standard deviation of the NLOS noise. The standard deviation of the LOS noise of all of the sensors was assumed to be identical. In addition, the single and omni-directional source was assumed to be in the stationary state. The MSE average was calculated as follows:

$$\text{MSE average} = \frac{\sum_{i=1}^{30}\sum_{k=1}^{500}\left[\left(\widehat{x}^k(i) - x(i)\right)^2 + \left(\widehat{y}^k(i) - y(i)\right)^2\right]}{30 \times 500} \tag{21}$$

where $\widehat{x}^k(i)$, $\widehat{y}^k(i)$ is the estimated position of the source in the ith position set and kth iteration and $x(i)$ and $y(i)$ indicate the ith true position of the source. Figure 1 illustrates a deployment of sensors, in which the radius of the sensor network was set to 10 m. The localization accuracy as a function of the standard deviation of the NLOS noise is shown in Fig. 2. In Fig. 2a, the contamination ratio (ϵ) was 20%, the standard deviation of the LOS noise (σ_1) was 0.01 m, the bias of the NLOS noise (μ_2) was 4 m, sensors 5, 6, and 7 were the LOS/NLOS sensors, and the remaining sensors were LOS sensors. The step size (μ) was set to 0.99 in the WBN algorithm. Also, the initial step size ($\mu^{(0)}$) was 0.99, ρ was 0.1, and $\mu_{\max}$ and $\mu_{\min}$ were one and 0.01 in the VSSWBN algorithm. It is clear that the MSE averages of the VSSWBN method are lower than those of other

Table 1 Comparison of the computational complexity

Algorithm	Computational complexity
Block Newton	$O\left(M^2N\right)$
WBN	$O\left(2M^2N\right)$
VSSWBN	$O\left(2M^2N + M^2\right)$
WCBN	$O\left(M^2\right)$

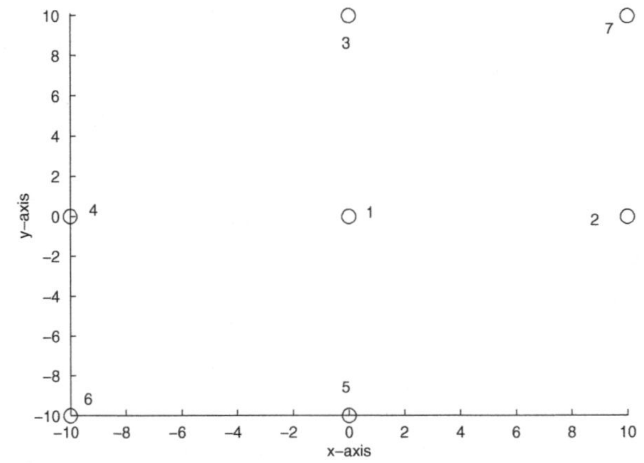

Fig. 1 Deployment of sensors

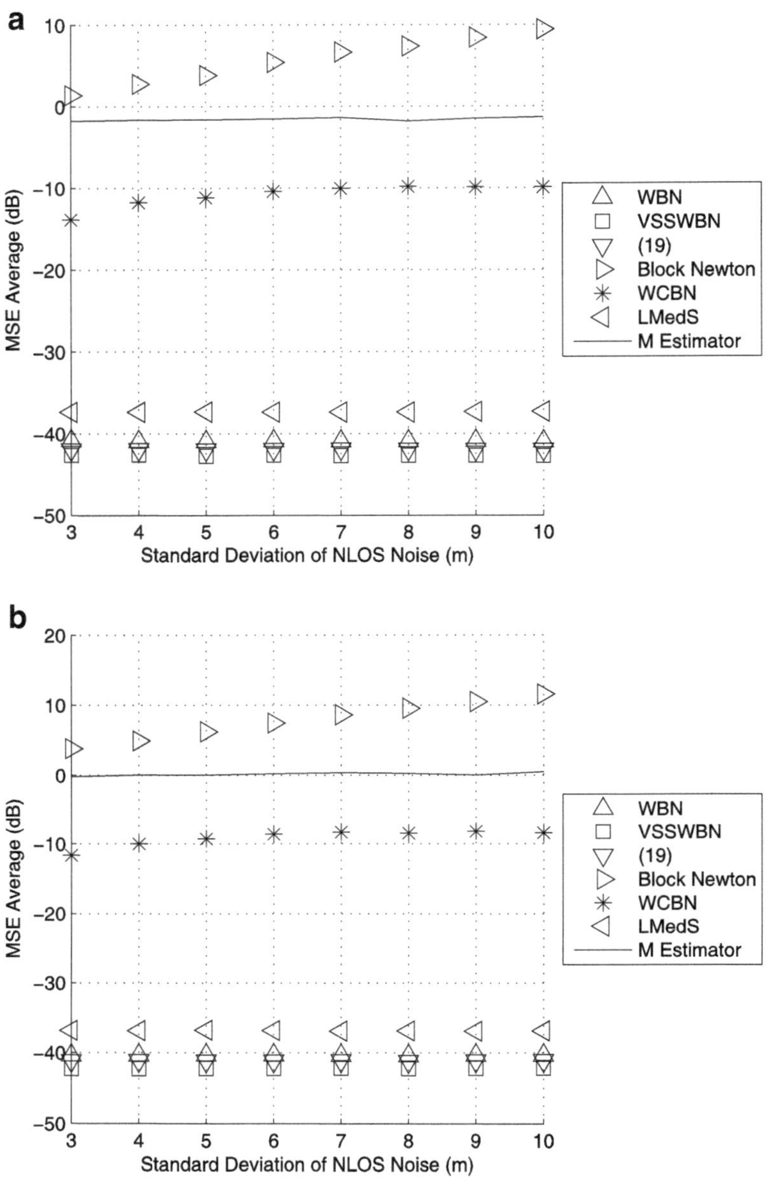

Fig. 2 Comparison of MSE averages of the proposed estimators with those of the existing methods when the sensors 5, 6, and 7 are the LOS/NLOS mixture sensors and the remaining sensors are the LOS sensors. **a** Contamination ratio (ϵ): 20%, the bias of NLOS noise (μ_2): 4 m, standard deviation of LOS noise (σ_1): 0.01 m. **b** ϵ: 30%, σ_1: 0.01 m, μ_2: 4 m

methods and nearly constant with respect to the standard deviation of NLOS error. This observation is caused because the weighting matrix attenuates the effect of the large variance of LOS/NLOS mixture sensor. In Fig. 2b, the contamination ratio was 30% and the remaining conditions are the same as those in Fig. 2a. Figure 2b shows that the MSE average performances of the VSSWBN method are much superior to those of the other methods. Figure 3 assumes the same condition as that in Fig. 2, with the exception that sensors 4, 5, 6, and 7 are the LOS/NLOS sensors in Fig. 3. Again, the proposed VSSWBN method outperformed the other methods, as shown in Fig. 3. Figure 4 shows the variation of the MSE averages as a function of the contamination ratio. The MSE averages gradually increase as the contamination ratio becomes larger, and the VSSWBN method was superior to the other methods. Figure 5 shows the MSE averages with respect to the standard deviation of LOS noise. The MSE averages of all robust methods increase as the LOS noise increases. Figure 6 illustrates MSE averages as a function of bias. As can be seen from Fig. 6, MSE averages do not change as the bias is varied and this is due to the weighting matrix

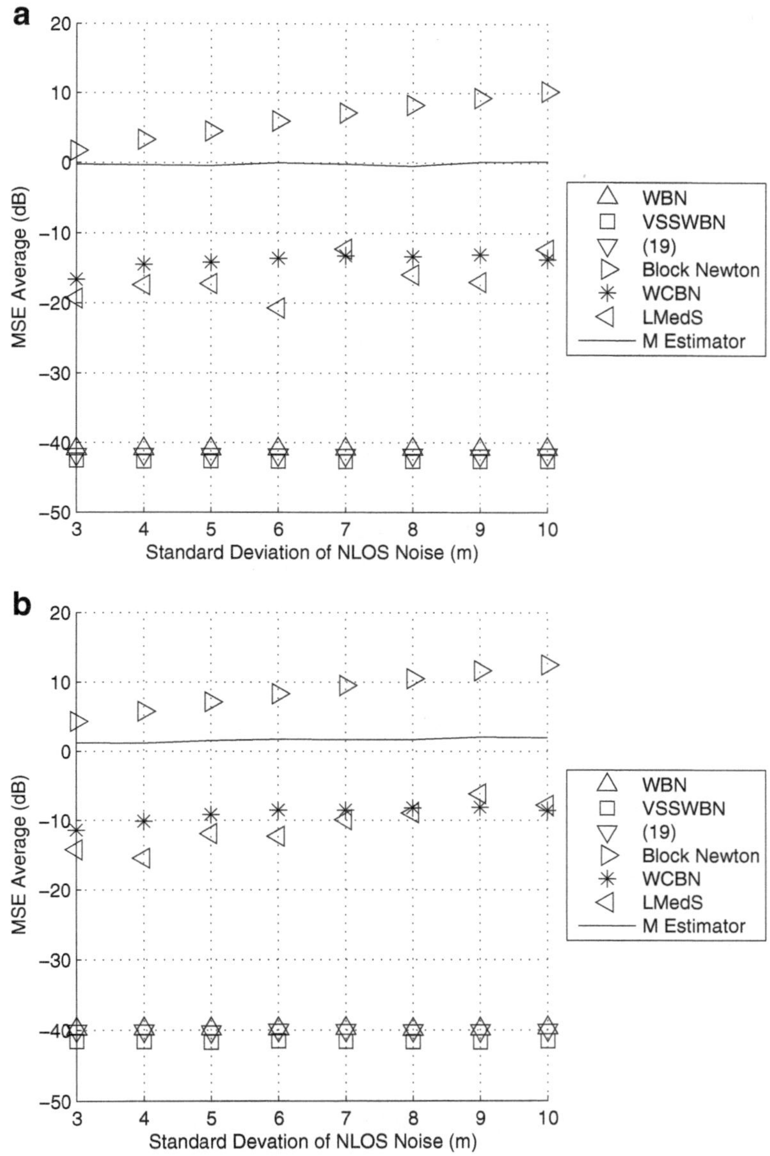

Fig. 3 Comparison of MSE averages of the proposed estimators with those of the existing methods when the sensor 4, 5, 6 and 7 are the LOS/NLOS mixture sensors and the remaining sensors are the LOS sensors. **a** Contamination ratio (ϵ): 20%, the bias of NLOS noise (μ_2): 4 m, standard deviation of LOS noise (σ_1): 0.01 m. **b** ϵ: 30%, σ_1: 0.01 m, μ_2: 4 m

which attenuates the effect of large variance caused by the bias ($\mu_2^2\sigma_2^2$ in (20)) because the corresponding weight is far small (squared residual is much large). Figure 7 shows the adaptation error as a function of the iteration number when the impulsive noise exists in the 10th iteration. The WBN and VSSWBN methods converged in two iterations; thus, the additional samples which are required in the transient period are negligible. Additionally, the proposed WBN and VSSWBN algorithm accurately tracked the abrupt change caused by the impulsive noise. Figure 8 shows the MSE average of the proposed algorithms as a function of the number of sensors. In this case, the number of sensors increases from 5 to 9 and the number of LOS/NLOS sensors is fixed to two and the number of LOS sensors is increased. The standard deviation of the LOS noise was 0.01 m, that of the NLOS noise was 10 m, the bias was 4 m, and the contamination ratio was 30%. We can see that the MSE averages of the localization decrease as the number of sensors increases. Meanwhile, Fig. 9 shows the MSE averages of the proposed methods as a function of the number of sensors when the number of the LOS/NLOS sensors increases. The number of the

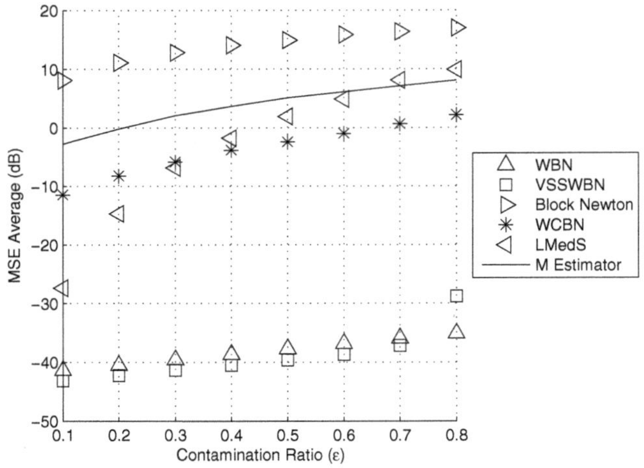

Fig. 4 MSE averages of the localization algorithms as a function of contamination ratio (bias of NLOS noise (μ_2): 4 m, standard deviation of LOS noise (σ_1): 0.01 m, standard deviation of NLOS noise (σ_2): 10 m)

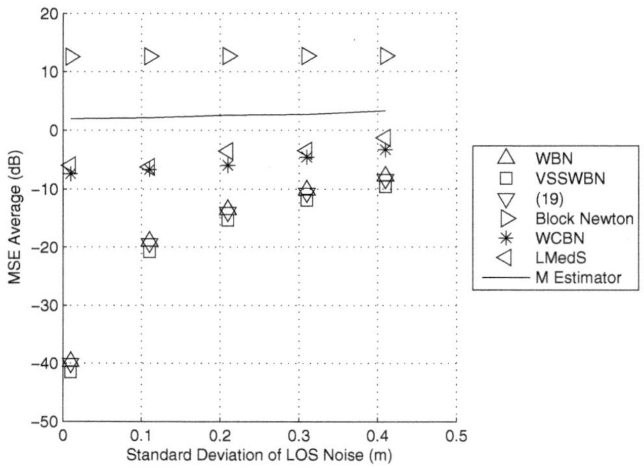

Fig. 5 MSE averages of the localization algorithms as a function of standard deviation of LOS noise (bias of NLOS noise (μ_2): 4 m, contamination ratio: 30%, standard deviation of NLOS noise (σ_2): 10 m)

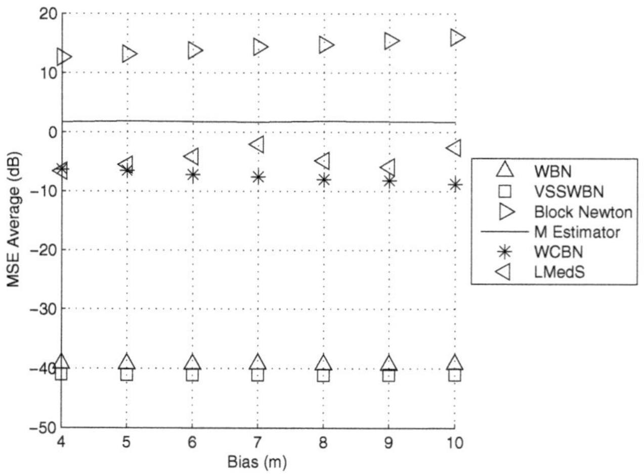

Fig. 6 MSE averages of the localization algorithms as a function of bias (contamination ratio: 30%, standard deviation of LOS noise (σ_1): 0.01 m, standard deviation of NLOS noise (σ_2): 10 m)

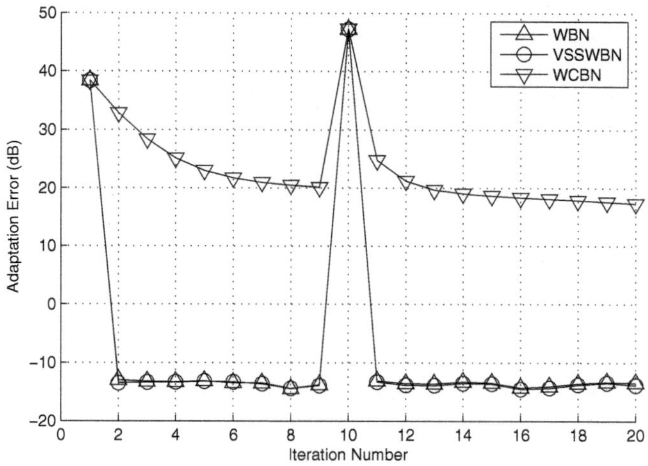

Fig. 7 Adaptation error of adaptive localization algorithms (contamination ratio: 30%, standard deviation of LOS noise (σ_1): 0.01 m, standard deviation of NLOS noise (σ_2): 10 m, bias (μ_2): 4 m)

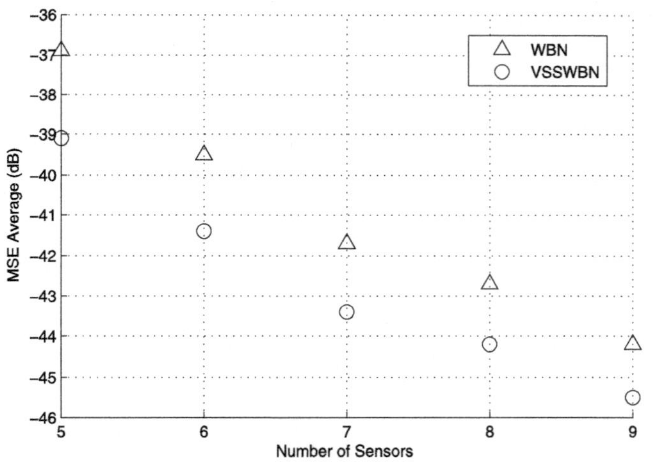

Fig. 8 Comparison of MSE averages of the proposed estimators as a function of the number of sensors (when the number of LOS sensors increases)

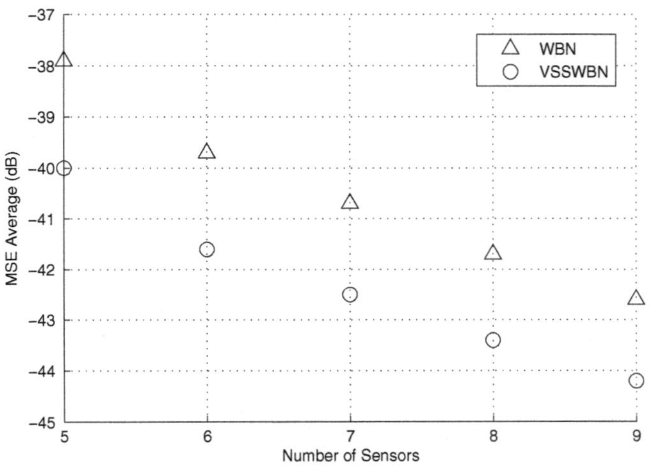

Fig. 9 Comparison of MSE averages of the proposed estimators as a function of the number of sensors (when the number of LOS/NLOS sensors increases)

LOS/NLOS sensors is one when the number of sensors is five and then increases in parallel with the number of sensors. The MSE averages of the proposed methods decrease as the number of LOS/NLOS sensors increases, and the decreasing rate of the MSE averages is lower compared to the case in which the number of LOS sensors increases.

7 Conclusions

The WBN algorithm was developed by modifying the block LMS algorithm to make it robust to outliers and the proposed method employed a weighting matrix. Furthermore, the VSSWBN method was proposed to improve the MSE performances of the WBN algorithm. We also analyzed the MSE of the WBN algorithm. In the simulation results, the MSE averages of the proposed methods were smaller than that of the other adaptive localization methods and robust positioning algorithms.

Acknowledgements
This research was supported by Basic Science Research Program through the National Research Foundation of Korea (NRF) funded by the Ministry of Science, ICT, and Future Planning (No.2014R1A2A1A10049735).

Competing interests
The authors declare that they have no competing interests.

Authors' contributions
In this research paper, the authors proposed a robust localization algorithm. All authors read and approved the final manuscript.

Author details
[1]Department of Electronics and Computer Engineering, Hanyang University, Seoul 133-791, Republic of Korea. [2]Department of Electronic Engineering, Hanyang University, Seoul 133-791, Republic of Korea.

References

1. CH Park, KS Hong, Block LMS-based source localization using range measurement. Digit. Signal Process. **21**(2), 367–374 (2011)
2. Y Sun, J Xiao, X Li, F Cabrera-Mora, in *Proc.of GLOBECOM*. Adaptive source localization by a mobile robot using signal power gradient in sensor networks, (New Orleans, 2008), pp. 1–5
3. S Zhong, W Xia, Z He, in *Proc.of IEEE China Summit and International Conference on Signal and Information Process*. Adaptive direct position determination of emitters based on time differences of arrival, (Beijing, 2013), pp. 230–234
4. S Venkatesh, RM Buehrer, in *Proc. of IEEE International Symposium on Information Processing in SensorNetworks (IPSN)*. A linear programming approach to NLOS error mitigation in sensor networks, (Nashville, 2006), pp. 301–308
5. X Wang, Z Wang, BO Dea, A TOA based location algorithm reducing the errors due to non-line-of-sight (NLOS) propagation. IEEE Trans. Veh. Technol. **52**(1), 112–116 (2003)
6. S Venkatesh, RM Buehrer, NLOS mitigation using linear programming in ultrawideband location-aware networks. IEEE Trans. Veh. Technol. **56**(5), 3182–3198 (2007)
7. Y Feng, C Fritsche, F Gustafsson, AM Zoubir, EM- and JMAP-ML based joint estimation algorithms for robust wireless geolocation in mixed LOS/NLOS environments. IEEE Trans. Signal Process. **62**(1), 168–182 (2014)
8. H Shen, Z Ding, S Dasgupta, C Zhao, Multiple source localization in wireless sensor networks based on time of arrival measurements. IEEE Trans. Signal Process. **62**(8), 1938–1949 (2014)
9. G Wang, H Chen, Y Li, N Ansari, NLOS error mitigation for TOA-based localization via convex relaxation. IEEE Trans. Wirel. Commun. **13**(8), 4119–4131 (2014)
10. YT Chan, WY Tsui, HC So, PC Ching, Time-of-arrival based localization under NLOS conditions. IEEE Trans. Veh. Technol. **55**(1), 17–24 (2006)
11. J Riba, A Urruela, in *Proc. of IEEE International Conference on Acoustics, Speech, and Signal Process*. A non-line-of-sight mitigation technique based on ML-detection (ICASSP, Quebec, 2004), pp. 153–156
12. Y Qi, H Kobayashi, H Suda, Analysis of wireless geolocation in a non-line-of-sight environment. IEEE Trans. Wirel. Commun. **5**(3), 672–681 (2006)
13. I Enosh, AJ Weiss, Outlier identification for TOA-based source localization in the presence of noise. Signal Process. **102**, 85–95 (2014)
14. A Abbasi, H Liu, Improved line-of-sight/non-line-of-sight classification methods for pulsed ultrawideband localisation. IET Commun. **8**(5), 680–688 (2014)
15. M Crocco, A Del Bue, V Murino, A bilinear approach to the position self-calibration of multiple sensors. IEEE Trans. Signal Process. **60**(2), 660–673 (2012)
16. I Dokmanić, R Parhizkar, J Ranieri, M Vetterli, Euclidean distance matrices: Essential theory, algorithms and applications. IEEE Trans. Signal Process. Mag. **32**(6), 12–30 (2015)
17. T-K Le, N Ono, Closed-form and near closed-form solutions for TOA-based joint source and sensor localization. IEEE Trans. Signal Process. **64**(18), 4751–4766 (2016)
18. T-K Le, N Ono, Closed-form and near closed-form solutions for TDOA-based joint source and sensor localization. IEEE Trans. Signal Process. **65**(5), 1207–1221 (2017)
19. F Gustafsson, F Gunnarsson, Mobile positioning using wireless networks. IEEE Signal Process. Mag. **22**(4), 41–53 (2005)
20. U Hammes, E Wolsztynski, AM Zoubir, Robust tracking and geolocation for wireless networks in NLOS environments. IEEE J. Sel. Top. Signal Process. **3**(5), 889–901 (2009)
21. Y Feng, C Fritsche, F Gustafsson, AM Zoubir, TOA-based robust wireless geolocation and Cramer-Rao lower bound analysis in harsh LOS/NLOS environments. IEEE Trans. Signal Process. **61**(9), 2243–2255 (2013)
22. P Huber, *Robust statistics*. (Wiley, Hoboken, 2009)
23. R Casas, A Marco, JJ Guerrero, J Falco, Robust estimator for non-line-of-sight error mitigation in indoor localization. EURASIP J. Adv. Signal Process. **Article ID 43429**, 1–8 (2006)
24. S Haykin, *Adaptive filter theory*. (Pearson, Upper Saddle River, 2013)
25. JL Moschner, *Adaptive filtering with clipped input data*. Ph.d.thesis. (Stanford University, Stanford, 1970)
26. RH Kwong, EW Johnson, A variable step-size LMS algorithm. IEEE Signal Process. **40**(7), 1633–1642 (1992)
27. VJ Mathews, Z Xie, A stochastic gradient adaptive filter with gradient adaptive step size. IEEE Trans. Signal Process. **41**(6), 2075–2087 (1993)
28. B Farhang-Boroujeny, *Adaptive filters: theory and applications*. (Wiley, Chichester, 2013)
29. X-W Chang, Y Guo, Huber's *M*-estimation in relative GPS positioning: computational aspects. J. Geod. **79**(6), 351–362 (2005)

15

Endcast: mobile stateless data delivery in MANETs

Sulochana Sooriyaarachchi* and Chandana Gamage

Abstract

MANETs are networks of mobile devices for multihop communication without a separate infrastructure for routing and control. If robust protocols are available, MANETs can exploit the true ubiquity of everyday mobile devices for application scenarios such as collaborative learning, multiplayer gaming, adaptive advertising as well as in disaster management and battlefield communication. Traditional schemes such as reactive, proactive and hybrid routing incur costs of often $O(n^2)$ for discovering and maintaining routes or node hierarchies for networks with n number of nodes. These routes or node structures become obsolete rapidly due to MANET dynamics resulting from mobility, limited battery life, and impairments of wireless links. These traditional approaches are based on a paradigm that we refer to as *fixed-stateful routing*. The routing paradigm that best matches the ubiquity of devices in a MANET is *mobile-stateless routing*. In mobile-stateless routing, nodes do not keep global network states or play special roles as control nodes. We propose flooding as an end to end packet delivery mechanism and introduce the concept of *endcast* that allows the communication from a single source to a single destination in which the data is carried by controlled flooding and the data propagation is inhibited after reaching the destination.

Problem statement: Simple flooding causes excessive packet redundancy resulting in collisions and contention which is known as broadcast storm problem. We introduce the term *broadcast flood problem* to refer to the propagation of data beyond the destination in flooding operation. This research is aimed at modeling endcast and developing an endcast scheme to enable MANET nodes to perform end to end communication in mobile-stateless manner.

Solution: We built a theoretical model to analyze the endcast concept. This model includes methods to quantify broadcast storm problem and broadcast flood problem. Based on this model, we proposed an endcast scheme that controls the storm situation using counter-based flooding and the flood situation using negative acknowledgement packets. The storm control scheme is inspired by the chalone mechanism that regulates growth of biological organs. The proposed scheme is in the mobile-stateless routing paradigm because the nodes keep only a set of local states in the form of a log of received data packets and the nodes do not act as infrastructure nodes. We apply the theoretical model on the proposed endcast scheme to validate its correctness and simulations were done to verify the performance.

Keywords: MANET, Multihop, Broadcasts, Flooding

1 Introduction

MANETs can be exploited as a ubiquitous computing architecture especially when a dense population of mobile nodes are connected with near field communication. However, the ad hoc mode of operation shows limited or negligible usage compared to infrastructure mode though it has been in existence in IEEE 802.11 specifications alone for about 20 years. The main cause that hinders true potential of MANETs is the lack of robust multihop communication protocols.

According to Abolhasan et al. [1], asymptotic costs of the proactive, reactive, and hybrid protocols have a memory overhead of magnitude $O\left(N^2\right)$ and control overhead of magnitude $O(N)$ (N—number of nodes). However, MANETs are dynamic networks and the routes thus discovered become obsolete rapidly, even before they

*Correspondence: sulochanas@cse.mrt.ac.lk
Department of Computer Science and Engineering, University of Moratuwa, Moratuwa 10400, Sri Lanka

converge in a time of magnitude $O(D)$ (D—diameter of network in number of hops). Similarly, the reactive protocol overheads also become higher because the routes become invalid rapidly due to dynamic topology of MANETs. Hybrid protocols require nodes to organize into zones, which also incurs costs apart from the costs of discovering and maintaining routes. Therefore, it is infeasible to route data packets in dense MANETs based on topological states.

Blind rebroadcasting is the simplest mechanism to send a data packet in fully stateless manner such that any node pair in the network can reach each other. However, blind rebroadcasting suffers from severe performance degredation due to massive redundant rebroadcasts resulting in *broadcast storm problem*. An improved version of blind rebroadcasting is the simple flooding in which a node rebroadcasts a packet only when it is received for the first time. The received packets are checked using a log of packet IDs where each entry comprises of source address and sequence number. There are flooding schemes that evaluate additional variables for making the rebroadcast decision such as counter-based flooding and distance-based flooding. In these flooding schemes, only a subset of nodes that receive the packet will rebroadcast and hence they are known as *efficient flooding schemes*.

Flooding schemes are conventionally used to disseminate a message to all the members of a network, and this communication mode is known as *broadcast*. The aim of routing is for a single source to reach a destination using a multihop path, and this communication is referred to as *unicast*. In this research, we leverage flooding for end to end data delivery where there are no preset point to point paths and only with rebroadcasting at intermediate nodes. We coined the term *endcast* to reflect this communication mode.

Viswanath and Obraczka [2] analyze flooding in its conventional use of broadcast. Their model includes node-to-node communication model of Wu and Varshney [3], multihop propagation of flooding modeled as a graph, and a model for probabilistic flooding. Our work extends these research by modeling efficient flooding schemes such as counter-based flooding. We incorporated mobility and time domain aspects such as random assessment delay (RAD) in our model.

We model endcast using a graph-based representation that we name *flooding graph* to make insights into the concept. Message propagation in flooding operation is perceived as a *flooding wave*. We geometrically estimate the area covered by the gradual propagation of a flooding wave. The number of newly reached nodes is estimated by considering node distribution. The characteristics of physical and MAC layer are captured by a probability (P_S) of node-to-node successful transmission using Markov chain models. The effects of multihop wireless communication such as hidden terminal problem are accounted for in calculating P_S. The successful reception of a rebroadcast is modeled by another probability (P_b). The properties of efficient flooding schemes are modeled by a probability (P_{tx}) of rebroadcasting, for example P_{tx} for counter-based flooding is calculated using the counter value and the threshold. Our theoretical model for endcast is capable of (i) modeling the effect of node mobility, (ii) modeling time domain aspects of flooding schemes such as RAD, (iii) modeling efficient flooding schemes, and (iv) quantifying broadcast storm and flood problems.

An endcast protocol should consist of a broadcast storm control mechanism and a broadcast flood control mechanism. We proposed an endcast scheme that controls the storm situation using counter-based flooding and the flood situation using negative acknowledgement packets. The storm control scheme is inspired by the chalone mechanism that regulates the growth of biological organs. An operational model was derived for the scheme using cellular automata with totalistic rules. The protocol was evaluated for reachability using the developed theoretical model, and its performance was analyzed using extensive simulations.

2 Endcast

We define *endcast* as the communication from a single source to a single destination in which the data is carried by controlled flooding and the data propagation is inhibited after reaching the destination. When any kind of flooding is used for end to end data forwarding, waste of network resources occur due to

- Broadcast storm: ideally, only one message should propagate from the source to the destination whereas flooding causes excessive number of messages propagating all over the network. Tseng et al. [4] define this situation as *broadcast storm problem*.
- Broadcast flood: even after reaching the destination, the messages continue to propagate and get accumulated in the network unless explicit mechanisms are implemented to stop the message propagation and remove the messages. We term this situation as *broadcast flood problem*.

Traditional approaches of routing in MANETs require global network information or at least neighbor information. Moreover some nodes may act as infrastructure nodes. We consider these approaches as *fixed-stateful routing*. We name the routing approaches that do not require nodes to maintain global network states or act as control nodes as *mobile-stateless routing*. Endcast is designed to be in mobile-stateless routing paradigm.

3 Related work

Ho et al. [5] argued that flooding is a possible alternative for reliable multicast in highly mobile ad hoc networks as traditional multicast solutions are not suitable. We propose that flooding is also an alternative for end to end data delivery as it allows to eliminate the cost of discovering and maintaining routes. Rahman et al. [6] also experimentally prove that flooding with tunable heuristics can be adopted as a reliable and efficient routing scheme in MANETs and these flooding-based schemes are not inferior to sophisticated point-to-point forwarding schemes. However, we did not come across any other work that studies utilizing of flooding for end to end communication in MANETs, particularly with a strong theoretical basis.

Viswanath and Obraczka [2] model the propagation of a data packet in a MANET via flooding as a graph rooted at the source. We refer to resulting graph as *flooding graph* hereafter and use this graph model for theoretical model of endcast. Further, flooding model in [2] consists of node-to-node communication model, multihop communication model, and probabilistic flooding model as explained below.

3.1 Node-to-node transmission

Wu and Varshney [3] propose a three-state Markov chain model for node-to-node wireless transmission in the presence of hidden terminal problem and it has been used by Viswanath and Obraczka [2] also. The three states are *idle state* (I), *successful transmission state* (S), and *collision transmission state* (C) as shown in Fig. 1.

Operational details of the MAC layer protocol specify the conditions for a successful node-to-node transmission. Four conditions must be fulfilled for a successful transmission from x to y in Fig. 2 according to CSMA. The notations used for describing these conditions are when r is the distance between nodes x and y, $C(r)$ is the area of the region $N(x) \cap N(y)$, $B(r)$ is the area of the region $N(y) - N(x)$. Takagi and Kleinrock [7] derived Eqs. 1, 2, and 3 for estimating the areas $C(r)$ and $B(r)$.

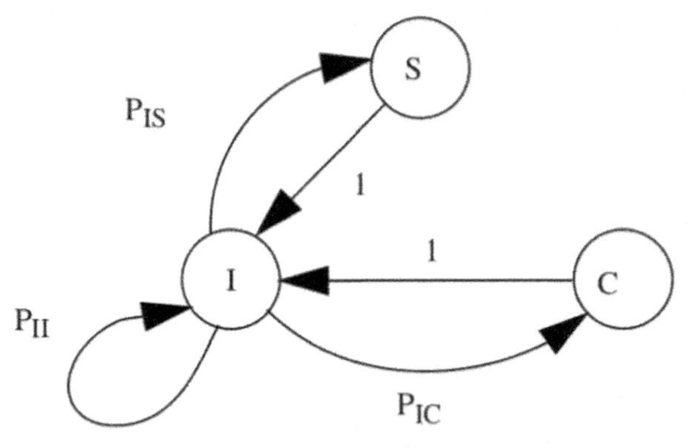

Fig. 1 Three-state Markov chain model of a node [3]

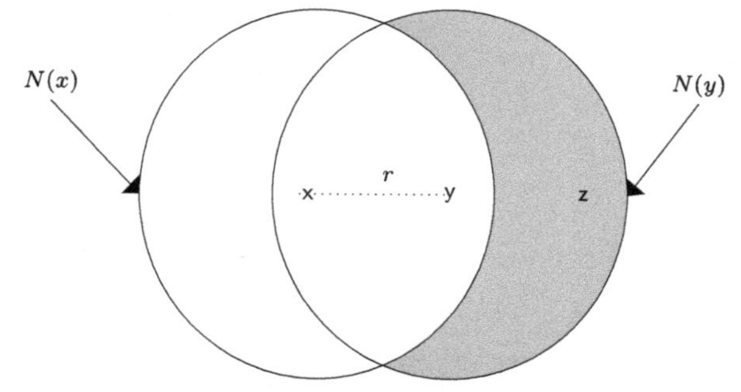

Fig. 2 Illustration of hidden node problem [3]

$$C(r) = 2R^2 q\left(\frac{r}{2R}\right) \quad (1)$$

$$B(r) = \pi R^2 - 2R^2 q\left(\frac{r}{2R}\right) \quad (2)$$

$$q(t) = arccos(t) - t\sqrt{1-t^2} \quad (3)$$

Equation 4 gives the conditions for a successful node-to-node transmission in the presence of hidden terminal problem in CSMA.

$$P_{IS}(r) = \begin{cases} \text{Prob}(x \text{ transmits in a slot}). \\ \text{Prob}(y \text{ does not transmit in the same slor}|r). \\ \text{Prob(nodes in } C(r) \text{ region do not transmit in the same slot}|r). \\ \text{Prob(nodes in B(r)do not transmit in } 2\tau+1 \text{ slots}|r) \end{cases} \quad (4)$$

3.2 Multihop flooding

Multihop flooding model in [2] is explained using Fig. 3. Accordingly, S is the source and N is the average number of nodes in the S's transmission region. The probability of successful transmission by S to any one of its neighbors is P_S from node-to-node transmission model where subscript "S" denotes *successful transmission state*. The

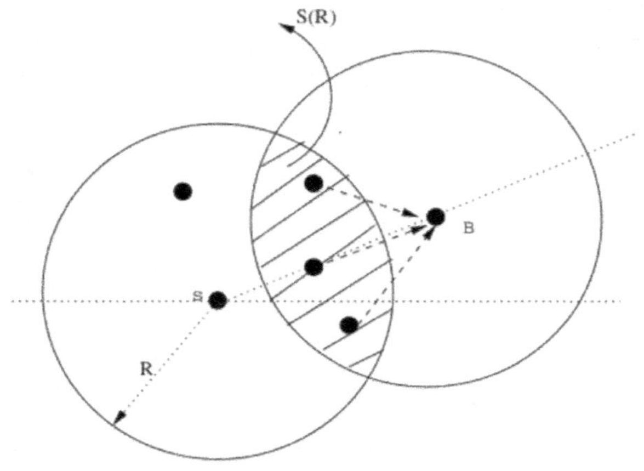

Fig. 3 Second level retransmission [2]

event of exactly N_S number of neighbors successfully receiving the transmission from S is modeled by Binomial probability distribution [2].

$\bar{N_S}$ is the number of neighboring nodes that will retransmit the received data. Out of these $\bar{N_S}$ nodes, only the retransmissions of nodes in the region $S(R)$ will reach node B. Number of nodes in the region $S(R)$ is denoted by N_b.

$$P_b = P\left(\text{B receives at least one copy of data}/\bar{N_b} \text{ nodes transmit}\right) \tag{5}$$

The probability of successful reception at any retransmission level is P_b.

It is required to estimate the number of newly reached nodes when a packet is flooded into a network by a single source. The number of nodes N_1 reached by the original broadcast of the source is given by:

$$N_1 = P_S N \tag{6}$$

Number of nodes N_2 that are newly reached by the rebroadcast of N_1 nodes is:

$$N_2 = \beta P_S P_b N^2 \tag{7}$$

Here, β in [2] is the additional coverage due to rebroadcast of receiving nodes at any level. However, we modify β as a variable dependent on the level i of rebroadcasting and each additional coverage β_i is estimated using a geometric method.

3.3 Modeling probabilistic flooding

In contrast to simple flooding, a message received for the first time by a node is rebroadcast with a probability P_{tx} in probabilistic flooding. The probability of successful reception at any retransmission level is P_b which is derived as follows:

$$P_b = 1 - P(\text{B receives no copy}/\bar{N_b}P_{tx}) \tag{8}$$

$$P_b = 1 - (1 - P_S)^{\bar{N_b}P_{tx}} \tag{9}$$

Comparison of our work with existing work is summarized in Table 1.

Table 1 Comparison of our work with Viswanath and Obraczka [2]

Aspect	[2]	Our work
Probability of successful transmission	From [3]	From [3]
Probability of successful reception	Original	From [2]
Probabilistic flooding model	Original	From [2]
Graph representation of flooding	Original	From [2]
Efficient flooding models	×	Original
Flooding wave propagation	×	Original
Effect of mobility on flooding	×	Original
Time domain representation in flooding graph model	×	Original

4 Theoretical model for endcast

4.1 Flooding graph

In modeling, we first extract the transmission graph for a given node placement. According to Rajaraman [8], in a transmission graph $G = (V, E)$ in Euclidean space, the set V represents all the nodes and set E contains an edge from node u to v if u can directly transmit to v. The ability to directly transmit is determined by the path loss equation and signal-to-noise ratio formula [8]. The transmission graph is interpreted as an adjacency matrix. A simple example is given in Fig. 4 for elaborating the graph-based modeling process of endcast. In the flooding graph in Fig. 4c, a broadcast of node A is received by the nodes that are adjacent to A (the neighbors of A). The number of frames accumulated in the network is equal to the number of edges in the graph. Number of edges in turn depends on MANET parameters such as network topology, neighborhood and wireless transmission characteristics.

A generalized flooding graph is given in Fig. 5 where N_{xy} denotes the nodes that are x hops away from the source and y is the index for nodes at that level. Further, x represents the rebroadcast level, for example the immediate neighbors (one hop neighbors) of the source will rebroadcast the packet for the first time and thus become level 1 rebroadcasters. Each edge (u, v) represents the reception of a packet by a node v due to a broadcast of a node u.

Equation 7 gives the number of nodes reached by each level of rebroadcast. We modify Eq. 7 by replacing β with β_i resulting in follows expressions.

N_1 is the number of nodes reached by the transmission of the source, given by:

$$N_1 = P_S N \tag{10}$$

N_2 is the number of nodes newly reached by the second transmission, given by:

$$N_2 = \beta_1 P_{tx} P_b N N_1 \tag{11}$$

N_l is the number of nodes newly reached by the lth retransmission, given by:

$$N_l = \beta_{l-1}(P_{tx}P_b N)^{l-1} P_S N \tag{12}$$

Therefore, N_T is the reachability, given by:

$$N_T = P_S N + P_S N \sum_{i=1}^{l-1} \beta_i (P_{tx} P_b N)^i \tag{13}$$

4.2 Geometric analysis of message propagation in flooding

We perceive the propagation of the data frame via flooding as a *flooding wave* as shown in Fig. 6a. When node A initiates a flooding operation, A's neighbors rebroadcast such that the nodes in the dotted area receive the frame. These newly reached nodes will rebroadcast at the next level to reach another set of new nodes. This process can

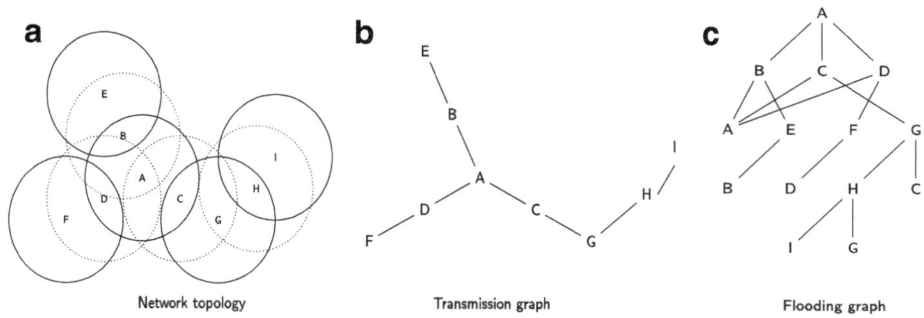

Fig. 4 Extracting the flooding graph of a MANET. **a** Network topology. **b** Transmission graph. **c** Flooding graph

be visualized as similar to how a drop of water fallen on a static pool of water causes a water ripple that propagates outwards as in Fig. 6b.

Consider a scenario in which a node A broadcasts a message and another node B that receives it decides to rebroadcast. Figure 7 represents this scenario and the shaded area will be the additional area that can benefit from the transmission of B. If S_A and S_B are the circular areas covered by A and B respectively, then S_{B-A} (the shaded area in Fig. 7) is given by $S_{B-A} = S_B - S_{A \cap B} = \pi r^2 - INTC(d)$, where $INTC(d)$ given by Eq. 14 is the intersection of S_A and S_B as derived in [4].

$$INTC(d) = 4 \int_{\frac{d}{2}}^{r} \sqrt{r^2 - x^2} dx \tag{14}$$

If B is randomly located in the transmission region of A, the expected additional coverage (EAC) is given by Eq. 15 where r is the transmission range of each node. By solving Eq. 15, EAC is obtained as 41% [4].

$$EAC = \int_0^r \frac{2\pi x \left[\pi r^2 - INTC(x)\right]}{\pi r^2} dx \tag{15}$$

4.2.1 Nodes covered by the first level of rebroadcast

We have to estimate average increment of area b by rebroadcast of all the neighbors of A (dotted area in Fig. 6a). The average additional area covered by the rebroadcast of a single neighbor is 41% of radio range [4] (dark gray area in Fig. 8). The total effective additional coverage areas by the rebroadcast of all neighbors of A equal to b. We estimate b by considering the number of radio range circles needed to cover the circumference of the radio range circle of A. An arc of length l will be covered by a single neighbor. Then the number of arcs required to cover radio range circle equals $\frac{2\pi R}{l}$. Here, $l = R\theta$.

From Fig. 8,

$$\theta = 2\cos^{-1}\frac{d}{2R} \tag{16}$$

$$l = 2R\cos^{-1}\frac{d}{2R} \tag{17}$$

Therefore, the number of arc lengths k that will cover the circumference of a circle of radius R is given by $2\pi R/l$ which will be simplified to Eq. 18;

$$k = \frac{\pi}{\cos^{-1}\frac{d}{2R}} \tag{18}$$

Since the average additional area covered by rebroadcast of a single neighbor is $0.41\pi R^2$ [4] the average additional area b covered by all neighbors of A will be given by $0.41k\pi R^2$ only if the maximum number of neighbors (highest node degree) $deg_{max} \geq k$, which simplifies to Eq. 19;

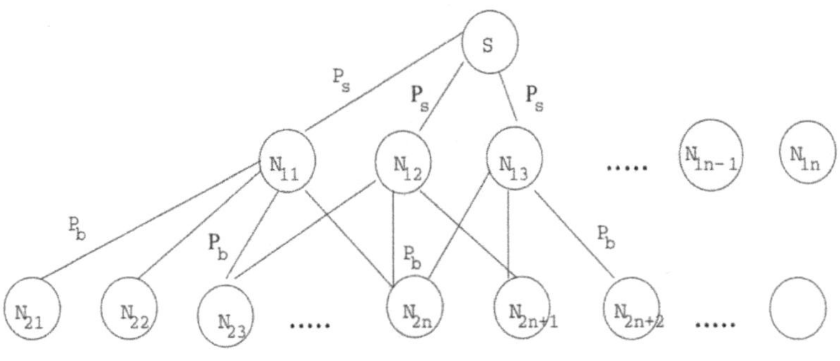

Fig. 5 First two retransmissions of flooding [2]

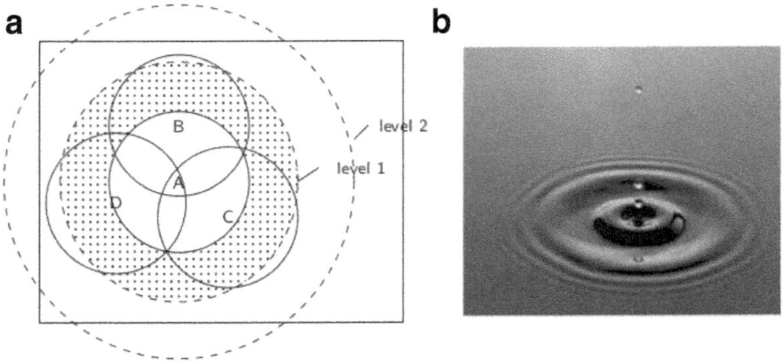

Fig. 6 Geometric estimation of reached nodes at each rebroadcast level. **a** Illustration of flooding wave. **b** Water ripple

$$b = \begin{cases} 0.41k\pi R^2, deg_{max} \geq k \\ 0.41deg_{max}\pi R^2, deg_{max} < k \end{cases} \tag{19}$$

The term d in Eq. 18 denotes the distance between A and any neighbor B and can vary from 0 to R. The lower bound of b is when $d = 0$ whereas the upper bound is when $d = R$. Average distance d_{av} corresponds to the case that the transmission ranges of A and B overlap by 59% of area as EAC is computed to be 41%.

According to Fig. 9b total area of the wedge is $\frac{R^2\theta}{2}$ and $A_1 = \frac{R^2\theta}{2} - A_2$.

$$A_2 = \frac{Rd\sin\frac{\theta}{2}}{2} \tag{20}$$

$$A_1 = \frac{R^2\theta - Rd\sin\frac{\theta}{2}}{2} \tag{21}$$

Intersection of transmission ranges of A and B denoted by $b' = 2A_1$.

$$b' = R^2\theta - Rd\sin\frac{\theta}{2} \tag{22}$$

$$\cos\frac{\theta}{2} = \frac{d}{2R} \tag{23}$$

$$\sin\frac{\theta}{2} = \sqrt{1 - \frac{d^2}{4R^2}} \tag{24}$$

Using $\cos\theta = 2\cos^2\frac{\theta}{2} - 1$;

$$\cos\theta = \frac{d^2}{2R^2} - 1 \tag{25}$$

Substituting in Eq. 22;

$$b' = 2R^2\cos^{-1}\left(\frac{d}{2R}\right) - \frac{d}{2}\sqrt{4R^2 - d^2} \tag{26}$$

Equation 22 should be solved for $b' = 0.59\pi R^2$ to calculate average distance, d_{av};

$$0.59\pi R^2 = 2R^2\cos^{-1}\left(\frac{d_{av}}{2R}\right) - \frac{d_{av}}{2}\sqrt{4R^2 - d_{av}^2} \tag{27}$$

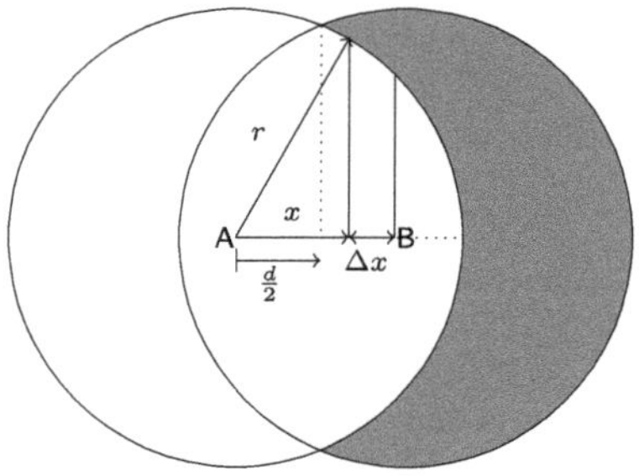

Fig. 7 Additional area covered by a rebroadcast [4]

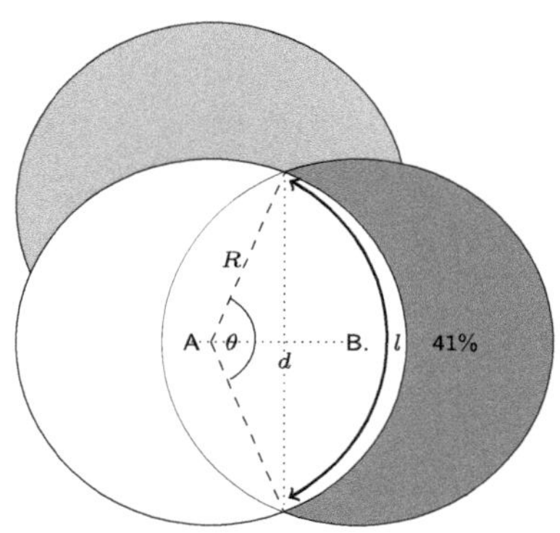

Fig. 8 Estimating the average area covered by rebroadcast level 1

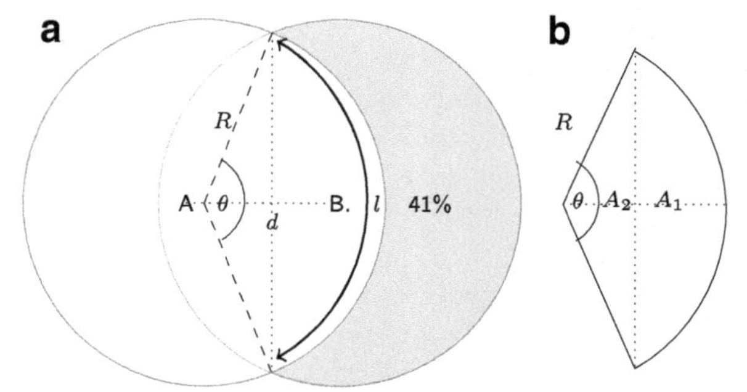

Fig. 9 Estimating average distance d_{av} between source and a neighbor. **a** Intersection between transmission ranges. **b** Calculating area of wedge

The average number of nodes newly reached at rebroadcast level 1 by the flooding wave denoted as the discrete random variable n was calculated using Eq. 28 and $p(n)$ is the probability that there exists n number of nodes in a considered region. The probability $p(n)$ is calculated assuming Poisson node distribution given by Eq. 29.

$$E[n] = \sum_{n=0}^{\infty} np(n) \tag{28}$$

$$P\{N(a+b) - N(a) = n\} = e^{-\lambda b}\frac{(\lambda b)^n}{n!}, n = 0, 1, \ldots \tag{29}$$

4.2.2 *Nodes covered by next levels of rebroadcast*

The additional area covered by rebroadcast level 1 is added to the coverage of the source to get the area covered by level 1 circle in Fig. 6a. This gives Eq. 30 where b is given by Eq. 19.

$$A_{\text{level1}} = \pi R^2 + b \tag{30}$$

$$R_{\text{level1}} = \sqrt{\frac{A_{\text{level1}}}{\pi}} \tag{31}$$

$$C_{\text{level1}} = 2\pi R_{\text{level1}} \tag{32}$$

$$k_{\text{level2}} = \frac{\pi R_{\text{level1}}}{R\cos^{-1}\frac{d}{2R}} \tag{33}$$

The circumference C_{level1} of level 1 circle is calculated from the area of the circle using Eqs. 30, 31, and 32. Then, the number of arc lengths of radius R that covers the circumference referred to here as k_{level2} is estimated using Eq. 33. The additional area b_2 that the second rebroadcast will cover is estimated using Eq. 19.

The probabilities that there are 1 through n_{total} number of nodes residing in the area b_2 are calculated using Eq. 29 to obtain the probability mass function (PMF). Thereafter, the number of newly reached nodes by the second rebroadcast is obtained using Eq. 28. This process is repeated for all the other levels of rebroadcasts until all the nodes have finished rebroadcasting or discarding the message for the first time.

4.3 Model validation

To validate the model, an analysis was done on a sample network consisting of 20 nodes that were randomly placed on a 85 $m \times 85$ m playground with each node having a transmission range of 40 m. The analytical results shown in Table 2 were obtained after applying the graph-based model with geometric analysis on this network in ideal wireless conditions of no collisions. The number of messages accumulated in the network was also calculated by building the flooding graph shown in Fig. 4c. The graph was built using actual neighborhood sizes that were obtained from the simulation.

The total number of packets accumulated in the network by the time each rebroadcast level finishes is shown in Fig. 10. As evident from the graph, this number grows exponentially. Hence, both the theoretical results and the simulation results show the same behavior. Theoretically estimated number of packets grows at a lower rate than shown in simulation results. This outcome is due to our use of average node degree $\overline{deg}$ in place of actual node degrees in the theoretical model.

4.4 Modeling efficient flooding schemes

In efficient flooding schemes, only a subset of receiving nodes will rebroadcast a packet after evaluating a set of conditions. For example, in counter-based flooding, a node counts the number of times the same packet is received and checks it against a threshold. The relative

Table 2 Theoretical results for flooding propagation for 20-node sample network

Level i	Area covered A_i (m^2)	Rebroadcast radius R_i (m)	No. of arc lengths on circumference k_i	No. of new nodes n_i	No. of total edges E
0	5027	40	2.53	3($= d_{src}$)	3
1	10,249	57	3.62	6	22.2
2	17,706	75	4.77	6	61.5
3	27,507	94	5.93	4	100.8

Transmission radius = 40m; average distance from a transmitting node to a neighbor = 26 m; θ when average additional coverage is 41%=2.48 radians (142°); $\overline{deg}$ average node degree=6.4

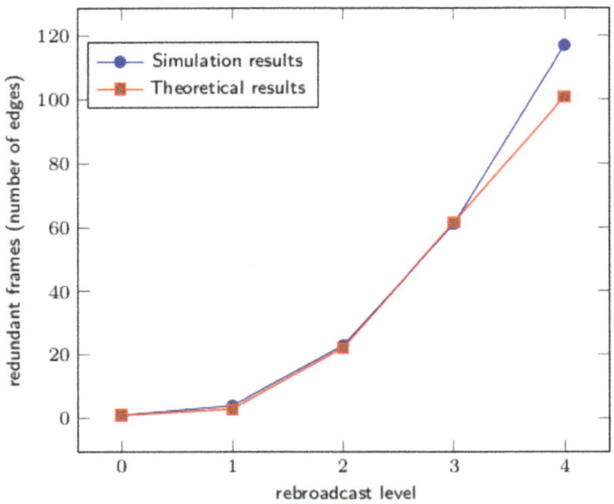

Fig. 10 Redundant rebroadcast frames with rebroadcast levels in simple flooding for a sample 20-node network

distance from the sender nodes against a threshold value is used in distance-based schemes. The assessment of variables representing conditionals for rebroadcast decision requires a time interval for the nodes to gather required neighbor information. This time interval, termed Random Assessment Delay (RAD) by Williams and Camp [9] is denoted as T_{RAD}.

The nodes at each level in the simple flooding tree are actual rebroadcasters. In efficient flooding schemes, these nodes are tentative rebroadcasters and would rebroadcast depending on their physical location relative to the source.

In contrast to the case of simple flooding, the graph model for efficient flooding requires the incorporation of time dimension. We index the set of edges with the time a packet is received by the receiving node so that we can determine which set of neighbors will rebroadcast upon the expiry of T_{RAD}. The set of nodes that actually rebroadcast is decided by the assessed variable against given criteria within T_{RAD} of that node. The order in which T_{RAD} expires for nodes in the same level determines the different configurations of the tree.

We represent parameters governing rebroadcast decisions in efficient flooding schemes in terms of P_{tx}. For example, according to the specification of counter-based flooding it is required that $C < K$ for a node to rebroadcast where C denotes the counter and K denotes the threshold. Therefore, $P_{tx} = P\{C < K\}$ for counter-based flooding. In order to estimate P_{tx}, the assessed variable such as counter in counter-based flooding is interpreted as a probability mass function.

4.5 Modeling node mobility

An impact of mobility is that the wireless links exist as long as the transmitter and the receiver do not move away from the communication ranges of each other. The probability of completing a transmission of a data packet between two moving nodes is derived by Cho and Hayes [10]. This probability should be incorporated in calculating probability P_S of successful node to node transmission for analyzing impact of mobility at a micro level.

However, as we model mobility at macro level, it is done by considering the impact of mobility on the variation of the variable assessed for making the rebroadcast decision. For example, the probability mass function of counter in counter-based flooding is analyzed for impact of mobility and P_{tx} is calculated as a mobility-dependent parameter.

5 Proposed endcast scheme

We started with blind rebroadcasting, which is the simplest mobile-stateless routing mechanism and cell proliferation was identified as an analogy to blind rebroadcasting. A grown organ in human body is made out of a dense population of biological cells. The organs originate from a small number of *stem cells* and these cells divide themselves repetitively in a process called *cell proliferation* [11–13]. The mapping we have created between cell proliferation and re-broadcasting of packets in a MANET is given in Table 3.

It is an interesting question to ask, "On what basis does our ears end up growing into a particular size?". The nature has its regulatory mechanisms for controlling the cell division in an organ whereby individual cells sense the growth in terms of number of accumulated cells. This is achieved using the concentration of some secreted molecules known as *chalones* [11]. Bullough [14] coined the term "chalone" and hypothesized it as the inhibitors of cell growth by negative feedback to control the size of a tissue [15–18].

Table 3 Comparison of cell proliferation with blind rebroadcasting

Cell proliferation	Blind rebroadcasting
It is local to the micro environment of the cells and uses information either by direct signaling or by indirect signaling via the micro environment	Each node has no state information about the topology or the neighborhood of the nodes
Cells are identical to each other and goes through the same life cycle	Every node performs the same simple task of forwarding a received data packet
Each cell divides into two so that the number of cells should grow exponentially if there are no regulatory mechanisms	Total number of packets in the network grows exponentially
Cells are organized into a dense population in a biological organ	Large number of interacting nodes in the network

Our endcast scheme is inspired by this chalone mechanism and is targeted to operate at link layer with a *frame* as the unit of transmission.

5.1 Biological system to MANET mapping

We map a biological cell to a data frame in a node as shown in Table 4. While cells increase in number due to proliferation, the frames in a MANET increase in number due to rebroadcasts. However, while the number of cells in the organ is represented by the chalone concentration, the total number of accumulated frames cannot be known using a similar local mechanism, although increase in the frame count depends on the neighborhood of the rebroadcasting nodes. Therefore, the growth of the number of frames is indirectly estimated using the number of neighbors.

A conventional method to obtain neighborhood information in MANETs is to periodically exchange beacon frames. However, beaconing is undesirable in dense MANETs due to increased collisions and contention. Therefore, the received frame count acts as the chalone concentration to indicate the total accumulated frame count in the network. For instance, if a node receives the same frame from many neighbors, that node may not reach new nodes by rebroadcasting the same frame.

A node decides to rebroadcast a received frame depending on the number of redundant copies of the same frame received from elsewhere. If the count of redundant frames reaches a threshold the node decides not to rebroadcast. This is similar to chalone mechanism and is already known as the counter-based flooding mechanism in the context of communication networks. We borrow the term *chalone* for the actions regarding the neighborhood in MANETs to denote the inspiration from chalone mechanism in cell proliferation.

5.2 Operational model of the bio-inspired system

In general, the functional behavior of cell biological systems are modeled in terms of cellular automata (CA) as in [19–22]. Similarly, the operation of biologically inspired rebroadcasting model for MANETs is interpreted based on cellular automata in this section.

Table 4 Mapping from biological system to MANET

Biological system	MANET
Biological cell	Data frame in a node
Cell division	Frame rebroadcasting
Chalone	The received data frames from the neighbors. Neighbors are the nodes within the transmission range of a node
Chalone concentration	Number of neighboring nodes who have already broadcast the data frame

According to [23], a two-dimensional CA consists of a regular lattice of sites. Each site takes on k possible values. Each site is updated in discrete time steps according to a set of rules ϕ that depends on the values of sites in some neighborhood around it at the previous time step. Packard and Wolfram [23] address the special class of ϕ known as *totalistic rules* in which the values of a site depends only on the sum of values of the neighbors.

Our CA model of MANET is also viewed as a regular lattice on which mobile nodes are randomly placed. Each site takes only two values: 1 to denote that the node has already rebroadcast a specified data frame and 0 to represent that the node has not rebroadcast. Neighborhood for our CA is the circular region captured by the radio range of a mobile node. Any node within the range will respond as a neighbor. We apply *totalistic rule* in which we count the total number of neighbors who have rebroadcast a specific data frame. The number of neighbors who have rebroadcast (state 1) is given by the number of frames received from the neighborhood. If this number exceeds a predefined threshold a node can decide not to rebroadcast a data frame in hand and we name this predefined value as *chalone threshold*. This will be the cellular automata rule.

We designed the simulation model for the protocol in a modular architecture by considering the cellular automata-based representation of the protocol. Each node independently runs the same simple set of actions (Fig. 11): if a frame is received for the first time then buffer it; wait for a T_{RAD}; count the number of frames received subsequently; when T_{RAD} elapses evaluate the rule "counter is less than the threshold?"; if less then rebroadcast (state 1) else discard (state 0).

5.3 Protocol architecture and design

The protocol consists of two phases: storm control mechanism and the flood control mechanism.

5.3.1 Storm control mechanism

Our conservative frame rebroadcast scheme termed the *storm control mechanism* operates as follows: when a node receives a data frame for the first time, it enqueues the frame and creates a record in its frame log. Then, it waits for a time period termed *chalone wait* listening for the neighbor information. If the same data frame is received subsequently during the chalone wait then a counter incremented for that frame. Thus, the log record includes the number of neighbors who have already rebroadcast the frame. When the chalone wait time elapses, the node evaluates the number of rebroadcast neighbors to see whether it exceeds a pre-defined threshold that we name *chalone threshold*. If the number of rebroadcasting neighbors does not exceed the threshold, then the node dequeues the frame by rebroadcasting it. Otherwise, the

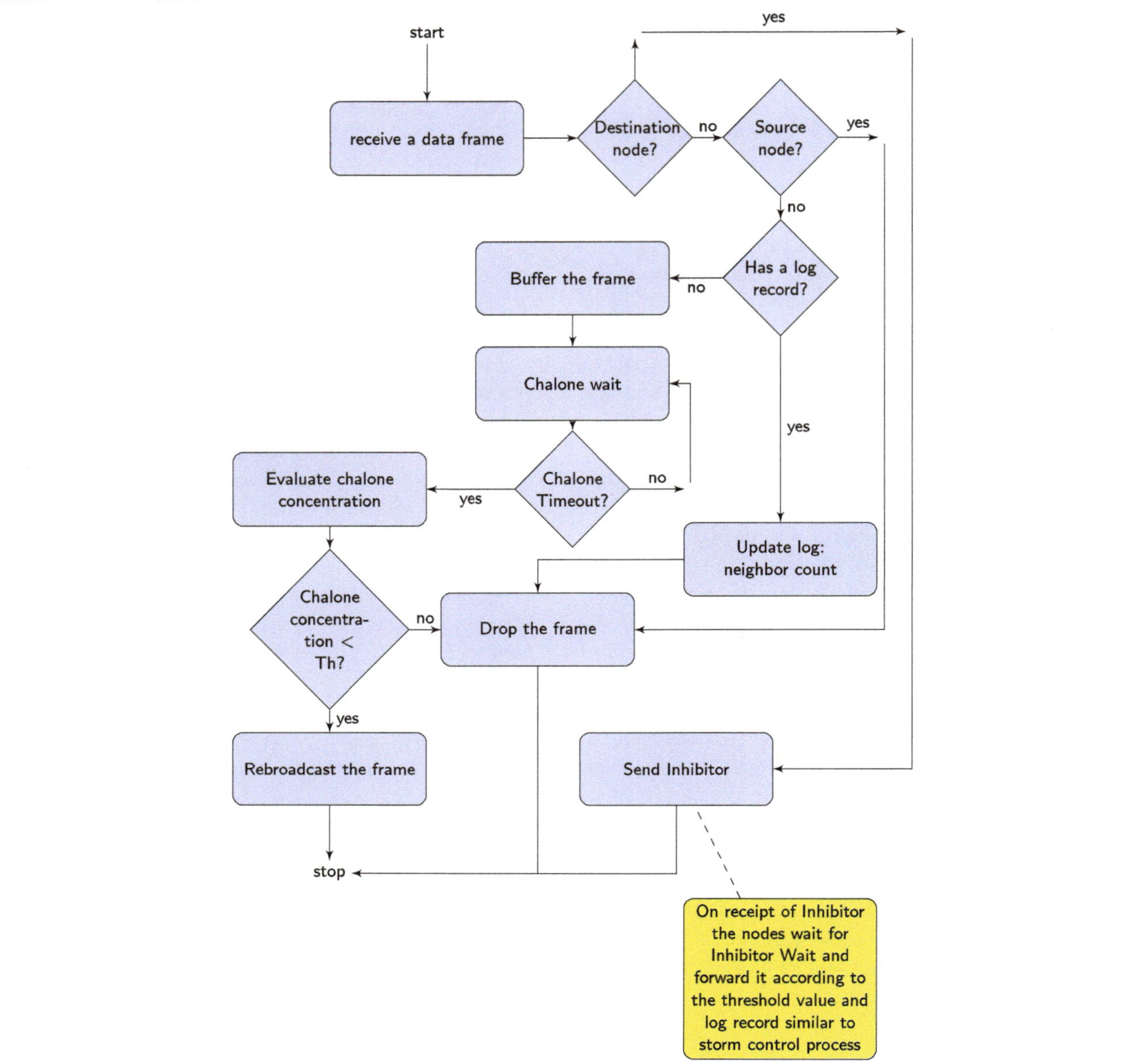

Fig. 11 Event flow at a receipt of a data frame by a node

frame gets discarded. According to the specification of counter-based scheme in [4], our approach can also be characterized broadly as counter-based flooding.

The forwarding decision is made per frame basis by referring to the frame log of the form shown in Fig. 12. A data frame is uniquely identified by the combination of source MAC, destination MAC, and a sequence number generated by the source which altogether is considered as the *Frame ID*. *Data state* carries the information such as whether the data is discarded or rebroadcast. *Control state* is to record any control message sent or received about the frame.

5.3.2 Flood control mechanism

Although the above threshold-based mechanism rebroadcasts the data frame in a conservative manner, the frame

time stamp	source MAC	destination MAC	seq num-ber	data state	inhibitor state	inhibitor count	neighbor count

Fig. 12 Log format

will continue to flood outwards in the MANET starting from the source node. Therefore, when it is received by the intended destination, the flooding wave must be stopped.

Our adaptive frame inhibition scheme which we call the *flood control mechanism* operates as follows: a node sends a control frame named *inhibitor* as soon as it receives a data frame destined to itself. Any other node that receives an inhibitor forwards it only if there exists a record of the data frame that the inhibitor corresponds to. In order to reduce the number of inhibitors forwarded by nodes, the storm control mechanism above is followed. However, the wait time referred in this case as *inhibitor wait* is considerably low compared to chalone wait time. Thus, the proposed inhibition scheme is initiated by the destination and the inhibitors propagate faster than the flooding wave only for the purpose of catching the flooding wave and not to propagate beyond the flooded area. Figure 11 illustrates the storm control mechanism and initiation of flood control. The set of actions are the same for both the data frames and the inhibitor frames.

5.4 Variables and justifications

Following variables may be set with default values when the endcast scheme is installed into devices by assuming or by letting the user select a MANET scenario, which is described by the approximate number of devices and the physical area.

5.4.1 *Chalone wait*

Chalone wait can be recognized as a random assessment delay (RAD) defined in [9]. Common practice of setting a RAD is to specify a maximum delay ($T_{\max}$) for a node to receive rebroadcast of neighbors and RAD is chosen randomly from the uniform distribution of delays within 0 to $T_{\max}$ [9]. This randomization avoids collisions. However, chalone wait should be large enough to receive as many neighbor responses as possible and should be small enough to give an acceptable end to end delay for the frame.

$T_{\max}$ is set to 0.001, 0.01, 0.05 and 0.1 s in the simulation study to compare broadcast protocols in [9]. $T_{\max}$ is set to 0.01 s in the simulation setup in [24]. Delay and backoff intervals in a Rene mote testbed for flooding experiments are randomly picked from the range 0.006 to 0.01 s. In [25] the maximal transmission latency is calculated by dividing the transmission range from transmission speed. IEEE 802.11 standard specifies the speed of radio waves propagating in free space as 300 $m/\mu s$. We set the $T_{\max}$ by dividing the maximum distance between a node and it's neighbors by the lowest transmission speed. This results in the worst case time duration for an immediate neighbor of a node to communicate with it.

5.4.2 *Chalone threshold*

Chalone threshold should be an optimal value so as not to cause too many redundant rebroadcasts and for the data frame to reach the destination. This threshold depends on the node density, and an adaptive counter-based flooding mechanism is proposed in [26] where the threshold is a function of the size of neighborhood (n). According to [26], the lowest possible threshold is 2 and the highest possible is $n + 1$. When n is small the threshold should be larger and vice versa. The threshold should be set to an initial value and then it should automatically adapt to the frame count according to a suitable function $C(n)$. It should also have an upper limit, and according to [26], only a few rebroadcasts can be saved by setting the threshold to be greater than 6 in sparse networks. An abstract shape for $C(n)$ is proposed, and the exact shape is derived using an experiment in [26].

5.4.3 *Inhibitor wait*

Inhibitor wait should be less than the *chalone wait* to ensure that the inhibitor has the necessary speed to catch the flooding wave that propagates beyond the destination. This is similar to the *recall* protocol proposed by Gao et al. [27]. It is also advisable to predict and initiate the inhibition process earlier rather than to wait until the destination is reached.

5.4.4 *Data buffer*

According to the protocol shown in Fig. 11, data frames should be buffered until a decision is made whether to rebroadcast or discard it. Time duration for buffering the frame and how the buffer should be maintained depends on a number of factors. For example, a frame should be kept in the buffer at least for chalone wait time added with some processing overhead time.

5.4.5 *Frame log*

A log record of a frame should reside in a log maintained by a node for a maximum period of time equal to the lifetime of the frame. Lifetime of a frame in the case of end to end data delivery is only until it is delivered to the destination. Therefore, the longest lifetime of a log record is the time taken by a frame to travel from one end to the other end of a network with all the delays included. Hence the lifetime of a log record depends both on the diameter of the network in terms of the number of hops and waiting times at each hop.

The cost of maintaining a frame log is comparable to the cost of performing simple flooding. In simple flooding, a node has to record the source address and the sequence number in a log for checking whether subsequently received frames are duplicates. In the proposed endcast scheme a log record includes the destination address of the message and few more state bits in addition to the sequence numbers logged in simple flooding. Details of the fields of a log record shown in Fig. 12 are as follows:

Frame ID is the combination of source MAC, destination MAC, and the sequence number. An end to end data frame is uniquely identified by a Frame ID, and its corresponding inhibitor also carries the same Frame ID.

Data state carries one of the states from the set: *unknown, originated, buffered, rebroadcast, discarded, and accepted*. The state *unknown* is when the node has never received a particular data frame so that there is no record of that particular frame in the log. The state *originated* refers to the case that the frame is originated at the node in concern. An incoming data frame from the physical layer will be buffered when it is received for the first time and this state is identified as *buffered*. A buffered data frame will be discarded when more duplicates than the threshold are received subsequently. When redundant frames are received while the frame is in data buffer, the neighbor count in the log record will be updated before discarding the frame. When the frame is no longer in the buffer so that it is finished with processing, the redundant frames will be silently discarded. The discarding of a data frame results in *discarded* state. If the data was rebroadcast then the log record carries *rebroadcast* state. The data state in the log record becomes *accepted* if the particular data frame is destined to the node and received for the first time.

Inhibitor state carries one of the states from the set: *unknown, originated, buffered, rebroadcast, discarded,* and the explainations are similar to data states except that the states are relevant to inhibitor messages. Similar to the data frames, the inhibitors incoming from the physical layer are either buffered or discarded. While an inhibitor is already in the buffer subsequently received inhibitors with same data frame ID result in an increment in the inhibitor count before getting discarded. If an inhibitor is completely processed according to the log and no longer resides in the buffer, redundant inhibitors are silently discarded.

The important parameters of the proposed endcast scheme are summarized in Table 5.

Table 5 Summary of parameters in proposed Endcast scheme

Parameter	Description
Chalone wait	Time for a data message to wait in buffer for neighbor information
Chalone threshold	Threshold for the number of copies of the same message (data or inhibitor) to receive for taking the rebroadcast decision
Inhibitor wait	Time for an inhibitor message to wait in buffer for neighbor information
Message ID	Source address, destination address and sequence number
Message log	Fields: timestamp, message ID, data state, inhibitor state, neighbor count, inhibitor count
Data and inhibitor buffers	Temporary store of messages until a rebroadcast decision is made

6 Theoretically modeling the proposed endcast scheme

The probabilities of node-to-node transmission and multihop frame propagation are derived using the previously introduced process described in the Section 4. The proposed endcast scheme uses counter-based flooding for storm control. The edges of flooding graph are indexed by time of reception of frames to analyze the effect of the assessment delay T_{RAD} as previously explained in Modeling efficent flooding schemes. Depending on the assignement of values to T_{RAD}s, the structure of the flooding graph changes in this proposed endcast scheme.

6.1 Modeling counter-based storm control mechanism

Chalone concentration C is a discrete random variable, which can be described by a probability mass function $f(c)$. Hence, $P = P\{C < K\}$ is the cumulative distribution function $F(C)$ for the chalone threshold K. By the definition:

$$F(C) = \sum_{i=0}^{K} f(c)$$

In order to construct the probability mass function $f(c)$, it is required to find the probabilities that a node receive c number of copies of the same frame where $0 \leq c < N$ and N is the number of tentative rebroadcasting nodes. The probability that a node will record a particular counter value at a given time depends on the relative positions of the tentative rebroadcasters at that time. Consider the original broadcast of a frame by node A in Fig. 13. The neighbors of A, namely, B, C, and D will receive this frame. Let all of them rebroadcast the frame and then any node in the dark gray area will record a counter value of three as it receives from B, C, and D after receiving originally from A. Similarly, the white area in A's range receives only from A and records a counter value of zero, light gray area records a counter value of one and gray area records a counter value of two.

$$P\{C = c\} = \frac{A_c}{A_{reg}} \tag{34}$$

Probability $P\{C = c\}$ that a node records a particular counter value c is the set of points in the region

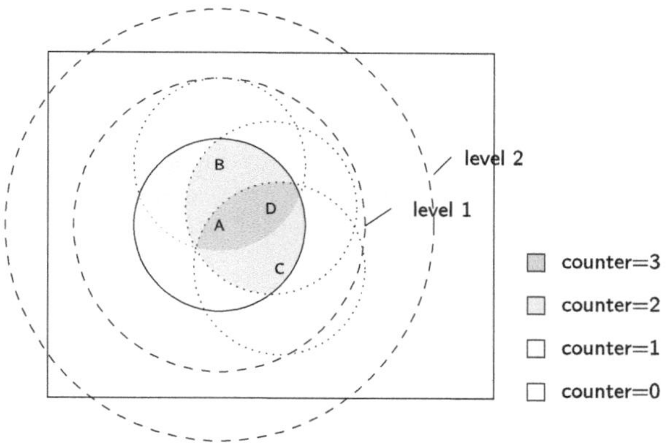

Fig. 13 Geometric interpretation of counter-based flooding

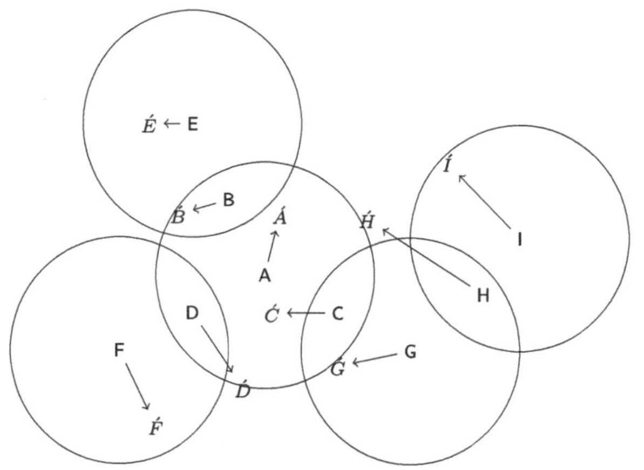

Fig. 14 Problem of endcast in the presence of mobility

that will be covered by c number of circles divided by all possible points in the region. This implies that the probability should be expressed in terms of areas as in Eq. 34 where A_c denotes the intersection area of c circles and A_{reg} is the total area of the region in which the circles are deployed. The centers of the circles are the potential rebroadcasters.

Calculating the area of intersection for more than three circles is considered practically unsolvable as a closed form expression because there are an infinite number of possible configurations for how the circles can intersect. Librino et al. [28] proposed an algorithm that iteratively computes the intersection area of an arbitrary number of circles in terms of matrix manipulations. This algorithm can be used to derive the PMF $f(c)$ for a specified location map of dense population of nodes in a MANET since the manipulation of large matrices is becoming practical with advances in high performance computing.

6.2 Modeling the effect of mobility on counter-based mechanism

The configuration of the radio range circles vary with the mobility of nodes as illustrated in the sample network in Figs. 14 and 15. Change of PMF $f(c)$ due to node mobility is used to analyze the effect of mobility on proposed endcast scheme. The Librino algorithm is run on the snapshots of the network to calculate $f(c)$ repetitively and thereby estimate a mobility dependent probability P_{tx} of rebroadcasting.

6.3 Modeling the effect of chalone wait and inhibitor wait

In order to analyze the protocol in time domain the graph model should be modified by indexing the edges with timestamp as shown in Fig. 16. Accordingly, every time a message is rebroadcast a new time index is added to the graph either on an existing edge or by inserting a new edge. Therefore, the total number of messages accumulated in the network is given by the total number of time indices. The number of indices plotted against time allows for storm and flood analysis.

Consider the 20-node sample network given in Fig. 17. Assume node h1 is the source and h17 is the destination. The original broadcast of the source will be received by all its neighbors inside the solid circle at time t_0. It will take the chalone wait time (t_{cw}) for any one of these nodes to rebroadcast or discard the message. However, h17 as the destination will immediately broadcast the inhibitor at time t_{i0}. The neighbors of h17 will receive this inhibitor (within the dotted circle). The nodes in the intersection between the solid and dotted circles are those who have received the data from h1 (source) and the relevant inhibitor from h17 (destination). These nodes will wait for the inhibitor wait time (t_{iw}) and rebroadcast the inhibitor. Node h6 does not rebroadcast the inhibitor because the relevant data is unknown for h6. According to Table 6, node h1 has the smallest t_{iw} and therefore h1 will be the first to rebroadcast the inhibitor. Since $t_{iw} << t_{cw}$, the nodes in the solid circle receive the inhibitor before t_{cw} expires for the level 1 rebroadcasting and thus all refrain from rebroadcasting. This way the inhibitor wave propagates only towards the area of the network that is already flooded with data and faster than the propagation of data.

Figure 18 shows the total number of data messages and inhibitor messages in the network against time assuming chalone wait is around 0.5 s and the inhibitor wait is around 0.1 s. Fast propagation of inhibitor succeeds in limiting the number of data messages to 10 that resulted from initial broadcast in this case. The number of inhibitors is 17 but the inhibitor is a small size message that consists only of a header to carry the ID of the data message being inhibited and own type as inihibitor.

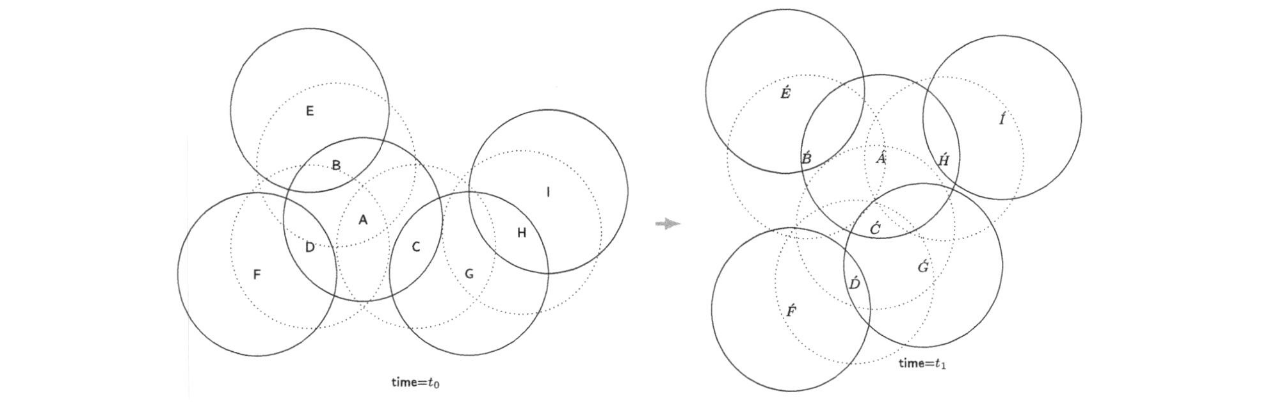

Fig. 15 Network configurations at two time instances while the nodes are moving

7 Theoretical results and discussion

$$N_T = P_S N + P_S N \sum_{i=1}^{l-1} (\beta P\{C < K\} P_b N)^i \qquad (35)$$

Reachability of the proposed endcast scheme was calculated according to Eq. 35 using the parameter set in Table 7. In Eq. 35, we do not consider variable coverage areas by different rebroadcast levels and thus we use β instead of β_i for simplifying the analysis.

The resulting plot of reachability against different retransmission levels is shown in Fig. 19. According to Fig. 19, a low chalone threshold such as one or two results in reachability that asymptotically reach a relatively small percentage of around 20% of the nodes. A threshold value of 3 and above is capable of delivering the messages to all the nodes in the network. When the threshold is increased from three, reachability curve becomes steeper so that the nodes are reached by few rebroadcast levels. This result can be used to estimate an optimal threshold value for a target MANET application scenario.

8 Simulation study of the proposed endcast scheme

We refer to our earlier proposed endcast scheme as CA protocol to denote its cellular automata-based design. This section presents the simulation study of the CA protocol. The experiments were conducted on OMNET++ (version 4.4.1) using INET (version 2.1) and MiXiM (version 2.3) frameworks [29].

The simulations were designed for evaluating the performance of CA protocol in comparison with Sequence Number Controlled Flooding (SNCF) or simple flooding. We selected or defined following metrics after a broad survey of available performance metrics:

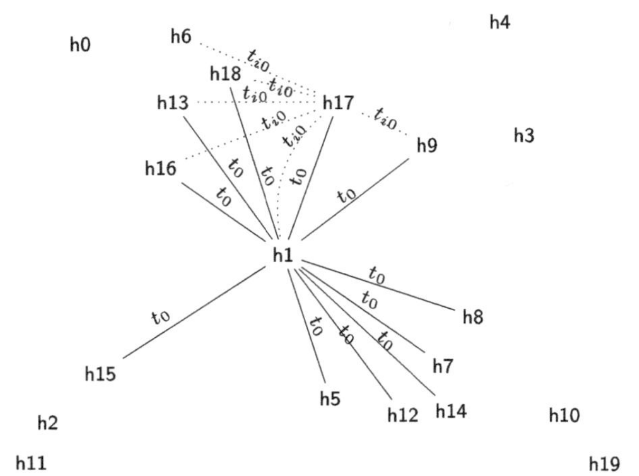

Fig. 16 Redundant rebroadcasting in flooding modeled as graph evolution with time

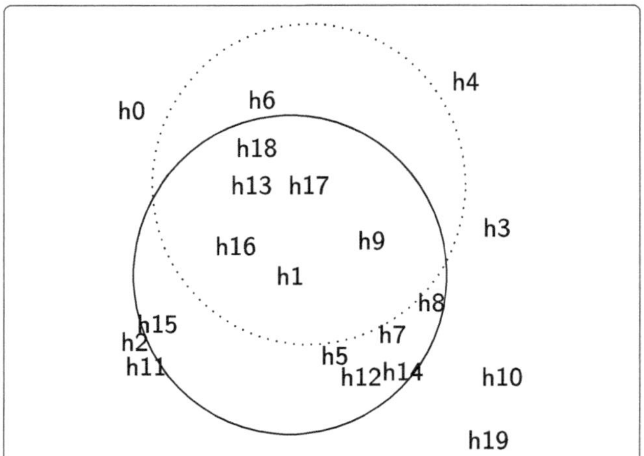

Fig. 17 Convergence of data message propagation and inhibition in proposed endcast scheme

Table 6 Order of expiring chalone wait and inhibitor wait (e.g., h0 has the smallest wait and h20 has the largest wait)

Node	h0	h1	h2	h3	h4	h5	h6	h7	h8	h9	h10	h11	h12	h13	h14	h15	h16	h17	h18	h19
RAD order	1	2	3	4	5	6	7	8	9	10	11	12	13	14	15	16	17	18	19	20

Redundancy overhead (ROH_{uf}): We refer to redundant copies of the same data frame accumulated in the network as *redundancy overhead for endcast* by modifying the metric *ROH* used by Reina et. al [30].

Saved ReBroadcasts (SRB): SRB is defined by Tseng et. al [26] as the ratio between the number of nodes that refrain from rebroadcasting and the number of nodes that receive a frame. Accordingly, SRB also measures the redundant rebroadcasts but in terms of number of nodes. $SRB = \frac{r-t}{r}$, where r is the number of receiving nodes and t is the number of transmitting nodes [26].

REachability (RE and RE_{uf}): Efficient flooding schemes impose a risk of not reaching the destination in the process of suppressing the rebroadcasts. Therefore, it is essential to evaluate flooding schemes in terms of reachability. Reachability is defined for network wide broadcasts as the ratio between the number nodes reached (r) by the flooding scheme and the number of reachable nodes (e) by topology. Thus, $RE = \frac{r}{e}$. However, we achieve endcast using efficient flooding by purposely curtailing the redundant rebroadcasts. The aim of the CA protocol is to reach the destination as opposed to reaching all nodes in the network in traditional use of flooding schemes. Therefore, we defined *Reachability for endcast* as the ratio between the number of reached node pairs (r_{uf}) by the flooding scheme and the number of all reachable pairs e_{uf}. Thus $RE_{uf} = \frac{r_{uf}}{e_{uf}}$. We use both of these definitions in our simulation study.

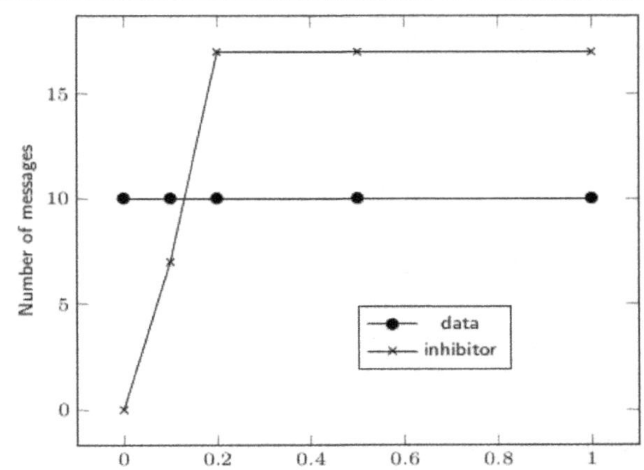

Fig. 18 Total number of data and inhibitor messages in the network with time

Flooding completion time (FCT): Flooding schemes involve a random assessment delay (RAD) [9]. For example a node that runs simple flooding waits a random period of time to avoid collisions. CA protocol waits for both to avoid collisions and to receive redundant frames to extract neighbor information. These waiting periods directly impact the end to end delay for a frame to travel from source to destination. Therefore, latency is an important metric to compare performance of flooding schemes especially when used for end to end data delivery. *Flooding completion time (FCT)* is the interval between the time at which the flooding is initiated and the time at which the final rebroadcast occurs or the final decision is made not to rebroadcast.

8.1 Experimental design

We conducted two sets of experiments: (i) based on ideal wireless conditions and (ii) to capture realistic network conditions. Ideal wireless conditions imply that there are no collisions, no interferences and even no propagation delays. Nodes are also assumed to be static. The protocols are tested in such a scenario in order to isolate protocol specific issues such as impact of the protocol logic on performance, for example, reachability. Further, in ideal wireless condition the frames are instantaneously delivered to all the neighbors. It results in maximum

Table 7 Parameter set for the reachability analysis of an example MANET

Parameter	Value
Playground	100 m×100 m
Total nodes	300
Transmission radius	40 m
Average neighborhood	100
Listening probability, p	0.0005
Slots per frame Tx time, τ	100
Rebroadcasting probabilities P_{tx}	0.04, 0.06, 0.08 and 0.11
Mobility	No

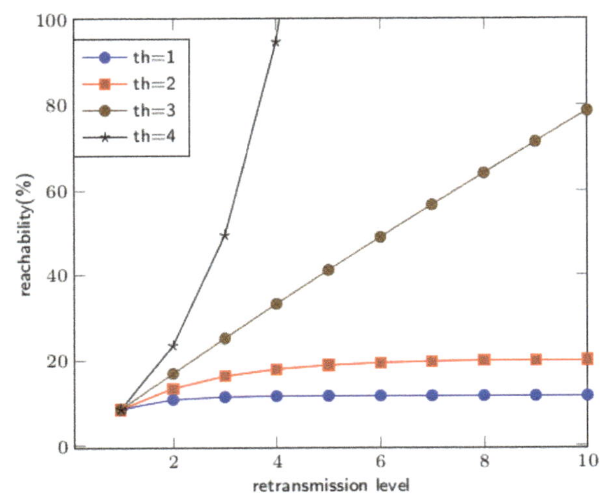

Fig. 19 Reachability with retransmission level in a MANET having density 100

frame counts whereas in realistic wireless conditions some frames may not be delivered, for example, due to collisions. Therefore, ideal wireless scenario results in the maximum reception of frames as well as maximum number of rebroadcasts so that it will give the worst performance in terms of redundant frames for any flooding scheme. As a first step of evaluating the protocols, we collected only the number of frames with varying network parameters such as number of nodes and neighborhood size.

Realistic wireless network conditions include collisions, interferences, and propagation delays of wireless links. In multihop networks, there are additional problems such as hidden terminal problem and exposed source problem. Furthermore, CA protocol when targeted at application scenarios, such as in a disaster situation, the protocol must be tested in the presence of node mobility. Hence, this set of experiments was designed to assess protocol performance in the presence of node mobility, realistic wireless conditions, and varying network densities.

8.1.1 Experiments based on ideal wireless conditions

Blind rebroadcasting causes a node to rebroadcast every time a message is received. This shows the closest analogy to uncontrolled biological cell proliferation and we based the CA protocol on blind rebroadcasting. However, in general, SNCF is the implemented version of blind rebroadcasting with an improvement such that the frame is rebroadcast only once by any node. Therefore, we compare CA protocol with both blind rebroadcasting and SNCF in these experiments.

These experiments correspond to the simulation parameters given in Table 8. We randomly place nodes on the same playground of size 600 $m \times 400$ m with each node having transmission radius of 200 m. This playground size and transmission range are comparable to number of similar simulation studies [5, 31, 32] and within the ranges in Table 9. Chalone timeout is set at a small value of 0.1 μs since the frames are delivered instantaneously in ideal wireless conditions.

8.1.2 Experiments based on realistic wireless conditions

The physical layer settings of commodity wireless devices conform to IEEE 802.11 wireless standards and we selected IEEE 802.11g implementation of MIXIM-INET for the following experiments. IEEE 802.11's medium access control (MAC) is called Distributed Coordinate Function (DCF) which is a CSMA/CA scheme according to [33]. We consider only the basic DCF in standard IEEE 802.11 MAC because broadcasts are handled using basic DCF and flooding schemes are broadcast based [34].

9 Simulation results and discussion

Redundancy, reachability and latency aspects of the CA protocol against SNCF were studied by configuring the simulation with ideal wireless conditions in Table 8, realistic wireless conditions in Table 10 and with network topologies described in Tables 11, 12, and 13.

9.1 Simulation results

9.1.1 Redundancy overhead

Experiment 1 We placed 20 to 1000 nodes on a playground of size 600 $m \times 400$ m with each node having a 200-m transmission range in ideal wireless conditions. We counted the total number of frames (ROH_{uf}) received by all the nodes in the network by the time the destination is reached, for the same source destination pair when the simulation was run for 10 different node placements under the ideal wireless conditions. We averaged the results over these 10 simulation runs and a plot of the results with 99% confidence interval is shown in Fig. 20. Note that the CA protocol was configured without an inhibition scheme to stop the propagation of flooding wave after the destination is reached and also the threshold is set to be one.

Table 8 Simulation parameters for experiments in ideal wireless conditions

Parameter	Value
Playground size	600 m × 400 m
Transmitter range	200 m
Chalone timeout	0.1 μs
Chalone threshold	1 to 3

Table 9 Surveyed simulation parameters

Parameter	Values	
Playground size	Minimum	250 $m \times 250\ m$
	Maximum	5500 $m \times 5500\ m$
Number of nodes	200 to 1000	
Network density (neighbors per node)	4 to 20	
Radio range (m)	88 to 500	
Data rate (Mbps)	1 and 2	
Packet size (bytes)	64, 100, 256, 280, 1024	

According to Fig. 20, the CA protocol outperforms both blind rebroadcasting and SNCF protocol for all numbers of nodes shown. For example, in a MANET with 100 nodes, blind rebroadcasting caused the network to receive the same frame around 2440 times on average, while SNCF and CA protocol caused 1870 and 1000 frames respectively. Thus the CA protocol saves nearly 60% of redundant rebroadcasts compared to blind rebroadcasting whereas the saving is nearly 45% compared to SNCF. Also for 1000-node network the frame count was more than 280,000 for blind rebroadcasting, around 210,000 for SNCF and only around 70,000 frames for CA protocol. CA protocol in small MANETs of 20 nodes also shows a saving of redundant rebroadcasts around 40% and 10% compared to blind rebroadcasting and SNCF protocol respectively.

Blind rebroadcasts are unacceptable for dense MANETs in terms of redundant rebroadcasts as the growth of the curve is steeply exponential. Though SNCF is slightly better compared to blind rebroadcasting, it also shows an exponential growth. CA on the other hand shows a slow growing curve thus making it a better protocol for all network sizes shown. Hence the CA protocol reduces storm situation better than SNCF.

Experiment 2 This experiment analyzed the redundancy overhead of CA protocol compared to SNCF in the presence of realistic wireless conditions in Table 10 with

Table 10 IEEE 802.11 parameters [146]

Parameter	Value	
	2.4 GHz band	5.0 GHz band
Slot time	Short = 9 μs, long = 20 μs	9 μs
SIFS	10 μs	16 μs
DIFS	Short = 28 μs, long = 50 μs	34 μs
Bitrate	2 Mbps	2 Mbps
CWmin	15	15
CWmax	1023	1023
Retry limit	7 (short)	
	4 (long)	

Table 11 Network densities of simulated topologies

Playgrond size	Average node degree	Median node degree
20 m×20 m	299.0	299
50 m×50 m	256.5	260
100 m×100 m	105.0	104
150 m×150 m	52.8	53
200 m×200 m	31.7	31
250 m×250 m	21.0	20
300 m×300 m	15.1	14
350 m×350 m	11.4	11
400 m×400 m	8.9	9

network densities in Table 11. We placed 300 nodes in playgrounds of sizes ranging from 20 $m \times 20\ m$ to 400 $m \times 400\ m$ so that the network density in terms of average neighbors per node varies from 10 to 300.

Figure 21 shows variation of the redundancy overhead in terms of total number of frames accumulate in the network (ROH_{uf}) for CA and SNCF protocols until the end of all the protocol events. Results are interpreted in two ways: redundant frames with the playground size and the same with node densities.

According to Fig. 18, the total number of accumulated redundant frames at the end of the simulation is low at about 1000 frames for CA protocol for all the network densities and playground sizes and it is a significantly better performance compared to SNCF, which records over 2000 to 50,000 frames for the same network configurations. The growth of redundant frames for SNCF is unacceptably high as it grows from around 2500 frames to 50,000 when the network density increases from 10 to 300 neighbors per node. However, CA protocol shows almost constant number of redundant frames for all node densities shown. This is because the destination is reached by the time around 1000 redundant frames are recorded in the network in all network configurations in CA protocol and thereafter, the inhibition scheme of the CA protocol effectively stops the propagation of frames throughout the network.

Experiment 3 We used the same test configuration in experiment 2 and obtained the number of nodes that

Table 12 Topologies with constant network density

Playground	Nodes	Density
20 m×20 m	11	10.00
85 m×85 m	20	9.61
150 m×150 m	50	10.70
220 m×220 m	100	10.55

Table 13 Simulation parameters for experiments in realistic wireless conditions

Parameter	Value	
Number of nodes	300	
Transmission range	40 m	
Playground sizes	Meeting room	Disaster site
	20 m×20 m	450 m×450 m
Chalone wait	0 to 6 s	
Flooding wait	0 to 6 s	
Threshold	1	

received the frame and the number of nodes that rebroadcast the frame in the networks. We calculated *SRB* as a percentage. The plot of SRB against both the playground size and the network density is shown in Fig. 22.

It was observed that in SNCF, all nodes receiving a frame did rebroadcast it thus giving no saving of rebroadcasts. CA protocol performs better for all node densities and all playground sizes with over 60% saving recorded for all these network configurations. Interestingly, SRB increases when the network density increases in CA protocol. The saving in rebroadcasts sharply increases by about 30% when the node density increases from 10 to 50.

9.1.2 Reachability

The performance gain in terms of redundant rebroadcasts is due to suppression of rebroadcasts using the threshold driven counter-based mechanism. However, suppression of rebroadcasts imposes a risk of not reaching the destination. Therefore, reachability was analyzed according to the original definition, $RE = \frac{r}{e}$ where r is the number of reached nodes and e is the number of reachable nodes. We find r using simulations whereas e is found using graph theoretic approach as follows:

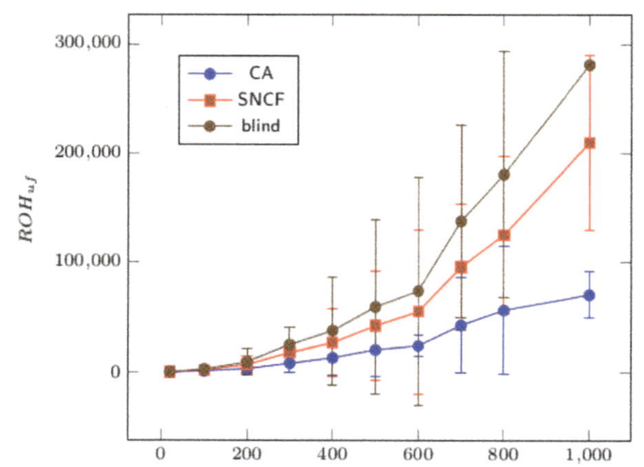

Fig. 20 Total number of redundant frames received in the network (ROH_{uf}) until the destination is reached for blind, SNCF, and CA protocols in ideal wireless conditions

We model the MANET as a graph in which vertices are the mobile nodes and the edges denote the availability of a direct wireless link between nodes at a given point in time. We first take the Euclidean distance between every node pair and then determine the immediate neighborhood by considering the transmission ranges. It is analogous to super imposing the communication ranges of nodes on the physical location map of the network. The resulting graph can be identified as a *transmission graph*. In order to evaluate the reachability of this graph, we use Floyd-Warshall algorithm that provides simplicity, parallel computation for all node pairs and low pre-processing needs. If a path can be found by the algorithm then we can conclude that a node can reach any other node by simple flooding.

Experiment 4 We configured the simulation as in experiment 2 and obtained number of reachable nodes using Floyd-Warshall all pair shortest path algorithm. Then we executed CA and SNCF protocols to obtain the number of reached nodes. After that we calculated *RE* and plotted the results against both the network densities and playground sizes. Figure 23 shows reachability as a percentage, achieved by CA protocol when the playground size and the node density varies. SNCF gave 100% reachability for all node densities and playground sizes considered and this data is not shown in the graph for the clarity of the data points. According to Fig. 23 the reachability of CA protocol decreases when the node density decreases or the playground size increases and this is for the smallest possible threshold value which is 1.

However, this reduction in RE should not be interpreted as a weakness of the CA protocol because it purposely curtails the rebroadcasts and is aimed at delivering end to end data. Therefore, we run experiment 5 to measure RE_{uf} to analyze the performance of CA protocol in end to end data delivery.

Experiment 5 In this experiment we maintained the average network density at around 10 neighbors per node by varying the number of nodes from 11 to 100 and varying the playground size from $20\ m \times 20\ m$ to $220\ m \times 220\ m$ as shown in Table 12. The experiment was on realistic wireless conditions from Table 10. When all node pairs are considered, 100-node network requires to be simulated with end to end data delivery for 9900 ($100 \times 100 - 100$) node pairs. These tests required heavy computational costs, therefore we had to semi automate and repeat the tests manually in batches. Therefore, only 3 data points were obtained as shown in Fig. 24.

It was observed according to Fig. 24a that the reachability was 100% for smaller networks of size 10 nodes for SNCF and CA protocol even with threshold of one. We get this complete reachability with CA when threshold is

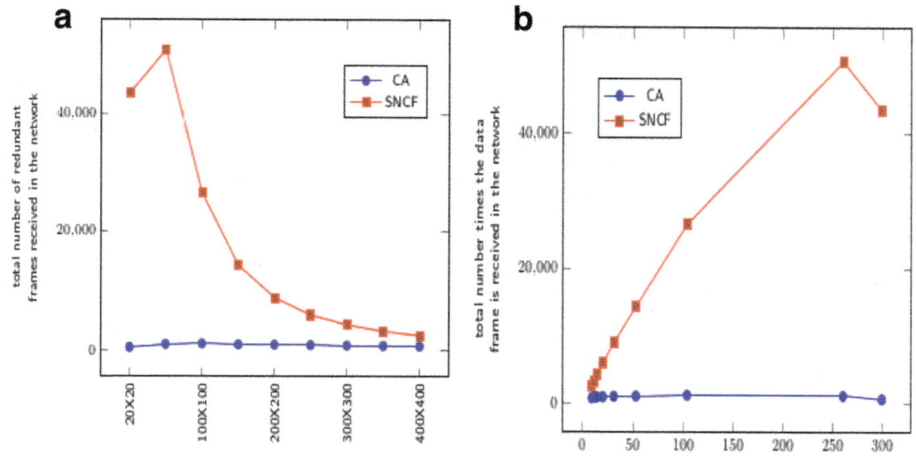

Fig. 21 Redundant frames until the end of events in SNCF and CA protocols. **a** Against playground size. **b** Against node density

one, while saving about 80% of redundant frames compared to SNCF as in Fig. 24b. However, when the network is large and nodes are dispersed in a larger area, the frame should propagate in a multihop manner to reach some node pairs. Depending on which nodes decide to rebroadcast in CA protocol, the destination may not be reached in some cases. This is visible in the 50-node network where reachability of CA is only about 75% when the threshold is one whereas SNCF shows around 90% reachability.

According to Fig. 24a reachability increases when the threshold increases in realistic network conditions also. For example, reachability improves from 75 to 90% in the 50-node network by increasing threshold from one to two. This significant reachability improvement is gained while still saving around 80% of redundancy overhead compared to SNCF.

The redundancy-reachability tradeoff can be analyzed using the ratio between RE and ROH_{uf} (RE/ROH_{uf}). For example, for the static 50-node network scenario in Fig. 24 this ratio is 0.34 for SNCF, 0.97 for CA protocol with threshold 1 and 0.88 for CA protocol with threshold 2. Accordingly when threshold increases the ratio decreases and ultimately CA protocol acts as SNCF when threshold is increased beyond the maximum neighborhood in the network.

9.1.3 Latency

Experiment 6 We configured the simulation as in experiment 2, where CA and SNCF protocols are simulated on realistic network conditions with 300 nodes placed on playgrounds of sizes 20 $m \times 20\,m$ to 400 $m \times 400\,m$. We obtained flooding completion time (FCT) by taking

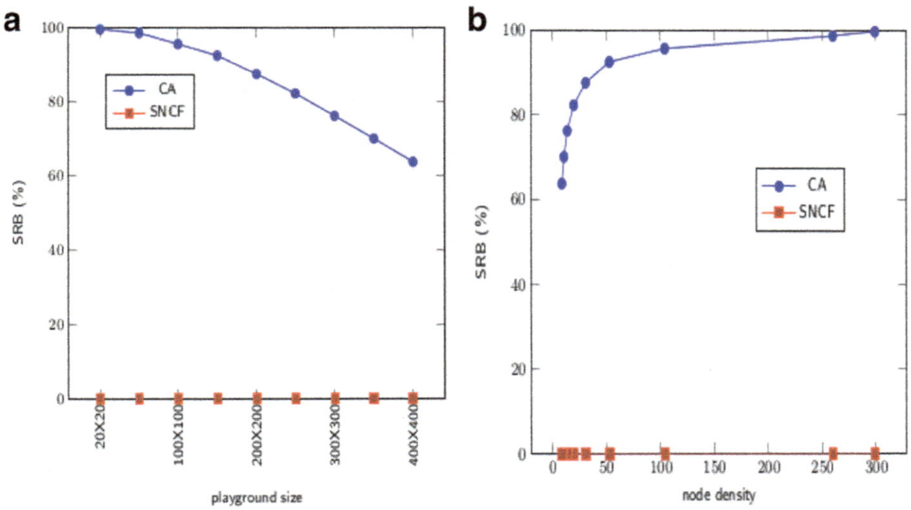

Fig. 22 Saved rebroadcasts (SRB) for CA and SNCF protocols for a 300-node network. **a** Against playground size. **b** Against node density

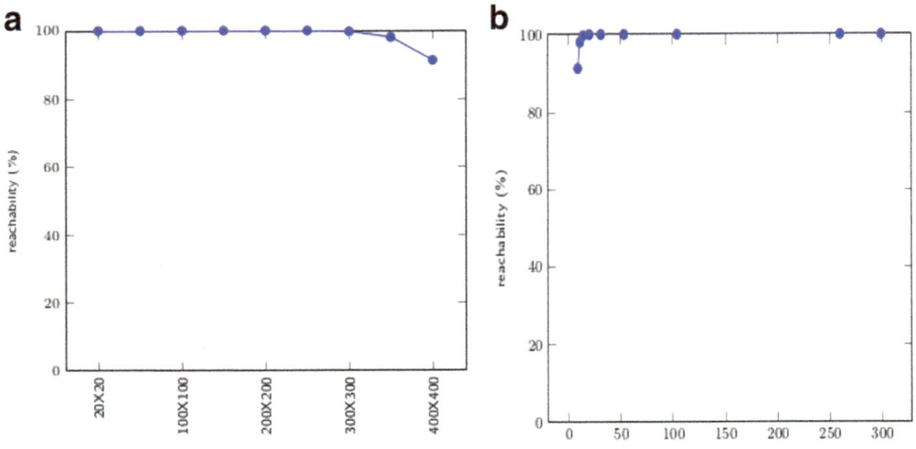

Fig. 23 Reachability of SNCF and CA protocols. Reachability is 100% for SNCF for all the considered network configurations. **a** Against playground size. **b** Against node density

the difference between the timestamp at which the frame gets originated and the timestamp at the end of the simulation. Accordingly, FCT consists of chalone wait periods in CA protocol and random wait periods in SNCF, backoff periods and propagation delays of IEEE 802.11 MAC and physical layer. The ACK timeouts do not apply in IEEE 802.11 MAC specifications as broadcasts are not acknowledged. Only single stage backoff at the transmitting node applies in detecting an idle channel. However, the MAC level delays are at microsecond scale whereas RAD times such as chalone wait in CA protocol and random delay in SNCF are much higher at 0.6 s. Therefore, the latency metrics are governed by RAD time.

The plot of FCT resulting from CA protocol is comparable with that resulting from SNCF according to Fig. 25 as the two curves are almost superimposed on one another. However, there is slightly higher FCT recorded by SNCF compared to CA protocol for highly dense networks according to Fig. 25b. For example, in networks of around densities of 250 and 300 neighbors per node, SNCF takes about 5 more seconds to finish its frame propagation than CA protocol. On the other hand for low density networks such as 10 neighbors per node, the latency is higher in CA protocol with a gap of about 25 s.

9.2 Discussion

9.2.1 Effect of storm control

CA protocol controls the redundant rebroadcasting of frames using a counter-based mechanism and this was termed storm control. According to simulation results, CA protocol saves over 45% of redundant frames compared to SNCF even when it runs without inhibition

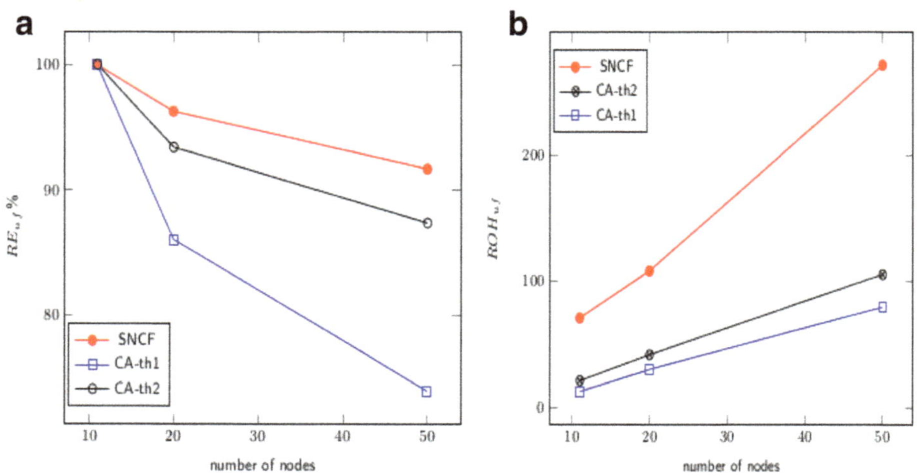

Fig. 24 Reachability (RE_{uf}) and redundancy overhead (ROH_{uf}) of endcast with SNCF and CA protocols when the network size varies with network density kept constant at 10. **a** Reachability. **b** Redundancy overhead

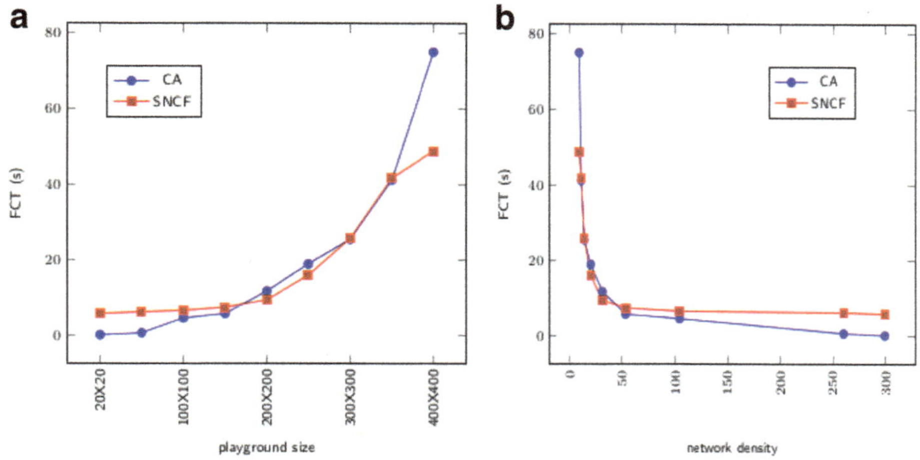

Fig. 25 Latency of CA protocol in terms of flooding completion time (FCT) in comparison with simple flooding. **a** Against playground size. **b** Against node density

scheme for flood controlling (Fig. 20). This saving is recorded when the threshold for storm control was one and the number of nodes were over 100. The nodes were placed on a playground of size 400 $m \times 600\,m$ with each node having 200 m transmission range. Further, when the number of nodes were increased to 1000, the saving of redundant frames increases to 65%.

With the inhibition scheme in operation the saving in redundant frames is over 50% (Fig. 18) for networks of densities 10 through 300 of average number of neighbors per node. These network densities were obtained by placing 300 nodes in playgrounds of sizes 20 $m \times 20\,m$ to 400 $m \times 400\,m$ and configuring each node to have power-saving radio range of 40 m. It was observed that the number of frames remain almost constant at around 1000 throughout the total range of network densities in CA protocol whereas the number increased from 2000 to 50000 in SNCF. This implies the saving in redundant frames in CA protocol increases from 50% at network density of 10 to about 98% at network density of 300 with respect to the redundant frames caused by SNCF. These statistics for variety of network configurations can be interpreted as evidence for the effectiveness in saving redundant frames in the network by the storm control using counter-based flooding and the flood control using negative acknowledgement.

9.2.2 Effect of inhibition scheme

The effectiveness of inhibition scheme is further evaluated by comparing the total number of redundant rebroadcasts with and without inhibition until all the protocol events are over. Figure 20 shows variation of number of redundant frames with varying number of nodes, when the CA protocol operates with and without inhibition scheme in ideal wireless conditions with simulation parameters configured according to Table 8.

Figure 26 shows almost constant growth rate of about eight frames per node in the plot for no inhibition. This growth is due to the continuous propagation of the frame throughout the network after reaching the destination. When the inhibition scheme is in place there is almost no growth in the number of frames when the number of nodes increases as the frame stops propagating soon after the destination is reached. The destination is reached in all networks considered within the total frame count of 100. For example in a 200-node network, the given destination was reached by the time the total number of frames was counted as 84 when the inhibition scheme was in operation. The frame count was 1405 when the inhibition was not used. This is around 15 times the recorded number of frames when the inhibition scheme is used.

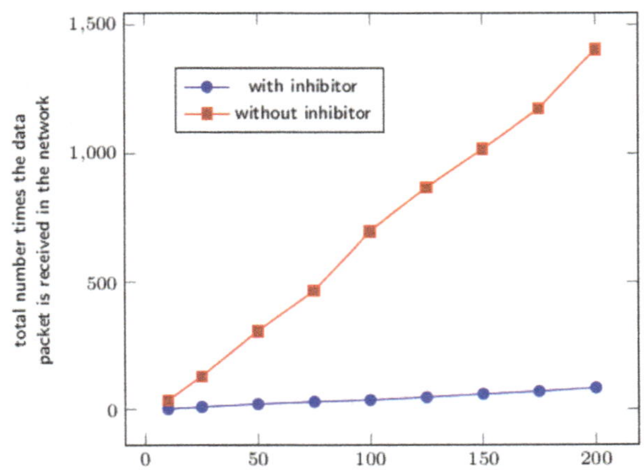

Fig. 26 Number of redundant rebroadcasts until the end of all events in CA protocol with and without inihibition scheme

This shows that the inhibition scheme significantly saves redundant frames accumulated after reaching the destination. Therefore, inhibition scheme satisfactorily controls flood situation.

9.2.3 Effect of mobility

It is essential to evaluate MANET protocols in the presence of mobility. Random Way Point (RWP) mobility is extensively applied in wireless network simulations [35–38]. In RWP mobility model, a node selects a random starting position and a random point in the area as the next waypoint. Then it moves towards the selected waypoint at a randomly picked speed between a minimum and a maximum speed. After reaching the determined waypoint it waits for a time duration, normally distributed between a minimum and a maximum, and repeats the process with next selected waypoint [36]. If the target application scenario is for human mobility as in the case of managing a disaster relief camp, minimum and maximum speeds are selected as 1 to 2 m/s and the pause time is between 0 and 0.5 s. These values are in line with similar studies that capture human mobility scenarios [35–38].

Figure 27 corresponds to a network density of 10 as defined in terms of average neighborhood per node. The simulation was run by increasing the number of nodes and the playground size while maintaining this network density. Transmission radius was also kept fixed at power saving range of 40 m. Therefore, the network diameter increases in terms of hop count when the number of nodes increases.

The simulations were run by configuring the network to be static as well as by allowing the nodes to move according to RWP mobility with speed from 1 to 2 m/s and pause time from 0 to 0.5 s. We tested SNCF protocol and CA protocol with both static and mobile network configurations for varying network sizes. CA protocol was tested for threshold values of one and two, which are denoted as *th1* and *th2* in Fig. 27.

Figure 27a shows the number of redundant frames accumulated in the network until the destination is reached when the network size is increased while keeping the average network density constant at 10. Accordingly, SNCF shows the highest frame counts and it grows faster than for CA protocol. For example, in a 50-node network SNCF caused accumulation of around 250 frames while CA protocol resulted in a number less than half of that. Furthermore, frame count increases from around 75 to 250 by SNCF in static network when the network size increases from around 10 to 50 nodes. The growth of frame count is even steeper for SNCF when the nodes are mobile as it increases from about 75 to 400. In contrast CA protocol shows only an increase from about 20 to 150 when the network size increases from 10 to 50 nodes.

In the presence of mobility, there is a higher frame count than in static networks for all protocol settings. However, the redundant frames caused by SNCF with mobility shows a considerable increase compared to the case of SNCF in static networks, for example, in a 50-node network, the mobility of nodes caused about 60% increase in redundant frames. In contrast, mobility of nodes does not cause considerable increase in redundant frames compared to static networks when CA protocol is in operation.

Figure 27b shows how the reachability varies with network size for static nodes and mobile nodes in SNCF, CA protocol with threshold 1 and CA protocol with threshold 2. Here, the reachability is according to the definition of RE, redefined for end to end data delivery by considering source-destination pairs.

As per Fig. 27b, mobility seems to assist frame propagation rather than hinder it because the reachability of SNCF

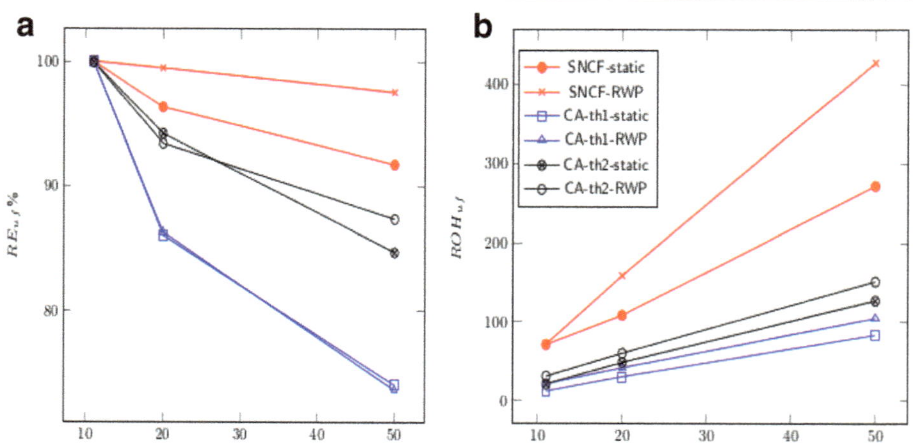

Fig. 27 Redundancy overhead and reachability of endcast until the destination is reached for SNCF and CA protocols with mobility when the threshold of CA protocol is set at 1 and 2 (*th1* and *th2*). **a** Reachability. **b** Redundancy

is significantly higher in mobile network compared to static network. Reachability is almost constant at around 100% for all the network sizes with mobile nodes for SNCF except for a slight decrease in higher network sizes. In contrast, in static networks reachability of SNCF drops from 100 to around 95% when network size increases from 10 to 50. A general perception is that mobility of nodes are disruptive in data communication in wireless networks and makes it challenging to device routing mechanisms. That may be true for routing schemes that set up virtual backbones and assign key nodes to act as routers. However, the flooding schemes considered here, can be considered to leverage mobility to enhance data delivery. This is in line with the argument supporting the branch of MANET study known as *opportunistic networks* in which mobility is considered an opportunity for delivering data using *store and carry* behavior of nodes. CA protocol with threshold of two, also shows a slightly higher reachability in mobile networks compared to static networks at least for higher network sizes. CA protocol with threshold of one does not show any difference in reachability when operating on static and mobile networks.

9.3 Effect of chalone threshold

In experiment 5, we evaluated RE_{uf} and ROH_{uf} in networks of different sizes but with constant network density of around 10 neighbors per node. When the two metrics are compared side by side as in Fig. 24, by slightly increasing the threshold, reachability could be improved considerably. For example, in a 50-node network the reachability could be increased by about 15% when the threshold was increased simply from one to two. Moreover, this slight increment in threshold does not increase the redundancy overhead in a considerable amount, for example, in the same 50-node network the number of frames increased only by about 6% when the threshold increased from one to two.

Accordingly, as chalone threshold becomes higher, for the CA protocol, the reachability-redundancy tradeoff becomes smaller and CA protocol acts similar to SNCF. Furthermore, mobility assists communication in flooding but at the expense of redundancy.

10 Conclusions

A key objective of this research was to build a theoretical basis to analyze the problem of routing in MANETs approached from a new paradigm that we term as *mobile-stateless paradigm*, which is completely different from traditional MANET routing. In this model, nodes do not keep any state information and they can become *stateless*. If they are allowed to blindly rebroadcast every received data frame then it results in unrestricted node *mobility*. However, *blind rebroadcasting* is unacceptably inefficient. Instead, we can use *simple flooding*, in which a node rebroadcasts a frame only when it is received for the first time. There are flooding schemes proposed in the literature that allow only a sub set of the nodes to rebroadcast and these schemes are referred to as *efficient flooding*. Conventional use of any kind of network flooding is to disseminate a message to all the nodes in a network. In contrast, we propose to use flooding for end to end data delivery. We did not come across any other work that attempts end to end data delivey using flooding with a theoretical analysis.

We leveraged efficient flooding as an end to end data delivery mechanism conforming to mobile-stateless routing paradigm and we term it as *endcast*. We also developed a model to analyze endcast schemes in terms of evolution of a graph $G(V, E)$, where the set of vertices V represent the nodes and an edge $(u, v) \in E$ is added when a frame is received by a node v due to rebroadcasting of another node u during the propagation of a frame via flooding. In our model, the number of edges are estimated using geometric analysis of the network.

Our model captures the major problems in flooding, namely, the *broadcast storm problem* and the situation that we call *broadcast flood problem* that is particularly observed in endcast. Broadcast storm problem refers to the situation where excessive number of redundant frames are transmitted all over the network causing heavy contention and collisions. This situation is already researched in the literature aiming at improving conventional use of flooding for network wide broadcast. When flooding is used for end to end data delivery, it is wasteful to leave the frames to propagate beyond the destination and we refer to this situation as broadcast flood problem.

The theoretical model that we built for analyzing endcast is capable of:

- Modeling the propagation of data via flooding schemes including simple flooding, probabilistic flooding, and counter-based flooding
- Analyzing flooding in networks with static nodes as well as mobile nodes
- Estimating the level of broadcast storm problem and broadcast flood problem
- Modeling the physical and MAC layer impacts such as hidden terminal problem

Our model extends the work of Viswanath-Obraczka [2] for analyzing multihop flooding in static networks as follows:

- We calculate the additional area covered by each level of rebroadcast geometrically so that β varies from level to level.
- We incorporate time domain in the graph to model flooding schemes that involve an assessment time before making a rebroadcast decision.

- We model node mobility using the time domain representation of the graph as well as the mobility-dependent variation of coverage areas.
- Instead of using a random P_{tx} we interpret P_{tx} in terms of different parameters in flooding schemes, for example we derive a P_{tx} for counter-based flooding based on the counter value and the threshold.

Additional file

Additional file 1: These are .dat files used for plotting graphs using pgfplot package. The commands for plotting are commented inside figure commands as the online editorial manager could not compile the commands. The data file for Fig. 10 (edgeGrow.dat); the data file for Fig. 19 (ReachabilityVsLevel.dat); a data file for Fig. 20 (CA.dat); a data file for Fig. 20 (SNCF.dat); a data file for Fig. 20 (blind.dat); a data file for Fig. 18 (CAUntilEnd.dat); a data file for Fig. 18 (sncfUntilEnd.dat); a data file for Fig. 18 (CApkts.dat); a data file for Fig. 18 (SNCFpkts.dat); a data file for Fig. 22 (CaSRB.dat); a data file for Fig. 22 (SncfSRB.dat); a data file for Fig. 22 (CaSRBnei.dat); a data file for Fig. 22 (SncfSRBnei.dat); a data file for Fig. 23 (ReachTopo.dat); a data file for Fig. 23 (CAReach.dat); a data file for Figs. 24 and 27 (eval.dat); a data file for Fig. 24 (eval-redframes.dat); a data file for Fig. 25 (DelayCAnSNCF.dat); a data file for Fig. 25 (DelayCAnSNCFwithDen.dat); the data file for Fig. 26 (withNwithoutInhib.dat); the data file for Fig. 18 (timeConverge.dat).

Funding
This research was conducted for the fulfillment of PhD in Computer Science and Engineering from the University of Moratuwa. The Department of Computer Science and Engineering provided necessary facilities and guidance for the research work.

Authors' contributions
The first author SJS conducted the research and prepared the manuscript under the supervision of the second author CDG. Both authors read and approved the final manuscript.

Competing interests
The authors declare that they have no competing interests.

References
1. M Abolhasan, T Wysocki, E Dutkiewicz, A review of routing protocols for mobile ad hoc networks. Ad Hoc Netw. **2**(1), 1–22 (2004)
2. K Viswanath, K Obraczka, Modeling the performance of flooding in wireless multi-hop ad hoc networks. Comput. Commun. **29**(8), 949–956 (2006)
3. L Wu, PK Varshney, Performance analysis of csma and btma protocols in multihop networks (i) - single channel case. Inf. Sci. **120**(1), 159–177 (1999)
4. YC Tseng, SY Ni, YS Chen, JP Sheu, The broadcast storm problem in a mobile ad hoc network. Wirel. Netw. **8**(2-3), 153–167 (2002)
5. C Ho, K Obraczka, G Tsudik, K Viswanath, in *The 3rd ACM International Workshop on Discrete Algorithms and Methods for Mobile Computing and Communications*. Flooding for reliable multicast in multi-hop ad hoc networks (ACM, 1999), pp. 64–71
6. A Rahman, W Olesinski, P Gburzynski, in *IEEE International Workshop on in Wireless Ad-Hoc Networks*. Controlled flooding in wireless ad-hoc networks (IEEE, 2004), pp. 73–78
7. H Takagi, L Kleinrock, Optimal transmission ranges for randomly distributed packet radio terminals. IEEE Trans. Commun. **32**(3), 246–257 (1984)
8. R Rajaraman, Topology control and routing in ad hoc networks: a survey. ACM SIGACT News. **33**(2), 60–73 (2002)
9. B Williams, T Camp, in *3rd ACM International Symposium on Mobile Ad Hoc Networking and Computing*. Comparison of broadcasting techniques for mobile ad hoc networks (ACM, 2002), pp. 194–205
10. S Cho, JP Hayes, in *IEEE Wireless Communications and Networking Conference*. Impact of mobility on connection in ad hoc networks, vol. 3 (IEEE, 2005), pp. 1650–1656
11. JC Lui, J Baron, Mechanisms limiting body growth in mammals. Endocr. Rev. **32**(3), 422–440 (2011)
12. GP Finkielstain, P Forcinito, JC Lui, KM Barnes, R Marino, S Makaroun, V Nguyen, JE Lazarus, O Nilsson, J Baron, An extensive genetic program occurring during postnatal growth in multiple tissues. Endocrinology. **150**(4), 1791–1800 (2009)
13. PJ Bryant, P Simpson, Intrinsic and extrinsic control of growth in developing organs. Q. Rev. Biol. **59**(4), 387–415 (1984)
14. WS Bullough, Mitotic and functional homeostasis:a speculative review. Cancer Res. **25**(10), 1683–1727 (1965)
15. H Onda, A new hypothesis on mitotic control mechanism in eukaryotic cells: cell-specific mitosis-inhibiting protein excretion hypothesis. Theor. Biol. **77**(3), 367–377 (1979)
16. SJ Lee, Regulation of muscle mass by myostatin. Annu. Rev. Cell Dev. Biol. **20**, 61–86 (2004)
17. K Elgjo, KL Reichelt, Chalones: from aqueous extracts to oligopeptides. Cell Cycle. **3**(9), 1208–1211 (2004)
18. D Bakthavatsalam, DA Brock, NN Nikravan, KD Houston, RD Hatton, RH Gomer, Chalones: from aqueous extracts to oligopeptides. Cell Sci. **121**, 2473–2480 (2008)
19. AR Kansal, S Torquato, GR Harsh, EA Chiocca, TS Deisboeck, Simulated brain tumor growth dynamics using a three-dimensional cellular automaton. J. Theor. Biol. **203**(4), 367–382 (2000)
20. Y Lee, S Kouvroukoglou, LV McIntire, K Zygourakis, A cellular automaton model for the proliferation of migrating contact-inhibited cells. Biophys. J. **69**, 1284–1298 (1995)
21. T Alarcóna, HM Byrneb, PK Mainia, A cellular automaton model for tumour growth in inhomogeneous environment. J. Theor. Biol. **225**(2), 257–274 (2003)
22. O Mireira, A Deutsch, Cellular automaton models of tumor development: A critical review. Advs. Complex Syst. **5**(2/3), 247–268 (2002)
23. NH Packard, S Wolfram, Two-dimensional cellular automata. J. Stat. Phys. **38**(5/6), 901–946 (1985)
24. A Mohammed, M Ould-Khaoua, L Mackenzie, *An efficient counter-based broadcast scheme for mobile ad hoc networks*. Formal Methods and Stochastic Models for Performance Evaluation. (Springer, 2007), pp. 275–283
25. S Pleisch, M Balakrishnan, K Birman, RV Renesse, in *The 7th ACM International Symposium on Mobile Ad Hoc Networking and Computing*. MISTRAL: efficient flooding in mobile ad-hoc networks (ACM, 2006), pp. 1–12
26. YC Tseng, SY Ni, EY Shih, Adaptive approaches to relieving broadcast storms in a wireless multihop mobile ad hoc network. IEEE Trans. Comput. **52**(5), 545–557 (2003)
27. J Gao, L Guibas, N Milosavljevic, J Hershberger, in *The 6th ACM International Conference on Information Processing in Sensor Networks*. Sparse data aggregation in sensor networks (ACM, 2007), pp. 430–439
28. F Librino, M Levorato, M Zorzi, An algorithmic solution for computing circle intersection areas and its applications to wireless communications. Wirel. Commun. Mob. Comput. **14**(18), 1672–1690 (2014)
29. OpenSim Ltd., OMNet++ (2001). https://omnetpp.org/. Accessed 19 Sept 2016
30. DG Reina, SL Toral, P Johnson, F Barrero, A survey on probabilistic broadcast schemes for wireless ad hoc networks. Ad Hoc Netw. **25**, 263–292 (2015)
31. Y Sasson, D Cavin, A Schiper, Probabilistic broadcast for flooding in wireless mobile ad hoc networks. IEEE Wirel. Commun. Netw. **2**, 1124–1130 (2003)
32. Y Yi, M Gerla, TJ Kwon, in *IEEE International Conference on Communications (ICC 2003)*. Efficient flooding in ad hoc networks: a comparative performance study, vol. 2 (IEEE, 2003), pp. 1059–1063
33. G Bianchi, Ieee 802.11-saturation throughput analysis. IEEE Commun. Lett. **2**(12), 318–320 (1998)

34. IEEE 802.11 Working Group and others, IEEE Standard for Information Technology–Telecommunications and information exchange between systems–Local and metropolitan area networks–Specific requirements–Part 11: Wireless LAN Medium Access Control (MAC) and Physical Layer (PHY) specifications Amendment 6: Wireless Access in Vehicular Environments. IEEE Std. **802**(11) (2010)
35. M Cornils, M Bahr, T Gamer, in *European Wireless Conference (EW)*. Simulative analysis of the hybrid wireless mesh protocol (HWMP) (IEEE, 2010), pp. 536–543
36. M Ikeda, T Honda, T Oda, S Sakamoto, L Barolli, in *IEEE 28th International Conference on Advanced Information Networking and Applications*. Analysis of WMN-GA simulation results: WMN performance considering stationary and mobile scenarios (IEEE, 2014), pp. 337–342
37. N Sabah, A Hocanin, Gamma random waypoint mobility model for wireless ad hoc networks. Int. J. Commun. Syst. **26**(11), 1433–1445 (2013)
38. O Younes, N Thomas, Analysis of the expected number of hops in mobile ad hoc networks with random waypoint mobility. Electron. Notes Theor. Comput. Sci. **275**, 143–158 (2011)

Femtocaching assisted multi-source D2D content delivery in cellular networks

Xiaoyan Zhao[1,2], Peiyan Yuan[2*], Yajun Chen[1] and Pei Chen[1]

Abstract

The influxes of diversified services and mass data lead to exponential growth of traffic load in mobile cellular networks. Cache-enabled device-to-device (D2D) communication provides a general framework to alleviate this situation. In contrast to previous single-source D2D models, this paper investigates a comprehensive content delivery framework based on a three-tier heterogeneous network (HetNet), where base station (BS), femtocaching auxiliary equipments (FAEs), and user terminals(UTs) are included. The cooperative D2D communication can be implemented from both FAEs and UTs to handle the ongoing explosive increase in ultra-dense scenario. Moreover, duplicate storage for requesting data in multiple neighbor nodes makes many-to-one D2D communication possible at the user layer. Considering the case that cellular users and D2D links reuse the same resources in the uplink period, the non-outage probability of the cellular communication is defined to guarantee the main communication quality. Under the constraints subject to cumulative interference, an optimization objective function based on multi-source D2D communication is deduced to achieve unprecedented average data rate. Numerical simulations show that our system yields network throughput exponentially while transferring traffic load of the BS reasonably.

Keywords: Device-to-device communication, Content delivery, Many-to-one, Non-outage probability, Heterogeneous network

1 Introduction

The latest report of the global system for Global Mobile Communication Systems Association (GSMA) has shown that three quarters of the world's population will be linked together through mobile terminals by 2020. Wireless data business is therefore expected to increase by a factor of 11 compared to the current, and each user needs to consume at least 1 Gb per day in the next 5 years [1]. This explosive growth is mainly caused by mobile multimedia traffic and other emerging content transmission such as popular videos and social networking services (SNS), which almost rise by a factor of 65 times [2]. The urgent demand for the thousandfold growth of mobile data inevitably leads to a slowdown, or even breakdown of the cellular systems. Thus, cellular networks are undergoing unprecedented paradigm shift in the way by which data is delivered [3].

Cache-enabled D2D communication is a key component in this shift, which can distribute content locally and transmit data directly. By pre-storing popular files and dispatching data via D2D, the system has a higher gain with the increase of file storage capabilities at the network edge [4, 5]. The reason behind this is that the most popular files are more likely to be accessed by the majority of users, although there are huge amount of traffic suffering by the network. Thus, a noteworthy development in this direction is to place as much popular contents as possible on a local storage device. In academia, there exist some possible solutions to support the cache-enabled D2D paradigm, such as [6–13]. These works discussed and elaborated the specific edge placement strategy to enhance the accessibility of data for requesting user terminal(RUT). In general, the existing D2D communication schemes can be classified into two categories by taking into account different devices to implement local-storage, as shown in Fig. 1. The first one is to deploy small cells and transfer data by auxiliary device, which can be seen as SBS. The other caches contents without special infrastructure and have the further advantage of mobile user terminals (UTs), such as smartphones, tablets, and laptops which are built in large

*Correspondence: peiyan@htu.cn
[2]School of Computer and Information Engineering, Henan Normal University, No. 46, Jianshe East Road, 453007 Xinxiang, China
Full list of author information is available at the end of the article

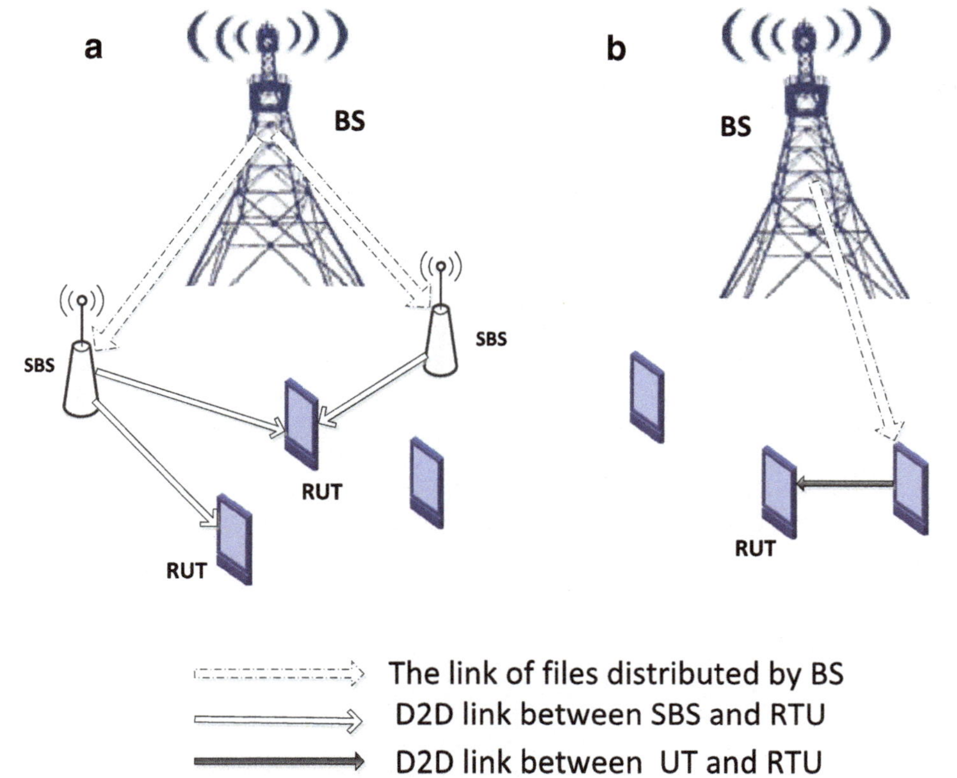

Fig. 1 Two main cache perspectives of D2D communication considering different placement devices. This figure shows the existing D2D communication schemes, which can be classified into two categories by taking into account different devices to implement local-storage, i.e., the mode of UTs or SBSs to carry out offloading. **a** Example of SBS offload, **b** Example of UT offload

storage devices. According to the special popularity distribution and storage capacity, a large percentage of video traffic can be offloaded through these two systems.

Note that available researches mainly focus on the single-source caching model, i.e., they either use femtocaching helper (i.e., SBS) or exploit UTs to carry on D2D communication. Although these two approaches were described separately, the measures were effective and brought about some enlightenment to us, e.g., these two strategies can also be mixed to some extent. Moreover, current wireless technologies such as time/frequency division multiple access and maximum ratio transmission (MRT) [14] support multiple users to transfer data simultaneously. For example, under MRT, each suitable UT and SBS in the cooperation communication radius can beamform to a RU such that the signals from the neighboring nodes and SBSs are coherently combined and ultimately result in diversity gain. That is to say, one RUT can receive contents from both the SBSs and other UTs, which can be called multi-source caching model. Here, the so-called multi-source caching, drawing a contrast with the single source(e.g., either SBSs or UTs), mainly refers to the collaboration between different types of storage devices(i.e., both SBSs and UTs). The heap offers us a feasible solution to the traffic offload problem from a technical and economical perspective. Accordingly, we propose a three-tier communication architecture in heterogeneous network incorporating femto-caching helpers and UTs to offload the backbone network traffic. This joint design offers not only two different caching gains and multiplexing gains but also diversity gains owning to many-to-one D2D communication. Many-to-one D2D communication means that each RUT may detect several mobile users storing the needed file within its scope of power coverage in the ultra-dense environment. In other words, three conditions are required for suitable UTs to take place many-to-one D2D: (1) they are within the transmission range of RU; (2) they have the file which is needed by RU; (3) there are more than one UTs matching both of criteria 1 and 2. In general, high data rate and low transmission delay can be obtained due to the proximity of paired devices in D2D links, thus introducing proximity gain. Nevertheless, it is unsurprisingly that the multi-source caching model proposed in this paper will introduce cumulative interference caused by multi-pair D2D links in the uplink transmission period. To achieve maximum transmission rate and to obtain high user satisfaction, non-outage probability of system is deduced to describe the influence of cumulative interference on the performance of cellular systems.

Compared with the existing work, we have the following contributions:

- Combining the storage advantages of fixed equipments and mobile terminals in ultra-dense environment, we model a three-tier heterogeneous architecture to distribute data. Furthermore, we introduce the concept of many-to-one D2D communication in the user layer.
- We present an optimization objection in terms of average transmission rate to discuss the performance characteristics of the heterogeneous network. Under the cumulative interference constrain due to multi-pair D2D links, non-outage probability of system is derived and simulated.
- We define the concept of user satisfaction to analyze the pareto solutions of the optimization problem. Moreover, we evaluate the impact of various parameters on user satisfaction, such as the request probability and D2D communication radius.

The remainder of the paper is organized as follows: after reviewing the main related works in Section 2, the three-tier content delivery mode is elaborated in Section 3. In Section 4, the average transmission rate and non-outage probability are derived. Numerical results in terms of network throughput and user satisfaction are discussed in Section 5. Finally, conclusive remarks are presented in Section 6.

2 Related work

D2D communication attracts a lot of attentions recently, and the potential research orientations include content delivery/dissemination [6, 7, 10], resource allocation [15–17], social- awareness video multicasting [18, 19], and so on. Existing researches for the content delivery can be mainly classified into two perspectives. The first one is to deploy SBS with mass storage device which does not require high-speed backhaul links. For example, in [6, 7], small base station called helper was introduced and placed in fixed position to serve user requests. Chou et al. [8] utilized mobile SBS to study the deployment problem and address the waste of resources in small cells. In [9], a cost-effective integration access technology between WiFi and cellular wireless was characterized for femto-based stations to fulfill time/space-varying traffic. In general, the scheme based on auxiliary devices in small cells, such as helper, femto-based station, and SBS can highly provide a performance boost to the network traffic. However, along with the growth in the number of helpers, each RUT can be covered by multiple helpers, and the hit rate will be increased as well as the interference and the investment cost. In addition, another potential deficiency of the small cell-based infrastructure is that, during peak traffic period, the backhaul link-capacity requirement to respond data is enormously high [4]. Obviously, if the storage capacity of existing devices (e.g., UTs) is considered, storage-investment efficiency will be further improved.

The second one is pushed even further by utilizing the storage ability of mobile UTs [7, 10, 13], which pre-store popular files in UTs through various storage strategies. The authors of [10] mainly focused on exploiting cognition to the cache-enabled D2D in the multi-channel cellular network, in which the number of users in a cell is formulated as mutually independent Poisson point processes (PPPs). Authors in [11] presented a paradigm for wireless video content dissemination based on caching popular files in mobile users with no additional infrastructure cost. Similarly, [12] put forward a systematic scheme in wireless content-centric networks. The storage capacity of users is adopted to maximize the content delivery capability with a fixed amount of wireless resources. Yang et al. [13] presented a comprehensive framework on D2D communication and advocated to proactively cache contents in partial users with caching ability when the network is off-peak. The studies have shown that pre-storing files in the UTs can increase spectral efficiency, thus introducing reuse gain. This scheme allows only one neighboring nodes to communicate with a RUT in the meantime, which can be described as one-to-one D2D communication. Nevertheless, the mobile devices like UTs cannot guarantee the effective transmission of popular files because of the unstable topology, limited storage capacity as well as constrained-energy in a sparse environment. Fortunately, this model can bring caching gains by taking advantage of multiple neighbor nodes to store the same required file. Therefore, we can activate multiple possible D2D links simultaneously in the dense or ultra-dense network named as many-to-one D2D communication.

In particular, the authors in [7] discussed about caching in these two ways. They introduced SBS placed in fixed position to serve user requests, and also adopted UTs' large hard disks built in to act as mobile helper stations. However, significant varies still exist between the reference [7] and our proposed system:

1. We mainly focus on the combination of the two approaches, in contrast, the reference [7] discussed about these two ways separately as well as other literatures. High data rate and system throughput can be obtained owning to caching gains, multiplexing gains, diversity gains, and proximity gains in our joint design.

2. We introduce the concept of many-to-one D2D communication in the user layer and make full use of the user redundancy in the dense environments . On the contrary, users in [7] were divided into smaller cluster based on their locations and at most one D2D transmission was allowed in each cluster at a time. That is to say, the authors

in [7] only allowed one-to-one D2D communication when they considered UTs for storage.

3. The interference of the proposed system is more complex than the reference [7]. There will inevitably be more interruptions for the cellular communication due to cumulative interference caused by many-to-one D2D communication. Thus, it is important to seek the trade-off at which interference and system throughput can supplement each other.

Therefore, efficient resource allocation and interference management have key impacts on the performance of content delivery in D2D and cellular coexisting system [15–17]. For theoretical derivations, only a single pair of D2D link reuse system resources is considered. Thus, it is always simplified to describe the interference, i.e., only one D2D pair is allowed to active within the power range of a user node or in a cluster. However, it is very critical to estimate the caused interference accurately when we calculate the system performance, such as maximize the data rate [16].

In short, the key study of this paper is to put forward a new strategy based on the study and the analysis of related literatures, which can not only take full advantage of the stability of SBS but also have the further numerical superiority and cost advantage of UTs. In other words, due to the different aspects of both traditional delivery methods, the joint design of these two approaches can be tried in practical project to offload the network traffic effectively and economically.

3 System model

Figure 2 illustrates the three-tier heterogeneous architecture with a single macro-cell including BS, femtocaching auxiliary equipments (FAEs), and multiple UTs. It is no doubt that the BS serves as the first layer. The BS can allocate and manage the channel state information (CSI) of the users and then distribute the files to FAEs and UTs relying on the rules of procedure during off-peak hours (at that time the network transmission is less costly, e.g., at night). FAEs and multiple UTs constitute the second and third layer of the network, respectively. FAE is a device that is similar to a proliferation of low-complexity small-base station with weak backhaul links and high storage capabilities. Considering the fact that users in similar geographical and social attributes are more likely to seek for common interest, we deploy FAEs in the center of hot spots. In addition, FAEs can further extend coverage area effectively, especially when they are placed in the edge of the cell. If a typical user initiates a file request, the propose cooperation scheme will combine FAEs and UTs to provide data flow through D2D communication, as well as through the BS in a traditional cellular (uplink/downlink) communication when the content cannot be found in nearby. The neighbor nodes and the nearest FAE caching the requested information will participate in the D2D communication. The neighbor node is defined as the UT in the communication distance of the requesting user. The maximum allowable distance r is determined by the power level of the RUT, and all of the UTs have the same power level. FAEs can have the same or different transmission power as the UTs, i.e., the same or different feasible communication range. To simplify the description, we assume that each RUT can only communicate to at most one FAE due to the investment efficiency. In fact, MRT and zero-forcing beam forming (ZFBF) technology in MIMO system [14, 20, 21] can make one FAE transmit multiple data stream to different users at the same time, and also support multi-devices in transmitting the same file to a receiver simultaneously. The complexity caching scheme for multiple access has been investigated deeply in [5, 14], by which the optimum file placement strategy can be found even if a RUT can connect several FAEs and we do not carry out further discussion.

The communication topology of the overall heterogeneous network may change over time. However, at each time instant, the transmission connection of network can be analyzed from the static perspectives, as shown in Fig. 3. It describes a scenario where four users request files at the same time, while FAE transmits the stored files concurrently to multiple users (to three users' three different files). If the FAE does not have the required files in its cache, the user can only establish a D2D link with the suitable neighbor, i.e., the neighbors who have store the needed files. If none of them have the required files, BS will serve the file request through cellular communication directly. Although the storage capacity of each user device is limited, even if they only store one file, the requesting user can find more than one suitable neighbor user with a certain probability in the dense environment. Many-to-one D2D relationship is thus formed when the appropriate neighbor nodes send data to the RUT simultaneously. For example, Tom can also form a local cooperative D2D pair with neighbor users (such as Bob and Alice) besides the FAE. As a consequence, FAE and the neighbor users that have the requested file in their cache beamform their content to the RUT, so that the data stream can be coherently combined at the receiver.

4 Problem formulation and analysis

4.1 Problem formulation

In Mobile World Congress 2017(MWC2017, Barcelona), Network equipment vendors have released a number of end-to-end network solutions for the future mobile communication and announced some experimental work on specific network technologies. For example, Nokia showed the new MIMO antenna (supporting 3.5, 4.5, 28, and 39 GHz spectrum), cloud residential access network (C-RAN), distributed core network and transmission

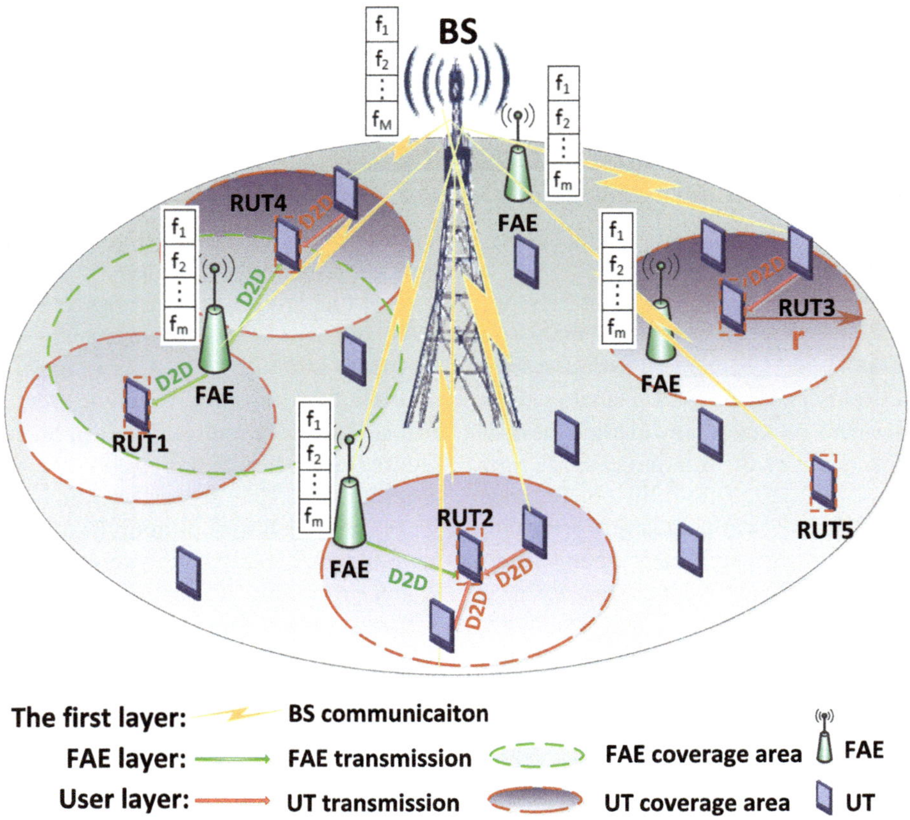

Fig. 2 The architecture of three-tier heterogeneous network equipped with FAEs in a single macro-cell. This figure describes the proposed three-tier heterogeneous architecture for content delivery in cellular networks, which includes BS, femtocaching auxiliary equipments (FAEs), and multiple users

scheme based on cloud architecture. Ericson announced to carry out a trial with Qualcomm, which was expected to build a standards-compliant 5G new gap system under 6 GHz. The trial was anticipated to use a variety of new technologies including large-scale MIMO, adaptive time division duplexing (TDD), beam forming, and extensible waveform technology based on OFDM. However, some problems related to network operation are also mentioned, such as mobile edge deployment which is suitable for video streaming and can reduce transmission delay. In this paper, we consider a joint caching to solve the problem of mobile edge deployment, which aims at maximizing the throughput, i.e., the sum of the effective data rate and subject to the interference constraint over all of the D2D links. The effective data rate refers to the total D2D transmission rate created by all requesting users at a steady-state moment. Consider a typical cell which serves N users and the coverage area is A. We assume that there are n users generating requests simultaneously. Each user requests a file independently from the library $\mathcal{F} = \{\mathrm{f}_1, \mathrm{f}_2, \ldots, \mathrm{f}_M\}$ of unit size M. The expected transmission rate obtained by the presented content delivery strategy can be denoted by $\mathcal{R}(n, x)$, where $x = [x_{ij}]$ represents the delivery strategy when the j-th user generates a request for the i-th file. Then, the optimization problem can be formulated as follows:

$$\mathrm{Max}\mathcal{R}(n, x) \tag{1}$$

Subject to

$$i \leq M$$

$$\gamma \geq \gamma_0, \quad \forall i, j$$

$$\gamma_d \geq \gamma_D, \quad \forall i, j$$

where γ is the signal-to-interference-plus-noise ratio (SINR) of the BS when D2D pairs reuse the given cellular resources. The performance of cellular communications should be guaranteed before D2D pairs are allowed to share the uplink resource [22]. Accordingly, the BS must be firstly satisfied as long as its SINR is no less than the threshold γ_0. And likewise, in order to ensure the reliability and stability of the D2D communication in the uplink transmission, the SINR of D2D receiver γ_d needs to exceed a given threshold γ_D.

Each request occurs independently and randomly according to the popularity distribution. Large quantities of researches certify that the popular distribution also can

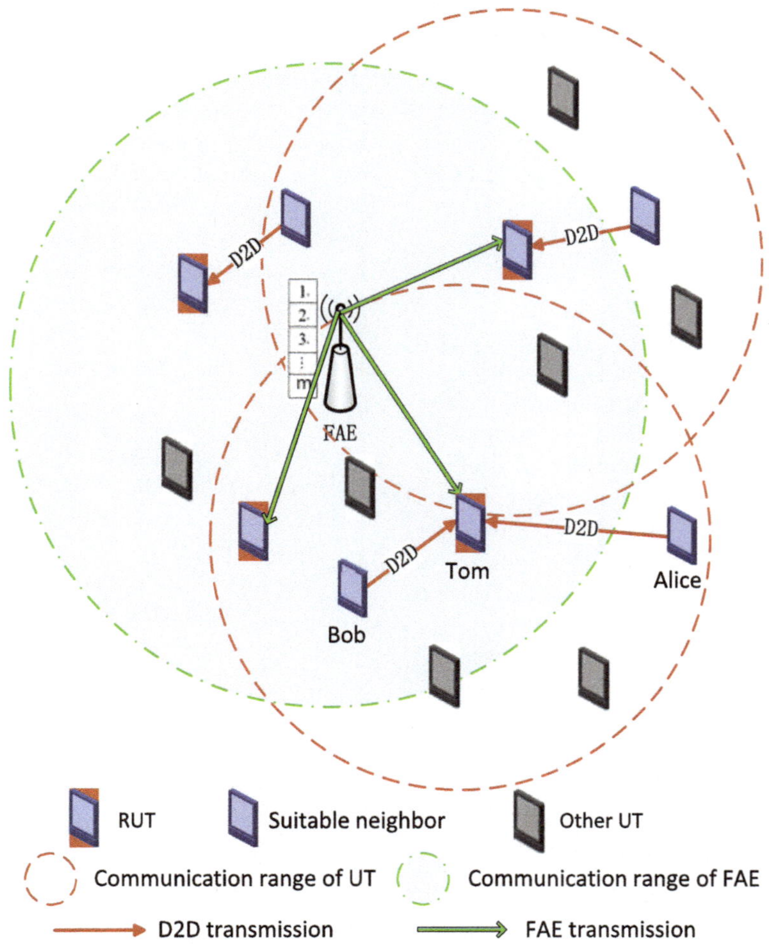

Fig. 3 Cooperation D2D communication for cache-driven multi-source caching model under MRT and ZFBF. This figure illustrates a transmission scenario under ZFBF and MRT when four users request files at the same time

be established as good as a Zipf distribution model to measure the user requesting probability for popular video files in D2D communication [22]. Under this model, the request probability of the i-th popular file, denoted by p_i, $1 \leq i \leq M$, $\sum_{i=1}^{M} p_i = 1$, is inversely proportional to its order:

$$p_i = \frac{1/i^{\gamma_r}}{\sum_{j=1}^{M} 1/j^{\gamma_r}} \quad (2)$$

The Zipf exponent γ_r characterizes the distribution function by controlling the relative popularity of files. Larger γ_r means higher content duplicate requests, i.e., the top few popular files account for the majority of requests.

An important aspect of the joint optimization is the storage allocation because hit ratio and transmission rate can be varied by caching different files in neighbor nodes. Caching multiple copies of a file can increase the chances to get diversity gains and multiplexing gains. For simplicity, we assume that all files have the same size, and one file is a basic storage unit. The library files are available in the BS, and users can directly connect via a cellular communication. Meanwhile, let C_f denotes the storage capacity of each FAE, which is measured by the maximum number of files it can store.

Using the collaboration storage capacity of UTs and FAEs is an important feature of this article. In a general random geometric graph, it is shown that finding the optimal deterministic file assignment is NP-hard even when terminals are static and interference is ignored [23]. However, in our three layers caching system, if we assume that each UT can access no more than one, we can find a simple centralized placement method for FAE. Given that there are m static memory units in each FAE, it should store the top m most popular files without repetition, while the storage capacity C_f is equal to the number of storage units in FAE, i.e., $C_f = m$. Each FAE can store on average up to m files with $m \leq M$.

Random caching can be implemented since users are highly mobile. Each UT caches files at random and independently obeys a caching distribution. Specifically, we assume that each UT stores exactly one file, though generalization to multiple files per user is trivial. On the basis of the study, this caching distribution can be also modeled as a Zipf distribution with exponent γ_c. Thus the storage probability of the i-th popular file , denoted by β_i, can be expressed as:

$$\beta_i = \frac{1/i^{\gamma_c}}{\sum_{j=1}^{M} 1/j^{\gamma_c}} \tag{3}$$

The exponent of caching distribution γ_c is one of our decision variables which is not necessarily equal to γ_r.

Note that under this storage policy, transmission rates related to the BS and FAEs are mainly influenced by the channel model. By contrast, the number of neighbor nodes determined by the D2D covering radius r has a greater influence on the system throughput. The larger the covering radius r is, the more the D2D links are to generate many-to-one D2D communication with a certain storage probability. Without loss of generality, we suppose all of the users are distributed randomly in the cell and can communicate with each other through D2D communications when the distance r_0 between them is less than the threshold r. Furthermore, define R_{ij} as the average transmission rate when the j-th user requests the i-th file. Considering the randomized request probability of the i-th file p_i, when there are n users generating requests, the sum of the effective data rate can be written as:

$$\mathcal{R}(n,x) = \sum_{j=1}^{n}\sum_{i=1}^{M} p_i R_{ij} \tag{4}$$

Therefore, the solution to the optimization problem can be obtained by solving the following binary linear program:

$$\text{Max} \sum_{j=1}^{n}\sum_{i=1}^{M} p_i R_{ij} \tag{5}$$

Subject to

$$\sum_{i=1}^{M} p_i = 1$$

$$0 < p_i < 1, \ \forall i \in M$$

$$C_f \leq M$$

$$r_0 \leq r, \ \forall i,j$$

$$\gamma \geq \gamma_0, \ \forall i,j$$

4.2 Transmission rate

When the j-th user requests the i-th file, network traffic is mainly generated by multi-mode communication, which means that the sender can be generally grouped into three categories in terms of the file storage locations: FAE, neighbor users, and BS. For the sake of analyzing the following derivations, we define the effective data rate as the transfer speed multiplied by the non-outage (transmission success) path existence probability (i.e., the probability that the channel can take place to support the data rate [24]). Following the view of graph theory, it means that the edge for the communication in the random geometric graph exists with a certain probability. The arguments in this section are similar, the distinction is that the subscripts B (b), F (f), C (c), D (d) are on behalf of the BS, FAE, cellular user and D2D user respectively.

- 1. FAE. Given that there are m storage units in a FAE, each FAE should store the top m most popular files without repetition. Define binary random variable α_i, such that α_i is equal to 1 if $1 \leq i \leq m$; otherwise, it is equal to 0. If the i-th file on demand is one of the top m popular files, the transmission rate that FAE can provide, denoted by $\overline{R_F}$, is equal to:

$$\overline{R_F} = \alpha_i R_f e^{-\eta_f} = \begin{cases} R_f e^{-\eta_f}, & 1 \leq i \leq m \\ 0, & m \leq i \leq M \end{cases} \tag{6}$$

  R_f refers to the ideal transmission rate of FAE and η_f indicates the attenuation coefficient of Rayleigh fading. Based on the dedicated bands frequency allocation strategy, the co-channel interference of FAEs can be ignored, and the SINR of its received signal is given by $\gamma_f = \frac{P_F x_{iF} d_{iF}^{-\sigma} H_{iF}^2}{N_0}$. Thus, the ideal transmission rate R_f can be obtained by Shannon theory [25].

$$R_f = Clb\left(1 + \frac{P_F x_{iF} d_{iF}^{-\sigma} H_{iF}^2}{N_0}\right) \tag{7}$$

  where P_F denotes the transmit power of FAE, d_{iF}, and $H_{i,F}$ represent the distance and the channel gain between the i-th user and the conterminous FAE, respectively. Path loss exponent is denoted as σ, and x_{iF} is the log normal shadow fading coefficient between the i-th D2D and the FAE. N_0 is white Gauss noise, and C is the available spectrum bandwidth.
- 2. Neighbor user. Each UT caches files at random and independently following the Zipf distribution with exponent γ_c. The number of neighbor nodes is usually referred to the binomial distribution with parameters N and P, i.e., $K = B(N,P)$, the probability of k neighbor nodes can be expressed as:

$$P_r(K=k) = \binom{N}{k+1} p^{k+1}(1-p)^{N-k-1} \tag{8}$$

  where $p = \frac{\pi r^2}{A}$, r is determined by the power level for each transmission, and A is the coverage area of the cell. Thus, the numerical probability of cooperating users for the transmission to the i-th file, that is to say, the numerical probability of neighbor nodes who have stored the i-th file is:

$$P(T_i = c) = \sum_{k=1}^{N} P\{c|K = k\}P_r(K = k) = \sum_{k=c}^{N} \binom{k}{c} \beta_i^c (1-\beta_i)^{k-c} P_r(K = k) \quad (9)$$

After a request to the i-th file, the average number of D2D links that one hit can be established as:

$$E(D) = \sum_{c=1}^{M} cP(T_i = C) \quad (10)$$

Taking into account the request probability of the i-th popular file, the mean value of any request for files can be summed up as:

$$\overline{E}(D) = \sum_{i=1}^{M} p_i E(D) = \sum_{i=1}^{M} p_i \sum_{c=1}^{M} cP(T_i = C) \quad (11)$$

$n\overline{E}(D)$ can represent the total number of D2D communication activated by n requesting users at the same time, i.e., $n\overline{E}(D)$ equals to the number of D2D pairs T_{total} mentioned in the later. An important theme to notice is that $P_r(K = k)$ is a function of r. Thus, $\overline{E}(D)$ and T_{total} are both functions of variable r and γ_c.

Then, the transmission rate provided by the neighbor users, denoted by $\overline{R_D}$, can be formulated as:

$$\overline{R_D} = \overline{E}(D)\, R_d e^{-\eta_d} = R_d e^{-\eta_d} \sum_{i=1}^{M} p_i \sum_{c=1}^{N} cP\left(T_i = c\right) \quad (12)$$

R_d refers to the ideal transmission rate of D2D communication. Due to reusing the same frequency resources with the cellular communication, the interference of D2D receiver comes from two aspects: white Gaussian noise and the cellular communication. Therefore, when uplink resources allocation is performed efficiently, the SINR of D2D users can be calculated by $\gamma_d = \frac{P_D x_{iB} d_{iB}^{-\sigma} {H_{iB}}^2}{P_C x_{CB} d_{CB}^{-\sigma} {H_{CB}}^2 + \mathrm{N_0}}$, and then R_d can be derived through the Shannon formula [25]:

$$R_d = Clb\left(1 + \frac{P_D x_{iB} d_{iB}^{-\sigma} {H_{iB}}^2}{P_C x_{CB} d_{CB}^{-\sigma} {H_{CB}}^2 + \mathrm{N_0}}\right) \quad (13)$$

where P_D, P_C denote the transmit power of D2D users and one cellular user, d_{iB} and d_{CB} represent the distance between the i-th D2D sender and cellular user to the BS, respectively. H_{iB} and H_{CB} denote the channel gain. σ and x_{iB} are expressed as the path loss exponent and the log normal shadow fading coefficient between the i-th D2D sender and BS.

- 3. BS. The base station will carry out data communication only when the required files are not stored in the neighbor nodes and FAE. Defined P_B as the transmit power of each resource block, the transmission rate can be calculated by:

$$\overline{R_B} = (1-\alpha_i)P\left(T_i = 0\right) \cdot R_b e^{-\eta_b} = \begin{cases} R_b e^{-\eta_b} \sum_{k=1}^{N} (1-\beta_i)^k P_r\left(K = k\right), & m < i \le M \\ 0, 1 \le i \le m \end{cases} \quad (14)$$

In this case, a file has to be transmitted from the BS and the RU can be thought of as an casual cellular user. Because D2D communications occur in the uplink direction, cellular users, as the receiver of the downstream link communication, are not affected by D2D communication. Then, the ideal transmission rate of BS R_b can be described as:

$$R_b = Clb\left(1 + \frac{P_B x_{iB} d_{iB}^{-\sigma} {H_{iB}}^2}{\mathrm{N_0}}\right) \quad (15)$$

In this optimization problem, self-request (i.e., the user stores the requesting file in its own cache) does not play any role. However, some users can achieve the required files without delay because that self-request needs not special link to transmit date. Based on the above theoretical calculation of transmission speed, the optimization formula can be rewritten as:

$$\mathcal{R}^*\left(\mathrm{n,x}\right) = \mathrm{Max}\left\{\sum_{j=1}^{n}\sum_{i=1}^{M} p_i \left[\alpha_i R_f e^{-\eta_f} + \sum_{c=1}^{N} c\, P\left(T_i = c\right) .R_d e^{-\eta_d} + (1-\alpha_i) R_b e^{-\eta_b} \sum_{k=1}^{N} (1-\beta_i)^k P_r\left(K = k\right)\right]\right\} \quad (16)$$

Notice that $P(T_i = c)$ is a function of γ_c and r, consequently, we can be informed that the optimization formula is a function of several variables, such as γ_c, γ_r, and r as well as the number of requesting user n. It can be generated by varying one or more of the parameters in a general form.

4.3 Noise and co-channel interference

In addition to the traditional interference in cellular systems, there are two kinds of interference in the D2D coexisting with cellular system. One is the interference from D2D pairs to the traditional cellular system; the other is the interference from the cellular system to the D2D receiver. For our theoretical derivations, we assume FAEs in the system utilize the dedicated mode which is assigned by the orthogonal resource, interference caused by FAE to BS and users can be negligible, or can be made negligible through an appropriate frequency reuse scheme. Furthermore, we presume that multiple D2D links to the same RUT can be operated on the different resources sharing bandwidth and each cellular wireless resource can be reused up to one at most in the uplink period. The main reason we reuse the frequency resources in the uplink is that BS, as the uplink receiver of cellular communication, has more powerful ability to suppress noise interference than UT (i.e., the receiver in the downlink). This ability is ideally suitable for our proposed system, which may introduce larger cumulative interference caused by multiple D2D links. For simplicity, we neglect inter-cell interference and consider one macro cell in isolation. Thus, the main purpose of this paper is to analyze the model taking into account co-channel interference, which means that the same resource block is spatially reused.

In order to guarantee D2D communication, the SINR of D2D users should be greater than its threshold when it is reusing with the cellular uplink resources, i.e., $\gamma_d \geq \gamma_D$. As shown in 4.2, the SINR of D2D users is a function related to D2D transmit power and the distance from the cellular user to the D2D receiver, and so it is best for D2D communication to reuse the resources of cellular user who is as far as possible from the D2D receiver. Moreover, even the parameters in this function are not associated with those in the optimized objective function, we can derive the minimum transmit power of D2D from this constraint.

To maintain the comparable fairness, cellular communication should be given priority. The maximum transmission rate of D2D links is restricted to guarantee the non-outage communication of the cellular network. Notice that BS is the receiver of cellular communication in the uplink period, it will then suffer from cumulative interference introduced by multi-pair D2D links. At this time, the signal received by the BS can fall into two categories:

- Cumulative interference of the base station caused by multiple D2D transmitters can be described by Ψ_D while T_{total} is the number of D2D links:

$$\psi_D = \sum_{i=1}^{T_{\text{total}}} \sqrt{P_D x_{iB} d_{iB}}^{\,-\sigma/2} H_{iB} \tag{17}$$

- Uplink signals Ψ_C to the BS transmitted by cellular user can be expressed as:

$$\psi_C = \sqrt{P_C x_{CB} d_{CB}}^{\,-\sigma/2} H_{CB} \tag{18}$$

Furthermore, we can approximate the SINR of the BS as:

$$\gamma = \frac{P_C x_{CB} d_{CB}^{-\sigma} H_{CB}^2}{\sum_{i=1}^{T_{total}} P_D x_{iB} d_{iB}^{-\sigma} H_{iB}^2 + N_0} = \frac{H_{CB}^2}{\frac{N_0}{P_C x_{CB} d_{CB}^{-\sigma}} + \frac{\sum_{i=1}^{T_{total}} P_D x_{iB} d_{iB}^{-\sigma} H_{iB}^2}{P_C x_{CB} d_{CB}^{-\sigma}}} \tag{19}$$

Under high SINR, $\frac{N_0}{P_C x_{CB} d_{CB}^{-\sigma}}$ can be approximated to zero. For the convenience of description, defining $\mu = \frac{1}{P_C x_{CB} d_{CB}^{-\sigma}}$ and $v_i = P_D x_{iB} d_{iB}^{-\sigma}$, then the SINR of the BS can be overridden as:

$$\gamma = \frac{H_{CB}^2}{\mu \sum_{i=1}^{T_{\text{total}}} v_i H_{iB}^2} \tag{20}$$

Since $\|H_{CB}\|^2$ follows exponential distribution and $\|H_{iB}\|^2$ is exponential random variables subjecting to independent identically distribution, according to [26, 27], the probability density function (PDF) of γ can be rewritten as:

$$F_\gamma(\gamma) = \sum_{i=1}^{T_{\text{total}}} \frac{A_i}{\mu v_i} \left(\gamma + \frac{1}{\mu v_i}\right)^{-2} \tag{21}$$

$$A_i = \prod_{j=1, j\neq i}^{T_{\text{total}}} \left(\frac{v_i}{v_i - v_j}\right) \tag{22}$$

Following the above derivations, the non-outage probability of the BS is defined as the probability that the instantaneous γ is greater than the threshold γ_0, which can be computed as:

$$\begin{aligned} P_{\text{Nout}} = P_\gamma(\gamma \geq \gamma_0) &= 1 - P_\gamma(\gamma < \gamma_0) = 1 - \int_0^{\gamma_0} f_\gamma(\gamma)\, d\gamma \\ &= 1 - \int_0^{\gamma_0} \left\{ \sum_{i=1}^{T_{\text{total}}} \frac{A_i}{\mu v_i} \left(\gamma + \frac{1}{\mu v_i}\right)^{-2} \right\} d\gamma \\ &= 1 - \sum_{i=1}^{T_{\text{total}}} \prod_{j=1, j\neq i}^{T_{\text{total}}} \left(\frac{v_i}{v_i - v_j}\right) \left(1 - \frac{1}{\mu v_i \gamma_0 + 1}\right) \end{aligned} \tag{23}$$

Thus, if more than one D2D links multiplex cellular uplink resources at the same time, the communication performance of the BS will be influenced by the system parameters such as transmit power, distance between D2D transmitter and BS, SINR threshold γ_0 as well as the number of D2D pairs T_{total}. T_{total} indicates the total number of D2D links when n users generate requests simultaneously and can be seen as a variable relate to r and γ_c.

4.4 User satisfaction

To analyze the impact of various arguments on the performance of the system, the concept of user satisfaction is introduced as follows:

$$\tau = \alpha P_{\text{Nout}}^* + (1 - \alpha) R_{\text{total}}^* \tag{24}$$

Here, α is the weighted coefficient, P^{*}_{Nout} and R^{*}_{total} are min-max normalization values. Finding the optimal solution analytically in closed form does not seem feasible, numerical solutions are possible with very low effort. Nevertheless, the pareto solutions of the optimization problem can be obtained using the above equation and numerical experiments based on the multi objective optimization model.

5 Numerical results and analysis

In the aforementioned references, parameters such as the popularity profile of data files and D2D communication radius are always assumed to be known perfectly. In practice, such an assumption cannot be dynamically adjusted and reasonably justified in different circumstances.

In the following, we provide some numerical results to investigate the effects of relevant parameters on the system throughput with the simple caching strategies. Unless specified otherwise, all figures set limits according to the following except those considering the effect of the parameters. The BS has a coverage area of $R = 500$ m with a storage capacity of $M = 100$ files. In the described system, FAEs use dedicated channels and only a linear impact on the network throughput. For the sake of simplicity, only four FAEs are installed in fixed position. Each FAE has a coverage area of 200 m, and its capacity is 30% of the total files. All channels are subject to path loss model where $\sigma = 3$. The noise power of the receiver is 70 dBm, and bandwidth of one channel is set to 1 MHz. We use the Monte Carlo experiment to prove the effectiveness of the proposed algorithm, and the number of Monte Carlo experiments is 10000. In the process of the experiment, we assume that the BS uses a transmission scheme similar to the fourth-generation long-term evolution (LTE) standard, based on an OFDM-TDMA physical layer. The FAEs are operating with dedicated spectrum resource and have multiple antennas, i.e., WiFi-like links. We envision that UTs have single-antenna although multi-antenna system is inevitable in the future.

The number of D2D links generated by user layer is an important criterion that affects the system throughput and cumulative interference. Thus, the first part of the experiment begins with the influence of each parameter on it.

Figures 4 and 5 show the number of D2D links for requesting the different files, which is named as well as the file sequence number (FSN) and ordered by the popular rankings. Intuitively, the number of D2D links will be relatively large when users request the first ten files, as in Fig. 4. However, looking closely at Fig. 5, we can observe that the number of D2D links rapidly descends and then becomes gradually lower if users request the remaining 80% of the files. The results will be slightly fluctuated but will not affect the the overall trend of the experiment.

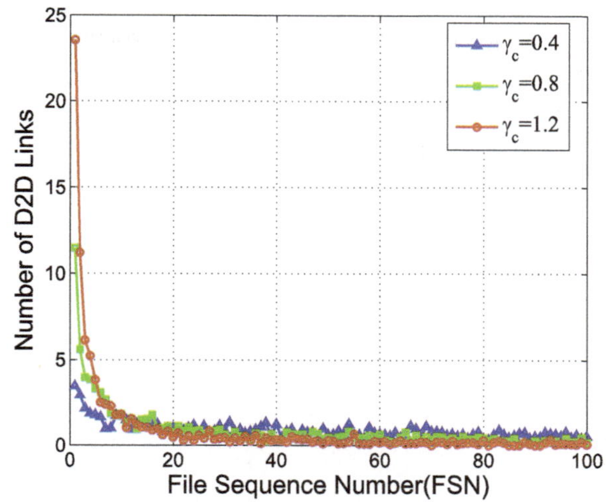

Fig. 4 The average number of D2D links for different File Sequence Number with r=100. This figure shows the number of D2D links established for requesting different files

This occurs because the selection of RU is random in the process of experiment. Meanwhile, the position of UTs is practically random although their storage probability is modeled as the Zipf distribution. Then, the Monte Carlo experiment results fluctuate by a number of random sampling variability. These fluctuations are almost all under 1, which is a good indicator that the remaining 80% of the content has limited contribution to the traffic. Moreover, the greater the γ_c value, the more obvious the change. We can conclude that γ_c determines the file storage probability; the larger γ_c means the greater probability that the most popular file is stored. This storage mode is more effective only when the request probability is considered simultaneously.

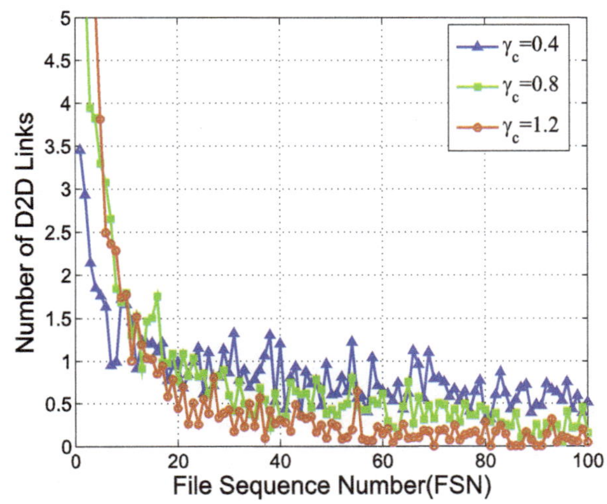

Fig. 5 The average number of D2D links with different ordinate scale. This figure shows the number of D2D links established with different ordinate scale

The average number of D2D links versus the collaboration distance r for different γ_c and γ_r is shown in Fig. 6. Different γ_r indicates different request probabilities. We can observe that the number of D2D links has at least twice as γ_r and γ_c increase by 0.4. This agrees with the following intuition: the larger γ_r and γ_c, the higher probability that a popular file will be requested, and it is just consistently stored in the neighbor nodes, especially with the increase of the cooperation distance. Furthermore, we can find that the impact of γ_r and γ_c on the number of D2D links is essentially the same, although they can take different values separately. In most of the case, γ_r can be obtained from the data statistics. Accordingly, in the following simulation experiments, we just need to think about the impact of γ_c by setting $\gamma_r = 0.8$, this value is based on a study conducted on the University of Massachusetts Amherst campus in 2008. Meanwhile, we can observe that in the case of γ_r invariant, to achieve the same number of D2D links, r will decrease with the increase of γ_c.

The total number of users and the number of users seeking files concurrent also affect the average number of D2D links in the cellular network. Figure 7 represents different curves by tracing the lines of different parameters. The x-axis refers to the number of users requesting files simultaneously. The bigger the N value is, the higher the user density is, and then the better the D2D connection performance is. We can also observe that the effect of N value on the curve is similar to that of r. This is consistent with the theoretical results. D2D connection capability has doubled and redoubled as users grow exponentially.

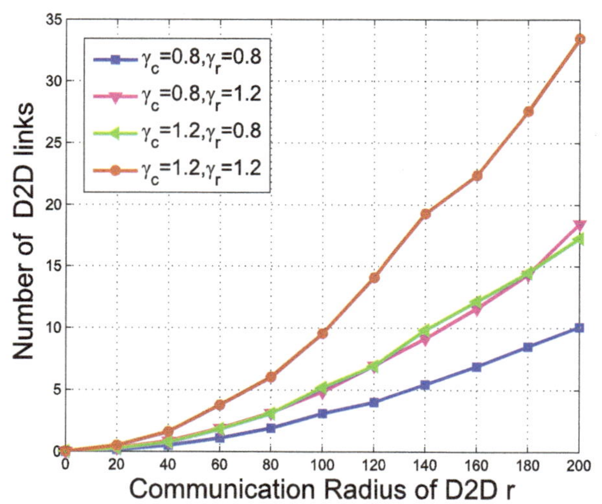

Fig. 6 The average number of D2D links for different request probability and storage probability with N =2000. Illustrates the average number of D2D links versus the collaboration distance r for different γ_c and γ_r

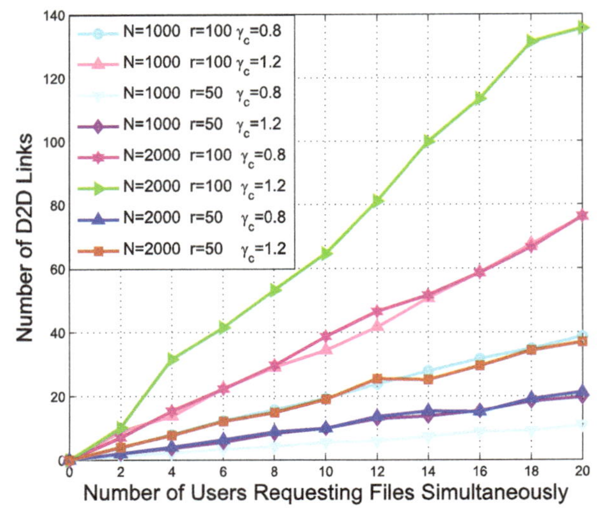

Fig. 7 The average number of D2D links with different number of request users. This figure represents the average number of D2D links by tracing the lines of different parameters, such as user density, communication radius, and request probability

Furthermore, we note that N value has lower impact on the average number of D2D links compared with γ_c. The reason is that user requests are based on the probability of request in the experimental process. Changing the popularity Zipf distribution exponent γ_c has greater impact on the popularity of the document. If the request is subject to the uniform distribution, N will have a greater impact on the chance of D2D connection with a fixed radius.

It can be observed that the theoretical simulation results are basically consistent with the simulation results using Monte Carlo simulation in Fig. 8. It means that the method for D2D users to reuse cellular wireless resources based on random distribution is effective. Moreover, with the same storage probability, i.e., the same γ_c, the numbers of D2D communication satisfying certain condition increases as the D2D transmission radius rises. Sometimes slight jitter will be produced due to the impact of random channel. In the remaining figures, except those we want to consider the effect of γ_c, we set it to 1.

From now on, in order to analyze the influence of the system interference, we just present the simulation results of non-outage probability in which the average number of D2D links is a crucial technical parameter.

When multiplexing the uplink resources, multi-pair D2D communications will give rise to cumulative interference. In the D2D sharing mode based on cellular system, we must give priority to ensure the quality of main chain link. Thus, the effect of multiple D2D links on the performance of cellular communication should be studied. Figure 9 shows the comparison of the non-outage probability between the reference [7] and the proposed system through Monte Carlo simulation. In simulating process,

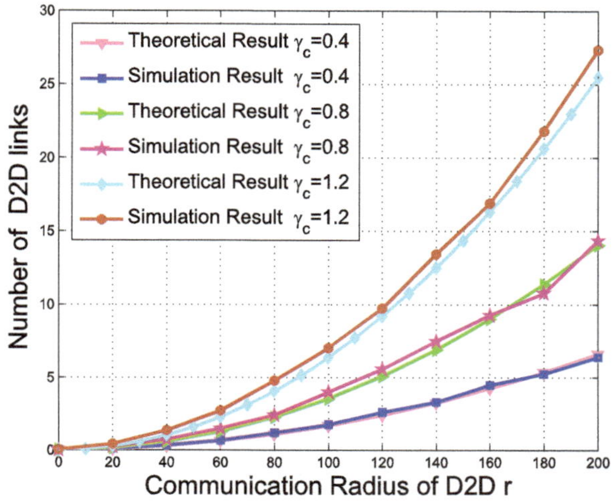

Fig. 8 The comparison between theoretical results and simulation results with N=2000. This figure describes the comparison result between Monte Carlo simulation and theoretical experiment under different parameters

the transmit power of cellular communication is set to 1 W and D2D transmit power is 10 mW, the SINR threshold of BS γ_0 is limited to 2. With the increase of RUs' number, the non-outage probability of the BS decreases. This is due to the cumulative interference caused by D2D communication to the BS, which makes the communication quality between cellular user and BS decrease. Moreover, as can be seen from the picture, the non-outage probability of reference [7] is slightly better than the proposed system. The reason behind is that the proposed system allows many-to-one D2D communication, which brings huge profits as well as interference. The number of D2D links activated under the same conditions is higher than the reference [7]. The system will even introduce cumulative interference while bringing higher throughput. Therefore, in order to ensure the quality of the cellular communication, the maximum number of D2D links allowed to reuse resources can be obtained through the non-outage communication constraints.

The average transmission rate versus r under four different transmission strategies is shown in Fig. 10 when a request occurs. The average rate increases with the increase of r. In fact, traditional D2D refers to one-to-one D2D communication, which uses random storage strategies and is shown in the studies pre-storing files in the UTs. For example, as described in [7], UTs act as mobile helper stations can be seen as traditional D2D. Considering SBS dissemination mode, single SBS indicates the case that just one SBS can be connected in the reference [7]. Intuitively, a larger r means that more qualified UTs, i.e., the neighbor nodes within the communication range as well as storing the required files, can be detected. This will almost certainly lead to higher frequency reuse. Therefore, multi-dimension diversity gain caused by FAE and multiple neighbor nodes can tremendously improve the transmission rate in the proposed strategy. In contrast, traditional D2D can connect neighbor node no more than just one. Although the connection probability increases as the D2D communication distance rises, when the probability reaches to 1, the channel attenuation becomes more serious with the increase of transmission distance. Therefore, the transmission speed decrease with the distance increasing after the connection probability is 1. Not surprisingly, the average speed transmitted through BS or SBS is relatively stable but smaller because of the lack of

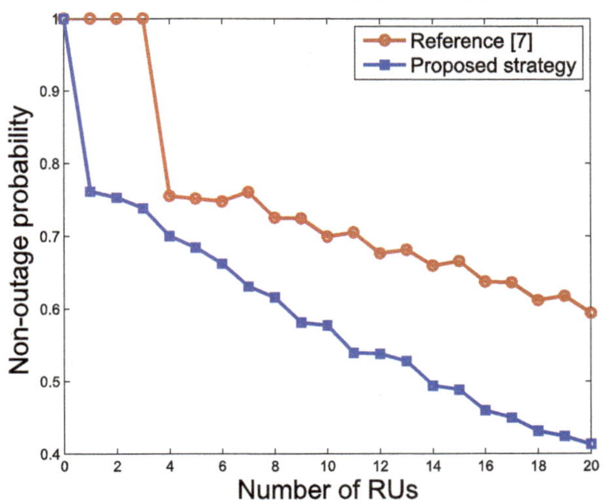

Fig. 9 The relationship between the non-outage probability of BS and the number of requesting users. This fuigure shows the comparison of the non-outage probability between the reference [7] and the proposed system through Monte Carlo simulation

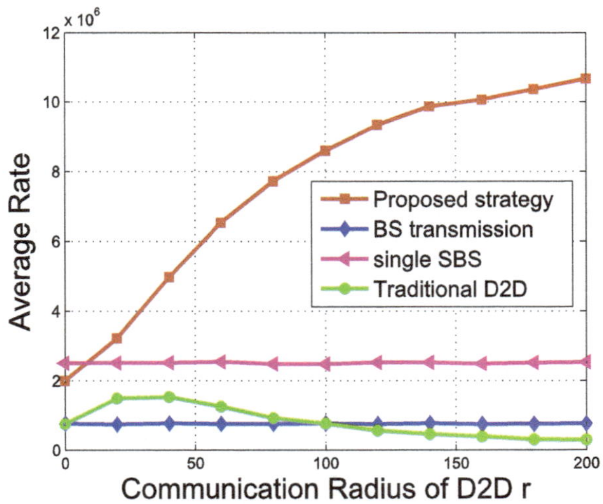

Fig. 10 The relationship between average rate and D2D communication radius under different transmission strategies. This figure illustrates the average rate and r under four different transmission strategies when a file request occurs

proximity gain. Moreover, we should keep in mind that the proposed transmission strategy not only improve the transmission speed and reduce the transmission delay but also unloads the transmission flow of the BS.

Considering the interference, user satisfaction is defined to measure the effect of system outage probability on user transmission rate. The effects of changing the popularity Zipf storage exponent γ_c on user satisfaction with different coefficient a are investigated in Fig. 11. As seen in the graph, the greater the non-outage probability weight, the faster decay of user satisfaction with the communication distance. It is because that the greater weight indicates the higher requirements of quality communication. User satisfaction fluctuates more and more slightly when a decreases with γ_c increasing. In addition, we can observed that its optimal distance r_{opt} decreases with the increase of γ_c. For the small γ_c, there is a little redundancy in the users' storage, i.e., the probability of finding files in the neighbor nodes is generally small. Thus, in order to increase the chance of having D2D communication and to achieve customer satisfaction, the collaboration distance r should increase. Hence, the optimal communication distance r_{opt} decreases for large γ_c when the number of requesting users is fixed.

User satisfaction versus r is shown in Figs. 12 and 13 with different number of requesting users. The two pictures tell us that the greater the non-outage probability weight, the faster the decay of user satisfaction with the certain communication distance. The reason behind is that the greater weight indicates the higher requirement for reliable communication. After the peak, the large D2D communication radius means higher number of D2D links, and then results in larger outage probability. User satisfaction decreases as well as the communication quality. The best D2D communication radius r_{opt} can be found from the figure when the user satisfaction reaches the top. Comparing Figs. 12 and 13, we can conclude that the optimal collaboration distance become smaller along with the number of requesting users increasing.

The influence of Zipf storage exponent γ_c on user satisfaction with $a = 0.8$ are investigated in Fig. 14. It shows the cases that different number of users request files under optimized distance. For small γ_c, there is a little redundancy in users' storage, i.e., the probability of finding files in the neighbor nodes is generally small. From the figure, it can be known that with the increase of γ_c, user satisfaction will rise accordingly. After reaching the peak,

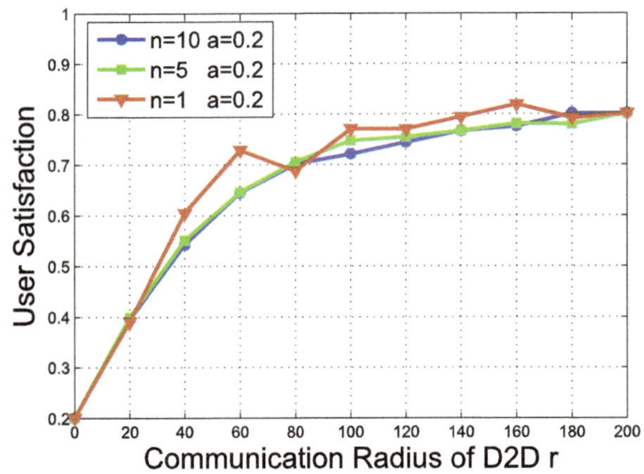

Fig. 12 The relationship between user satisfaction and D2D communication radius with $\gamma_c = 1$ and $a = 0.2$. This figure illustrates the effect of system outage probability and user transmission rate on user satisfaction

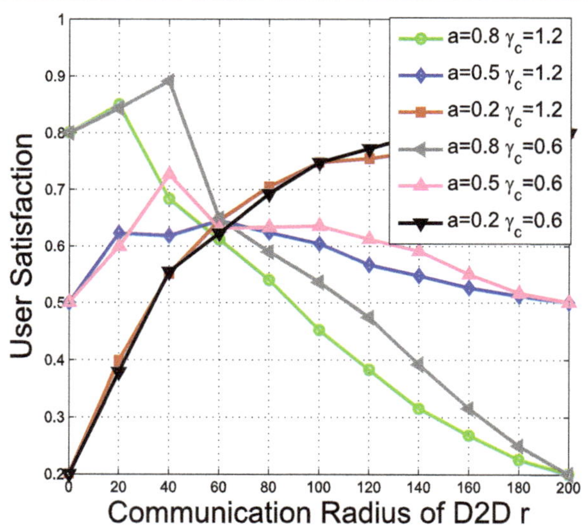

Fig. 11 The influence of γ_c on user satisfaction under different weight with $n = 5$. This figure illustrates the effects of changing the popularity Zipf storage exponent γ_c on user satisfaction with different coefficient a

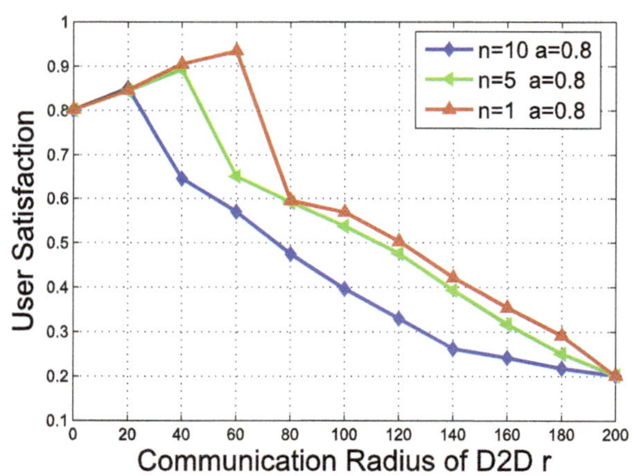

Fig. 13 The relationship between user satisfaction and D2D communication radius with $\gamma_c = 1$ and $a = 0.8$. This figure also illustrates the effect of system outage probability and user transmission rate on user satisfaction with different weight

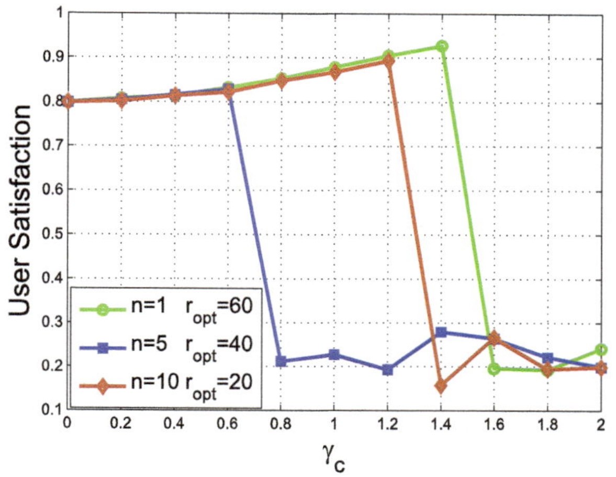

Fig. 14 The influence of γ_c on user satisfaction under optimized distance. This figure shows the influence of related parameters on user satisfaction

satisfaction decays rapidly. This is because the weight of outage probability is large, as already explained above. It is obvious that, we can achieve a set of pareto solutions for the optimization goal.

6 Conclusions

Content delivery strategy based on active storage in local is an effective way to settle the explosion of mobile data traffic. In this paper, we introduced cooperative transmission scheme by jointly using the storage capacity of UTs and FAEs. Aiming at the scenarios which have multiple-pair D2D links in the system, the network transmission rate is optimized on the premise of guaranteeing the main communication performance. Furthermore, the optimal D2D communication distance and storage probability exponent can also be found based on multi-objective optimization problem. The analytical and simulation results show that the proposed scheme can obviously improve the network throughput, can reduce the transmission delay, and can restrain the mutual interference among different links effectively. Meanwhile, the data traffic of the most popular files can be offloaded to D2D communication, which can provide high spectral efficiency and free up BS to provide other business.

In the future, we still have a lot of additional work to do, such as the question of storage optimization as well as the cooperative strategies of content allocation between UTs and FAEs. Furthermore, more complex D2D scenes need to be studied, e.g., multiple D2D users reuse the same CU resource simultaneously.

Acknowledgements

This work was supported by the National Natural Science Foundation of China under Grant No. U1404602 , the Young Scholar Program of Henan Province under Grant No.2015GGJS-086, the Science and Technology Foundation of Henan Educational Committee under Grant No. 14A510011, the Science and Technology Key Research Program of Henan Province with No. 172102210341, the Dr. Startup Project of Henan Normal University under Grant No.qd14136, and the Young Scholar Program of Henan Normal University with No. 15018.

Authors' contributions

XZ put forward the idea and wrote the manuscript. PY took part in the discussion and gave the original ideas, he also guided, reviewed, and checked the writing. YC and PC carried out the experiments and analyzed the experimental results. All authors read and approved the final manuscript.

Competing interests

The authors declare that they have no competing interests.

Author details

[1]National Digital Switching System Engineering and Technological Research Center, No. 62, Kexue Road, Zhengzhou, China. [2]School of Computer and Information Engineering, Henan Normal University, No. 46, Jianshe East Road, 453007 Xinxiang, China.

References

1. China Mobile, *Chinamobile technology vision 2020 + white paper*, (2016). [Online]. Available: http://www.gtigroup.org/
2. N Golrezaei, AG Dimakis, AF Molisch, in *Information Theory Proceedings (ISIT), 2012 IEEE International Symposium on*. Wireless device-to-device communications with distributed caching (IEEE, Cambridge, 2012), pp. 2781–2785
3. CVNI Cisco, *Global mobile data traffic forecast update. white paper*, (2016). [Online]. Available: http://www.cisco.com/
4. M Afshang, HS Dhillon, PHJ Chong, Fundamentals of cluster-centric content placement in cache-enabled device-to-device networks. IEEE Trans. Commun. **64**(6), 2511–2526 (2016)
5. BN Bharath, KG Nagananda, HV Poor, A learning-based approach to caching in heterogenous small cell networks. IEEE Trans. Commun. **64**(4), 1674–1686 (2016)
6. N Golrezaei, AG Dimakis, AF Molisch, et al., in *Signals, Systems and Computers (ASILOMAR), 2011 Conference Record of the Forty Fifth Asilomar Conference on*. Wireless video content delivery through distributed caching and peer-to-peer gossiping (IEEE, Pacific Grove, 2011), pp. 1177–1180
7. N Golrezaei, AF Molisch, AG Dimakis, et al., Femtocaching and device-to-device collaboration: A new architecture for wireless video distribution. IEEE Commun. Mag. **51**(4), 142–149 (2013)
8. SF Chou, TC Chiu, YJ Yu, et al., in *Global Communications Conference (GLOBECOM), 2014 IEEE*. Mobile small cell deployment for next generation cellular networks (IEEE, Austin, 2014), pp. 4852–4857
9. M Bennis, M Simsek, A Czylwik, et al., When cellular meets WiFi in wireless small cell networks. IEEE Commun. Mag. **51**(6), 44–50 (2013)
10. X Zhao, C Yang, Y Yao, et al., Cognitive and cache-enabled d2d communications in cellular networks. [Online]. Available: https://pdfs.semanticscholar.org/7359/2aa42bfc4f9d69b4a300049f46bd330ed628.pdf
11. G Negin, P Mansourifard, MF Andreas, DG Alexandros, Base-station assisted device-to-device communications for high-throughput wireless video networks. IEEE Trans. Wirel. Commun. **13**(7), 3665–3676 (2014)
12. H Liu, Z Chen, X Tian, X Wang, M Tao, On content-centric wireless delivery networks. IEEE Wirel. Commun. **21**(6), 118–125 (2014)
13. C Yang, Y Yao, Z Chen, B Xia, Analysis on cache-enabled wireless heterogeneous networks. IEEE Trans. Wirel. Commun. **15**(1), 131–145 (2016)
14. WC Ao, K Psounis, in *Proc. of ACM MobiHoc 2015*. Distributed caching and small cell cooperation for fast content delivery, (2015), pp. 127–136
15. Q Bodinier, A Farhang, F Bader, et al, in *Communications (ICC), 2016 IEEE International Conference on*. 5G waveforms for overlay D2D communications: Effects of time-frequency misalignment (IEEE, Kuala Lumpur, Malaysia, 2016), pp. 1–7

16. R Yin, C Zhong, Yu G, Z Zhang, KK Wong, X Chen, Joint spectrum and power allocation for d2d communications underlaying cellular networks. IEEE Trans. Veh. Technol. **65**(4), 2182–2195 (2016)
17. N Lee, X Lin, JG Andrews, RW Heath, Power control for d2d underlaid cellular networks, modeling, algorithms, and analysis. IEEE J. Selected Areas Commun. **33**(1), 1–13 (2016)
18. C Yang, T Jiang, X Chen, Z Junshan, Social-aware video multicast based on device-to-device communications. IEEE Trans. Mobile Comput. **15**(6), 1528–1539 (2016)
19. M Gregori, et al., Wireless distributed storage in socially enabled D2D communications. IEEE Trans. Wirel. Commun. **15**(6), 1–13 (2016)
20. HV Balan, R Rogalin, A Michaloliakos, K Psounis, G Caire, Airsync, Enabling distributed multiuser MIMO with full spatial multiplexing. IEEE/ACM Trans. Netw. **21**(6), 1681–1695 (2013)
21. E Antonio-Rodríguez, et al., Wideband full-duplex MIMO relays with blind adaptive self-interference cancellation. Signal Process. **130**, 74–85 (2017)
22. I tube, you tube, everybody tubes, analyzing the world's largest user (2007)
23. M Dehghan, A Seetharam, Jiang, in *Computer Communications (INFOCOM), 2015 IEEE Conference on*. On the complexity of optimal routing and content caching in heterogeneous networks (IEEE, Kowloon, 2015), pp. 936–944
24. V Sciancalepore, V Mancuso, A Banchs, S Zaks, A Capone. Enhanced content update dissemination through D2D in 5G cellular networks. IEEE Trans. Wirel. Commun. **15**(11), 7517–7530 (2016)
25. P Mogensen, W Na, IZ Kovács, et al., in *Vehicular Technology Conference, 2007. VTC2007-Spring. IEEE 65th*. LTE capacity compared to the Shannon bound (IEEE, Dublin, 2007), pp. 1234–1238
26. AM Mathai, SB Provost, *Quadratic forms in random variables*. (New York, Marcel Dekker, 1992)
27. A Papoulis, SU Pillai, *Probability, random variables and stochastic process*. 4th ed. (New York, McGraw-Hill, 2002)

Permissions

All chapters in this book were first published in EURASIP JWCN, by Springer International Publishing AG.; hereby published with permission under the Creative Commons Attribution License or equivalent. Every chapter published in this book has been scrutinized by our experts. Their significance has been extensively debated. The topics covered herein carry significant findings which will fuel the growth of the discipline. They may even be implemented as practical applications or may be referred to as a beginning point for another development.

The contributors of this book come from diverse backgrounds, making this book a truly international effort. This book will bring forth new frontiers with its revolutionizing research information and detailed analysis of the nascent developments around the world.

We would like to thank all the contributing authors for lending their expertise to make the book truly unique. They have played a crucial role in the development of this book. Without their invaluable contributions this book wouldn't have been possible. They have made vital efforts to compile up to date information on the varied aspects of this subject to make this book a valuable addition to the collection of many professionals and students.

This book was conceptualized with the vision of imparting up-to-date information and advanced data in this field. To ensure the same, a matchless editorial board was set up. Every individual on the board went through rigorous rounds of assessment to prove their worth. After which they invested a large part of their time researching and compiling the most relevant data for our readers.

The editorial board has been involved in producing this book since its inception. They have spent rigorous hours researching and exploring the diverse topics which have resulted in the successful publishing of this book. They have passed on their knowledge of decades through this book. To expedite this challenging task, the publisher supported the team at every step. A small team of assistant editors was also appointed to further simplify the editing procedure and attain best results for the readers.

Apart from the editorial board, the designing team has also invested a significant amount of their time in understanding the subject and creating the most relevant covers. They scrutinized every image to scout for the most suitable representation of the subject and create an appropriate cover for the book.

The publishing team has been an ardent support to the editorial, designing and production team. Their endless efforts to recruit the best for this project, has resulted in the accomplishment of this book. They are a veteran in the field of academics and their pool of knowledge is as vast as their experience in printing. Their expertise and guidance has proved useful at every step. Their uncompromising quality standards have made this book an exceptional effort. Their encouragement from time to time has been an inspiration for everyone.

The publisher and the editorial board hope that this book will prove to be a valuable piece of knowledge for researchers, students, practitioners and scholars across the globe.

List of Contributors

Fernaz Narin Nur, Selina Sharmin, Md. Ahsan Habib and Md. Abdur Razzaque
Green Networking Research (GNR) Group, Department of Computer Science and Engineering, University of Dhaka, Dhaka 1000, Bangladesh

Md. Shariful Islam
Institute of Information Technology, University of Dhaka, Dhaka 1000, Bangladesh

Ahmad Almogren, Mohammad Mehedi Hassan and Atif Alamri
Chair of Pervasive and Mobile Computing, College of Computer and Information Sciences, King Saud University, Riyadh 11543, Saudi Arabia

Wei Wang, Min Zhang, Dan Wang and Yu Jiang
Tianjin Key Laboratory of Wireless Mobile Communications and Power Transmission, Tianjin Normal University, Tianjin 300387, China

Antonio De Domenico and Vincent Heiries
CEA, LETI, MINATEC, F-38054, Grenoble, France

Mouhcine Mendil
CEA, LETI, MINATEC, F-38054, Grenoble, France
University Grenoble Alpes, G2Elab, F-38054, Grenoble, France

Raphael Caire and Nouredine Hadjsaid
University Grenoble Alpes, G2Elab, F-38054, Grenoble, France

Robin Gerzaguet
University of Rennes 1, IRISA, 38000 Grenoble, France

Laurent Ros and Jean-Marc Brossier
GIPSA-Lab, Image and signal department, 38400 Saint Martin d'Heres, France

Fabrice Belvèze
ST-Microelectronics, 38000 Grenoble, France

Pietro Savazzi and Anna Vizziello
Department of Electrical, Computer and Biomedical Engineering, University of Pavia, via Ferrata 5, Pavia, Italy

Kaito Funabiki, Takayuki Nishio, Masahiro Morikura and Koji Yamamoto
Graduate School of Informatics, Kyoto University, Kyoto, Japan

Daisuke Murayama and Katsuya Nakahira
NTT Network Innovation Laboratories, NTT Corporation, Yokohama, Japan

Zhenyu Na, Xiaotong Li and Zhian Deng
School of Information Science and Technology, Dalian Maritime University, Dalian, China

Xin Liu
School of Information and Communication Engineering, Dalian University of Technology, Dalian, China

BinWang, Zengshan Tian and Liangbo Xie
Chongqing Key Lab of Mobile Communications Technology, Chongqing University of Posts and Telecommunications, Chongqing 400065, China

Mu Zhou
Chongqing Key Lab of Mobile Communications Technology, Chongqing University of Posts and Telecommunications, Chongqing 400065, China
Tianjin Key Laboratory of Wireless Mobile Communications and Power Transmission, Tianjin Normal University, Tianjin 300387, China

Taehyun Kim and Jun Heo
The School of Electrical Engineering, Korea University, 145, Anam-ro, Seongbuk-gu, 02841 Seoul, Republic of Korea

Jonghyun Baik
LG electronics, 19, Yangjae-daero 11-gil, Seocho-gu, 06772 Seoul, Republic of Korea

Myeongwoo Lee
Samsung electronics, 1-1, Samsungjeonja-ro, 18448 Hwaseong-si, Republic of Korea

P. Muñoz, A. Gómez-Andrades, E. J. Khatib and R. Barco
Communications Engineering Dept., University of Málaga, Málaga, Spain

E. Cruz and N. Faour
Ericsson, 29590 Campanillas, Málaga, Spain

Zhuo Li
Tianjin Key Laboratory of Wireless Mobile Communications and Power Transmission, Tianjin Normal University, Tianjin 300387, China
College of Electronic and Communication Engineering, Tianjin Normal University, Tianjin 300387, China
Department of Computer Science, University of Arizona, Tucson, AZ 85721, USA

Yutong Chen
School of Microelectronics, Tianjin University, Tianjin 300072, China

Deliang Liu
Shijiazhuang Mechanical Engineering College, Shijiazhuang 050003, China

Xiang Li
College of computer and Information Engineering, Tianjin Normal University, Tianjin 300387, China

Pavel Prochazka, Tomas Uricar and Jan Sykora
Czech Technical University in Prague, Technicka 2, 16627 Prague, Czech Republic

David Halls
Toshiba—Telecommunications Research Laboratory, 32 Queen Square, Bristol, BS1 4ND, UK

Andreas Bollig, Martijn Arts and Rudolf Mathar
Institute for Theoretical Information Technology, RWTH Aachen University, Kopernikusstraße 16, 52074 Aachen, Germany

Anastasia Lavrenko
Institute for Information Technology, TU Ilmenau, 98684 Ilmenau, Germany

Chee-Hyun Park
Department of Electronics and Computer Engineering, Hanyang University, Seoul 133-791, Republic of Korea

Joon-Hyuk Chang
Department of Electronic Engineering, Hanyang University, Seoul 133-791, Republic of Korea

Sulochana Sooriyaarachchi and Chandana Gamage
Department of Computer Science and Engineering, University of Moratuwa, Moratuwa 10400, Sri Lanka

Yajun Chen and Pei Chen
National Digital Switching System Engineering and Technological Research Center, No. 62, Kexue Road, Zhengzhou, China

Peiyan Yuan
School of Computer and Information Engineering, Henan Normal University, No. 46, Jianshe East Road, 453007 Xinxiang, China

Xiaoyan Zhao
National Digital Switching System Engineering and Technological Research Center, No. 62, Kexue Road, Zhengzhou, China
School of Computer and Information Engineering, Henan Normal University, No. 46, Jianshe East Road, 453007 Xinxiang, China

Index